Studies in Computational Intelligence

Volume 953

The series "Studies in Computational Intelligence" (SCI) publishes new developments and advances in the various areas of computational intelligence—quickly and with a high quality. The intent is to cover the theory, applications, and design methods of computational intelligence, as embedded in the fields of engineering, computer science, physics and life sciences, as well as the methodologies behind them. The series contains monographs, lecture notes and edited volumes in computational intelligence spanning the areas of neural networks, connectionist systems, genetic algorithms, evolutionary computation, artificial intelligence, cellular automata, self-organizing systems, soft computing, fuzzy systems, and hybrid intelligent systems. Of particular value to both the contributors and the readership are the short publication timeframe and the world-wide distribution, which enable both wide and rapid dissemination of research output.

Indexed by SCOPUS, DBLP, WTI Frankfurt eG, zbMATH, SCImago.

All books published in the series are submitted for consideration in Web of Science.

More information about this series at http://www.springer.com/series/7092

Witold Pedrycz · Luis Martínez ·
Rafael Alejandro Espin-Andrade · Gilberto Rivera ·
Jorge Marx Gómez
Editors

Computational Intelligence for Business Analytics

Springer

Editors
Witold Pedrycz
Electrical and Computer Engineering
University of Alberta
Canada, AB, Canada

Rafael Alejandro Espin-Andrade
Accounting and Administration Faculty
Centre of Innovation Management
Research for Entrepreneurial and Regional Development (CIGIDER)
Autonomous University of Coahuila
Coah, Mexico

Jorge Marx Gómez
Department of Computing Science
Business Information Systems/VLBA
Carl von Ossietzky Universität Oldenburg
Oldenburg, Niedersachsen, Germany

Luis Martínez
Department of Computer Science
School of Computing
Universidad de Jaén
Jaén, Spain

Gilberto Rivera
División Multidisciplinaria de Ciudad Universitaria
Universidad Autónoma de Ciudad Juárez,
Chihuahua, Mexico

ISSN 1860-949X ISSN 1860-9503 (electronic)
Studies in Computational Intelligence
ISBN 978-3-030-73818-1 ISBN 978-3-030-73819-8 (eBook)
https://doi.org/10.1007/978-3-030-73819-8

This Springer imprint is published by the registered company Springer Nature Switzerland AG
The registered company address is: Gewerbestrasse 11, 6330 Cham, Switzerland

Preface

Business Analytics (BA) is reformulating corporate success. Transforming data in knowledge and using it harmonously in Decision Making is not, anymore, a luxury of the most prestigious companies and organizations, but an obligation to deliver competitive products and high-quality services. Approaches coming from different mathematical and computer sciences areas are deeply involved in this field. Nowadays, BA is probably the most important manifestation of the knowledge economy and knowledge society.

A growing number of companies are identifying innovative ways to take advantage of the developed solutions based on BA. These applications—when enriched with computational intelligence—are capable of generating information that can be vital for making strategic decisions in a more informed and reliable way even under high uncertainty environments. Consequently, research on computational intelligence for BA exhibits a substantial impact on our world today.

Eurekas is a Multinational Multidisciplinary Scientific Community joining professionals of Mathematics, Computer Sciences, Engineeging, as well as Administration, Economics and Social Sciences, towards theoretical and practical developments useful for Data and Business Analytics, in this way contributing to build the Knowledge Society and Knowledge Economy. It was founded in 2008 and it has been contributing by multiple projects to that purpose.

Computational Intelligence for Business Analytics is a new Eurekas editorial initiative which aims to collect the latest technological innovations in the field of BA to improve business models related to Group Decision-Making, Forecasting, Risk Management, Knowledge Discovery, Data Breach Detection, Social Well-Being, among other topics related to this field. This book consists of twenty-three chapters, organized into three main areas:

Part I. *Decision and Prescriptive Analytics.* In a nutshell, the nine chapters in this part are about ways of providing advice. They employ optimization, simulation, and decision-analysis algorithms to support decision makers in making future actions, according to their resulting outcomes. The original-research chapters in this part present approaches based on Computational Intelligence models—such as Compensatory Fuzzy Logic, TOPSIS, Fuzzy Preference Relations, SWOT Analysis, Ensemble Classifiers, and Causal Analysis—applied to the following domains:

Informal Trade, Low-Power Wide Area Networks, Consensus Reaching Process, Development of Women Entrepreneurs, Dermoscopy, Project Portfolio Selection, and Analysis of Socioeconomic Consequences of Debtors.

Part II. *Predictive Analytics*. The chapters in this part are all about understanding the future, providing companies with actionable insights based on data and delivered through machine learning, statistical models, and forecasting techniques. The eight chapters in this part address relevant problems by extending and adapting the following models: Compensatory Fuzzy Rough Predicates, Nonparametric Hypothesis Testing Methods, Deep Learning, Fuzzy Cognitive Maps, Linguistic Mathematical Morphology, Neural Networks, and others.

Part III. *Descriptive Analytics*. This part features six chapters introducing approaches to learn from data, understanding past behaviour that might influence the current business model. Here, the original-research contributions are inspired by computational models—e.g., Nonlinear Regression, Kernel-based Clustering, Metaheuristic Algorithms, Partial Square Minimums, Structural Equations Models, Interval-Valued Fuzzy Predicates and Fuzzy Mining—to succesfully deal with the following significant BA problems: Estimation of the Yield Curve, Students Satisfaction, Analysis of Economic and Social Conditions, and Social Well-Being Analysis.

Readers can also find the following studies providing a comprehensive literature review:

1. "Fuzzy logic-based approaches in supply chain risk management: a review" by A. Díaz-Curbelo, A. Gento and R. A. Espin-Andrade.
2. "A look at Artificial Intelligence on the Perspective of Application in the Modern Education" by A. Borges, R. Padilha, R. Arthur and Y. Iano.
3. "Nonparametric Tests for Comparing Forecasting Models" by Dmitriy Klyushin.

This book is expected to motivate readers to implement these technologies to form a Smart Business or Industry 4.0 environment, as well as encourage researchers to continue contributing to this field. *Computational Intelligence for Business Analytics* represents an interesting avenue fostering constructive discussions, conversations, and reflection about the impact and potential of BA for addressing everyday and emerging needs. Finally, we hope that readers will find this book (or any of its chapters) highly informative and useful, inspiring them to conduct productive research that benefits society through the production of new knowledge and deliver smart solutions that may impact not only the BA field but also impact other related disciplines.

Alberta, Canada — Witold Pedrycz
Jaén, Spain — Luis Martínez
Coahuila, Mexico — Rafael Alejandro Espin-Andrade
Chihuahua, Mexico — Gilberto Rivera
Oldenburg, Germany — Jorge Marx Gómez

Contents

Decision and Prescriptive Analytics

Multi-criteria Method for Evaluating the Impact of Informal Trade on the Mariscal de Puyo Market ... 3
Luis Eduardo Álvarez Cortez, Cynthia Paulina Cisneros Zúñiga, and Roberto Carlos Jiménez Martínez

Multicriteria Analysis for LPWAN Selection for Industry 4.0 Based on TOPSIS and a Model of Proportionality ... 15
Alberto Ochoa-Zezzatti, Roberto Contreras-Masse, and José Mejía

Comprehensive Minimum Cost Models Based on Consensus Measures ... 47
Álvaro Labella, Hongbin Liu, Rosa M. Rodríguez, and Luis Martínez

Strategies for the Development and Success of Women Entrepreneurs Through SWOT Analysis and Compensatory Fuzzy Logic ... 61
Magaly Oyervides Villarreal, Liliana Guerrero Ramos, Rafael Alejandro Espin-Andrade, and Israel Sánchez López

Fuzzy Logic-Based Approaches in Supply Chain Risk Management: A Review ... 79
Alina Díaz-Curbelo, Ángel Manuel Gento Municio, and Rafael Alejandro Espin-Andrade

Use of Fuzzy Logic in the Strategic Selection of Process Indicators ... 95
Israel Sánchez López, Rafael Alejandro Espin-Andrade, Liliana Guerrero Ramos, and Magaly Oyervides Villarreal

Evaluating Intelligent Methods for Decision Making Support in Dermoscopy Based on Information Gain and Ensemble ... 111
Newton Spolaôr, Rui Fonseca-Pinto, Ana I. Mendes, Leandro A. Ensina, Weber S. R. Takaki, Antonio R. S. Parmezan, Conceição V. Nogueira, Claudio S. R. Coy, Feng C. Wu, and Huei D. Lee

Modeling Performance of NP-hard Problems by Applying Causal Analysis for the VisTHAA Tool 129
Claudia Gómez-Santillán, Juan Gerardo Ponce-Najera, Luis Rodolfo García-Nieto, Laura Cruz-Reyes, Nelson Rangel-Valdez, Héctor J. Fraire-Huacuja, and Lucila Morales-Rodríguez

Fuzzy Logic to Measure the Legal and Socioeconomic Effect of the Debtors Declared in the Canton of Pastaza 155
Diego Vladimir Garcés Mayorga, Danilo Rafael Andrade Santamaría, and Luis Rodrigo Miranda Chávez

Predictive Analytics

A Look at Artificial Intelligence on the Perspective of Application in the Modern Education 171
Ana Carolina Borges Monteiro, Reinaldo Padilha França, Rangel Arthur, and Yuzo Iano

Knowledge Discovery by Compensatory Fuzzy Rough Predicates 191
Rafael Alejandro Espin-Andrade, Erick González, Rafael Bello, and Witold Pedrycz

Nonparametric Tests for Comparing Forecasting Models 213
Dmitriy Klyushin

Using an Innovative Model Based on Deep Learning to Determine Reduction of Habitats Associated with Arboreal Birds in Mexico 231
Alberto Ochoa-Zezzatti, Alberto Hernandez, Luis Alatorre, Luis Bravo-Peña, María Torres-Olave, and José Mejia

Method for Recommending Guardianship to Minors Based on Parental Responsibility Using a Fuzzy Cognitive Map 245
Hernán Patricio Castillo Villacrés, Mesías Elías Machado Maliza, and Diego Fabricio Tixi Torres

Linguistic Mathematical Morphology w-operators in Fuzzy Color Space 259
Juan I. Pastore, Virginia L. Ballarin, and Rafael Alejandro Espin-Andrade

Method for Treatment and Its Incidence in the Change of Social Rehabilitation Regime Using Compensatory Fuzzy Logic 273
José Rodolfo Calle Santander, Eduardo Luciano Hernández Ramos, and Klever Aníbal Guamán Chach

A Proposal for Data Breach Detection in Organizations Based on User Behavior 283
René Palacios and Victor Morales-Rocha

Descriptive Analytics

Estimation of the Yield Curve for Costa Rica Using Combinatorial Optimization Metaheuristics Applied to Nonlinear Regression 303
Andrés Quirós-Granados and Javier Trejos-Zelaya

Kernel-Based Clustering Driven by Density Index 317
Edwin Aldana-Bobadilla, Ivan Lopez-Arevalo, Ivan Mendez-Alvarez, Alejandro Molina-Villegas, and Hiram Galeana-Zapien

Students Satisfaction Description Based on Classical and Multivalent Discovery Techniques 345
Susana Beatriz Ruiz, Rafael Alejandro Espin-Andrade, and Myriam Beatriz Herrera

Is Economic Performance Affected by Social Conditions and Rights? The Case of the Central Region of San Luis Potosí, Mexico 367
Juan Carlos Yáñez-Luna and Leonardo David Tenorio-Martínez

Social Well-Being Analysis Using Interval-Valued Fuzzy Predicates 387
Diego S. Comas, Eugenio Actis Di Pasquale, Juan I. Pastore, Agustina Bouchet, and Gustavo J. Meschino

A New Plugin to Include FuzzyPred in KNIME 405
Orenia Lapeira, Ernesto Álvarez, René Cutie, Alejandro Prieto, Alejandro Rosete, and Taymi Ceruto

Decision and Prescriptive Analytics

Multi-criteria Method for Evaluating the Impact of Informal Trade on the Mariscal de Puyo Market

Luis Eduardo Álvarez Cortez, Cynthia Paulina Cisneros Zúñiga, and Roberto Carlos Jiménez Martínez

Abstract When people from different backgrounds offer products and services to the general population, without complying with the legal parameters that a business requires, they generate a phenomenon known as the Informal Market. In the Mariscal de Puyo market, there has been a flourishing of the informal market for the commercialization of basic necessities, which is gaining space in the internal supply chain of Ecuador. Quantifying its impact on society represents a task little tackled by science. This research proposes a solution to the problem posed by developing a method for evaluating the impact of informal trade. The proposed method bases its operation through a multi-criteria approach to evaluation. A case study is implemented with the aim of measuring the impact of informal trade on the Mariscal de Puyo market.

Keywords Multi-criteria method · Informal trade · Mariscal de puyo market

1 Introduction

The progress of the mercantile society is directly related to the levels of commerce with which economic transactions are carried out [1]. Economic transactions can be managed through a formal or informal market. The informal market has increased its forms of management, expanding in the different regions of Ecuador [2, 3].

After the expansion of informal markets, the community changes the perspective on this type of commercial form [4, 5]. Informal trade has within its characteristics [6, 7]:

- Informal management of your sales or services.
- A massive movement of cash and non-commercial bank transactions.
- Evidence of a relationship with poverty, lack of production in the region.
- The development of ingenuity from the need to find a lucrative way to work.

L. E. Á. Cortez (✉) · C. P. C. Zúñiga · R. C. J. Martínez
Universidad Regional Autónoma de Los Andes (UNIANDES), Puyo 160150, Pastaza, Ecuador
e-mail: up.luisalvarez@uniandes.edu.ec

W. Pedrycz et al. (eds.), *Computational Intelligence for Business Analytics*,
Studies in Computational Intelligence 953,
https://doi.org/10.1007/978-3-030-73819-8_1

Informal trade represents the performance of economic or service activities that are kept fundamentally hidden from the State administration, which brings a set of legal consequences and support for society since people do not have guarantees on the products obtained, among other characteristics, although it is evident that the level o f accessibility reaches a greater number of people [8, 9].

Based on the problems described above, this research aims to develop a method for evaluating the impact of informal trade. The proposed method bases its operation through a multi-criteria approach [10, 11] based on evaluative criteria using a multi-criteria method for evaluation.

2 Materials and Methods

This section describes the operation of the proposed method for evaluating the impact of informal trade. The general characteristics of the proposed solution are presented. The main stages and activities that make up the method are described.

The method for evaluating the impact of informal trade is designed low on a group of qualities [12]. The qualities that distinguish the method are:

- Integration: the method guarantees the interconnection of the different components in combination for the evaluation of the impact of informal trade.
- Flexibility: uses 2-tuples to represent uncertainty so as to increase the interoperability of the people who interact with the method.
- Interdependence: the method uses the input data provided by the process experts as a starting point. The analyzed results contribute to an experience base that forms the core of inference processing.

The method is based on the following principles:

- Identification through the team of experts of the indicators for evaluating the impact of informal trade.
- Definition and processing under a multi-criteria approach.
- The use of multicriteria methods in the evaluation.

The method for evaluating the impact of informal trade is structured to manage the workflow of the evaluation process based on a multi-criteria inference method; it has three fundamental stages: input, processing, and output of information [13]. Figure 1 shows a diagram illustrating the general operation of the method.

2.1 Description of the Stages of the Method

The proposed method is designed to ensure workflow management in the informal trade impact assessment process. It uses a multi-expert multi-criteria approach

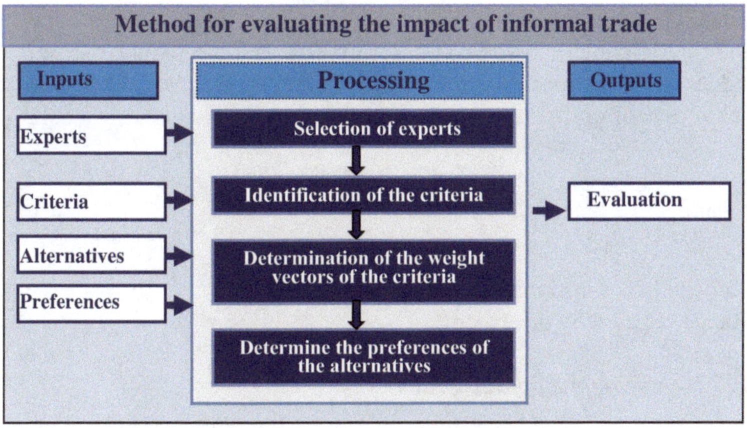

Fig. 1 General scheme of the method

where evaluative indicators are identified to determine the operation of the method's processing.

The processing stage is structured by four activities that govern the processing inference process. Figure 2 shows a diagram with the activities of the processing stage.

Figure 2 showed the activities of the processing stage. Its operation is detailed below:

Activity 1 Selection of experts

The process consists of determining the group of experts involved in the process. For its selection, the methodology proposed by Fernández is used [14, 15]. To start the process, a model is sent to potential experts with a brief explanation of the objectives of the work and the area of knowledge in which the research is framed. The following activities are carried out:

1. Knowledgeable experts are contacted and asked to participate in the panel. The activity results in the recruitment of the group of experts who will participate in the application of the method.

The process should filter out the experts with a low level of expertise participating in the process, those with the most knowledge and prestige in the area of knowledge

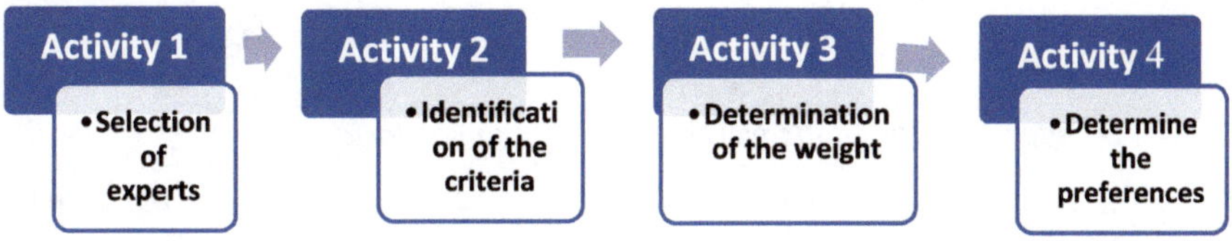

Fig. 2 Processing stage activities

that is the subject of the research study. To carry out the filtering process, a self-evaluation questionnaire is carried out for experts. he objective is to determine the knowledge or information coefficient (K_c), Eq. (1) expresses the method to determine the level of expertise

$$K_c = n(0, 1) \tag{1}$$

where:

K_c Knowledge or information coefficient,
n Range selected by the expert.

Activity 2 Identification of the criteria

Once the experts involved in the process have been identified, the evaluation criteria are identified. The criteria nurture the method; they represent input parameters that are used in the processing stage. From the group work of the experts, the following activities are carried out:

1. A questionnaire is sent to the panel members, and they are asked for their opinion for the selection of the evaluation criteria that support the research. From a previously prepared questionnaire, the set of criteria of the experts is obtained as a result.
2. The responses are analyzed, and the areas in which they agree and in which they differ are identified. The activity allows an analysis of the behavior of the answers issued by the experts, and the common elements are identified.
3. The summary analysis of all the responses is sent to the members of the panel; they are asked to fill out the questionnaire again and to give their reasons for the opinions in which they differ. The activity allows to obtain a new assessment from the group of experts on the knowledge collected and summarized.
4. The process is repeated until the responses stabilize. The activity represents the stopping condition of the method, from which the responses are stabilized, its application is concluded, considering this the general result.

The activity results in the set of evaluative criteria of the method. Employs a multi-criteria approach expressed as (Eq. 2).

$$C = \{c_1, c_2, \ldots, c_m\} \tag{2}$$

where:

$m > 1, \forall\, c_i \nexists\, \emptyset$

Activity 3 Determination of the weight vectors of the criteria

The group of experts involved in the process is used to determine the weights attributed to the evaluation criteria. They are asked to determine the level of importance attributed to the evaluation criteria identified in the previous activity.

The weights of the evaluation criteria are expressed by means of a domain of fuzzy values. Fuzzy sets give a quantitative value to each element, which represents the degree of belonging to the set [16, 17]. A fuzzy set A is an application of a referential set S on the interval [0, 1], such that:

$$A : S \rightarrow [0, 1],$$

and it is defined by means of a membership function:

$$0 \leq \mu_A(x) \leq 1. \tag{3}$$

Linguistic terms based on 2-tuples are used to increase interpretability in determining the weight vectors associated with the criteria [18, 19]. The use of linguistic labels in decision models involves, in most cases, performing operations with linguistic labels. Table 1 shows the set of linguistic terms with their respective values.

Once the weight vectors of the different experts involved in the process have been obtained, an information aggregation process is carried out using an average function, as shown in Eq. (4).

$$VA = \frac{\sum_{i=1}^{n} C_{ij}}{E} \tag{4}$$

where:

VA Value added,
E Number of experts involved in the process,
C_{ij} Weight vector expressed by experts for criteria C.

Table 1 Domain of values to assign weight to the criteria

Value	Linguistic terms
0.1	Without importance
0.2	Less important
0.3	Slightly important
0.4	Something important
0.5	Average importance
0.6	Important
0.7	Very important
0.8	Strongly important
0.9	Very strongly important
1	Extremely important

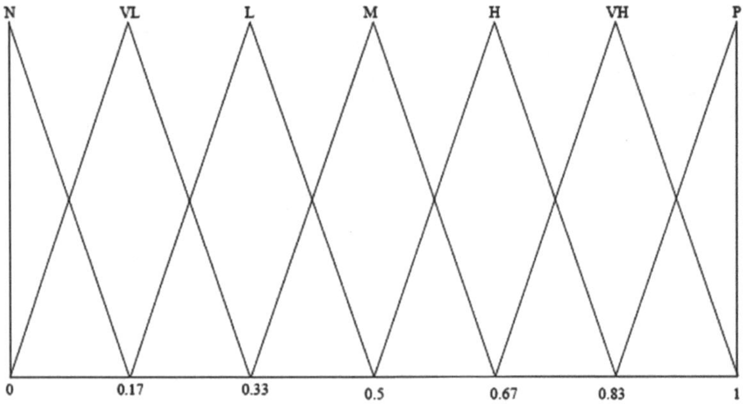

Fig. 3 Linguistic labels set

Activity 4 Determine the preferences of the alternatives

The activity for determining preferences consists of identifying the impact that evaluation criteria have on the study phenomenon. The evaluation process is carried out using a numerical scale so that the level of belonging of the indicators is expressed. Figure 3 shows a graph with the sets of linguistic labels used.

where:

N Null.
VL Very Low.
L Low.
M Medium.
H High.
VH Very High.
P Preferred.

For the evaluation of the impact of informal trade, the problem and the evaluation of each alternative from which the evaluation matrix is formed are described [20–22]. The matrix is made up of the alternatives, the criteria, and the evaluation of each criterion for each alternative.

From obtaining the preferences of each evaluative criterion on the object of study, the information inference process is carried out. The inference is guided by the use of information aggregation operators.

It is part of the set of alternatives A:

$$A = \{A_1, A_2, \ldots, A_m\} \tag{5}$$

To which the preferences P are obtained:

$$P = C_1, \ldots, C_n \tag{6}$$

A multi-criteria method is applied to the evaluation criteria to process the alternatives based on the weight vectors W defined by the experts on the evaluation criteria.

$$W = \{w_1, w_2, \ldots, w_n\} \tag{7}$$

The aggregation process is carried out with the use of information aggregation operators [23–25]. The fundamental objective is to obtain collective valuations from individual valuations through the use of aggregation operators. The OWA (Ordered Weighted Averaging) aggregation operator is used to process the proposed method [26–28].

The OWA operators work similar to the weighted average operators, although the values that the variables take are previously ordered in descending order and, contrary to what happens in the weighted averages, the weights are not associated with any specific variable [29–31].

Definition 1: Given a vector of weights $W = w_1, \ldots, w_n \in [0, 1]^n$ such that: $\sum_{i=1}^{n} w_i$, the operator (OWA) associated with W is the aggregation operator $f_n^w :\rightarrow$ R defined by:

$$f_n^w(u) = \sum_{i=1}^{n} w_i v_i \tag{8}$$

where v_i. is the i-th largest element of $\{u_1, \ldots, u_n\}$.

For the present investigation, the process of aggregation of the information used is defined, such as company (Eq. 9).

$$F(p_1, \ldots, p_2, \ldots, p_n) = \sum_{j=1}^{n} w_j b_j \tag{9}$$

where:

P Set of preferences obtained from the evaluation of the criteria for evaluating the impact of informal trade.

w_j Are the weight vectors attributed to the evaluation criteria.

bj Is the jth largest of the ordered p_n preferences.

3 Results

For the implementation of the proposed method, a case study has been carried out where an instrument focused on the specific case that is modeled is represented. The object of analysis was the Mariscal de Puyo market in Ecuador. The objective was to evaluate the impact of informal trade on the Mariscal de Puyo market.

Below are the ratings achieved for each activity:

Activity 1 Selection of experts

To apply the method, a questionnaire was applied in order to select a group of experts to intervene in the process. The disinterested commitment of 9 experts was achieved. The self-assessment questionnaire was applied to the 9 experts where the following results were obtained:

- 5 experts self-evaluate with a level of competence on the subject under study of 10 points.
- 1 expert self-assess with a competence level o f 9 points.
- 1 expert self-evaluates with a proficiency level of 8 points.
- 1 expert self-evaluate with a level of competence of 6 points.

The knowledge coefficient K_c represents an important parameter in the application of the proposed method. For the investigation, the K_c per expert are obtained as reported in Table 2:

Four questions were applied to the experts where the following results were obtained to identify the levels of knowledge on the subject:

- About question 1. Theoretical analyzes carried out by you on the subject: a self-evaluation of High for 5 experts and Average for 1 expert was obtained.
- About question 2. Study of works published by Ecuadorian authors: a self-evaluation of High for 5 experts, Average for 2 experts, and Low for 1 expert was obtained.
- About question 3. Direct contact with the informal market: a self-assessment of High was obtained for 5 experts, Medium for 2 experts, and Low for 1 expert.
- About question 4. Knowledge of the current state of informal trade: a self-assessment of High was obtained for 6 experts, Medium for 1 expert and Low for 1 expert.

Figure 4 shows a graph with the behavior of the experts' knowledge coefficients. From the analysis of the results, it is determined to use 7 of the 8 experts originally planned.

Activity 2 Identification of the criteria

For the activity, a survey was conducted of the experts involved in the process. The objective was to identify the evaluation criteria for evaluating informal trade. The indicators constitute the fundamental element on which the processing is carried out in subsequent stages.

Table 2 Knowledge coefficient by experts

1	3	4	5	6	7	8	9
1	0.90	1	0.60	0.80	1	1	1

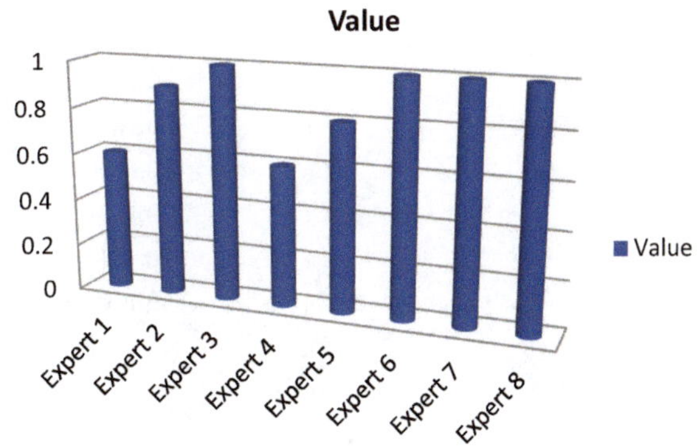

Fig. 4 Representation of the knowledge coefficient of the experts

Table 3 displays the evaluation criteria obtained from the activity.

Activity 3 Determination of the weight vectors of the criteria

To determine the weights on the criteria, a multi-expert approach was used, in which the 7 selected in Activity 1 participated. Using 2-tuples as proposed in Table 1, the work was carried out by the group of experts.

From the aggregation carried out using Eq. (4), the weights of the 7 experts are unified into an added value. Table 4 shows the result of the weight vectors resulting from the activity.

The consensus was reached in the second iteration of the process. From which it was taken as the stop value.

Table 3 Evaluative criteria obtained

Number	Evaluation criteria
C_1	Product quality certificate
C_2	Competitive prices
C_3	Contribution to the employment of personnel
C_4	Variety of products
C_5	Accessibility
C_6	Hygienic sanitary conditions

Table 4 Criterion weights based on expert criteria

Number	Weight vectors W for criteria C
C_1	0.1630
C_2	0.1956
C_3	0.1521
C_4	0.1847
C_5	0.1304
C_6	0.1739

Table 5 Result of evaluations obtained by experts

Number	W	Preference	$\sum_{j=1}^{n} w_j b_j$
C_1	0.1630	0.67	0.1956
C_2	0.1956	0.83	0.1533
C_3	0.1521	1	0.1443
C_4	0.1847	0.67	0.1092
C_5	0.1304	0.83	0.1019
C_6	0.1739	0.5	0.0652
Index			0.7697

Activity 4 Determine the preferences of the alternatives

For the proposed case study with the objective of evaluating the impact of informal trade on the Mariscal de Puyo market in Ecuador, an evaluation of compliance with the criteria was carried out. The weight vectors attributed to each evaluation criterion were taken as starting information. Compliance with the indicators in the Mariscal de Puyo market was evaluated using the set of linguistic labels proposed in Fig. 3.

As a result, a system with fuzzy values that are added as output values was obtained. Table 5 shows the result of the processing carried out.

Figure 5 shows the behavior of inferences about the evaluation criteria for the proposed case study.

From the data presented in Table 5, an impact index of informal trade for the Mariscal de Puyo market in Ecuador with an II 0.7697 is identified. The results obtained are valued as a high index of the impact of informal trade.

Fig. 5 Inference behavior

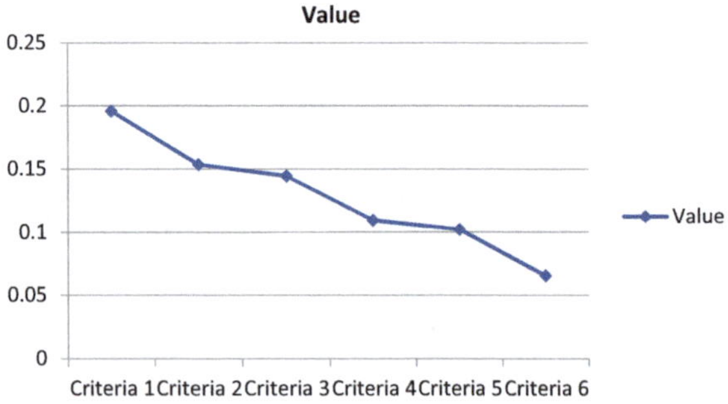

4 Conclusions

From the implementation of the proposed method, weights of aggregation are obtained for the evaluation of the evaluation criteria that represented the basis of the evaluation process for the informal market.

The disinterested participation of 9 experts was obtained as a result of the method, of which 7 were used based on their competence coefficient for the implementation of the proposed method.

The application of a case study to assess the impact of informal trade made it possible to determine the impact of informal trade on the Mariscal de Puyo market. It was found that informal trade gains space in the Ecuadorian trade network.

By applying the method in a real context, a method for evaluating the impact of informal trade is obtained from the use of information aggregation operators. The proposal can be extended to work with other multicriteria methods for inference processing, where the results obtained by the different methods can be compared.

References

1. Ricardo, J.E., et al.: Neutrosophic model to determine the degree of comprehension of higher education students in Ecuador. Neutrosophic Sets Syst. **26**(1) (2019). https://doi.org/10.5281/zenodo.3244297
2. Teltscher, S.: Small trade and the world economy: Informal vendors in Quito, Ecuador. Econ. Geogr. **70**(2), 167–187 (1994). https://doi.org/10.2307/143653
3. Middleton, A.: Informal traders and planners in the regeneration of historic city centres: the case of Quito, Ecuador. Prog Plann. **59**(2), 71–123 (2003). https://doi.org/10.1016/S0305-9006(02)00061-2
4. Waters, W.F.: The road of many returns: rural bases of the informal urban economy in Ecuador. Lat. Am. Perspect. **24**(3), 50–64 (1997). https://doi.org/10.1177/0094582X9702400304
5. Lanjouw, J.O., Levy, P.I.: Untitled: A study of formal and informal property rights in urban Ecuador. Econ. J. **112**(482), 986–1019 (2002). https://doi.org/10.1111/1468-0297.00067
6. Kromann, P., Montesdeoca, F., Andrade-Piedra, J.: Integrating formal and informal potato seed systems in Ecuador. In: Andrade-Piedra, J., Bentley, J., Almekinders, C., Jacobsen, K., Walsh, S., Thiele, G. (eds.) Case studies of roots, tubers and bananas seed systems, 2016–3 (2016)
7. Canelas, C.: Informality and poverty in Ecuador. Small Bus. Econ. **5**(4), 1097–1115 (2019). https://doi.org/10.1007/s11187-018-0102-9
8. Gamble, J., Puga, E.: Is Informal Transit Land-Oriented? Investigating the Links Be-tween Informal Transit and Land-Use Planning in Quito, Ecuador. Lincoln Institute of Land Policy (2019)
9. Gamble, J., Dávalos, C.: Moving with masculine care in the city: Informal transit in Quito, Ecuador. City **23**(2), 189–204 (2019). https://doi.org/10.1080/13604813.2019.1615796
10. Alava, M.V., et al.: Single valued neutrosophic numbers and analytic hierarchy process for project selection. Neutrosophic Sets Syst. **21**(1), 13 (2018). https://doi.org/10.5281/zenodo.1408761
11. Leyva, M.: Modelo de ayuda a la toma de decisiones basado en Mapas Cognitivos Difusos. Universidad de las Ciencias Informáticas (UCI), La Habana (2013)
12. Leyva-Vázquez, M., Smarandache F., Ricardo, J.E.: Artificial intelligence: challenges, perspectives and neutrosophy role.(Master Conference).” Dilemas Contemporáneos: Educación, Política y Valore, 6(Special) (2018)7

13. Ponce Ruiz, D.V., et al.: Softcomputing in neutrosophic linguistic modeling for the treatment of uncertainty in information retrieval. Neutrosophic Sets Syst. **26** (2019). https://doi.org/10.5281/zenodo.3244320
14. Fernández, S.H.D.M.: Criterio de expertos. Su procesamiento a través del método Del-phy. Histodidáctica. http://www.ub.edu/histodidactica/index.php?option=com_content&view=article&id=21:criterio-de-expertos-su-procesamiento-a-traves-del-metodo-delphy&catid=11:. (2016). Accessed 12 Aug 2009
15. Smarandache, F., et al.: Delphi method for evaluating scientific research proposals in a neutrosophic environment. Neutrosophic Sets Syst. **34**(1), 204–213 (2020). https://doi.org/10.5281/zenodo.3766597
16. Solís, P.Y.J., et al.: Compensatory fuzzy logic model for impact. Neutrosophic Sets and Systems, Book Series, vol. 26, p. 40: An International Book Series in Information Science and Engineering (2019)
17. Hernandez, N.B., Ruilova Cueva, M.B., Mazacón, B.N.: Prospective analysis of public management scenarios modeled by the Fuzzy Delphi method. Neutrosophic Sets Syst. **26**(1), 17 (2019)
18. Chen, Z.-S., Chin, K.-S., Tsui, K.L.: Constructing the geometric Bonferroni mean from the generalized Bonferroni mean with several extensions to linguistic 2-tuples for decision-making. Appl. Soft Comput. **78**, 595–613 (2019). https://doi.org/10.1016/j.asoc.2019.03.007
19. Giráldez-Cru, J., Chica, M., Cordón, O., Herrera, F.: Modeling agent-based consumers decision-making with 2-tuple fuzzy linguistic perceptions. Int. J. Intell. Syst. **35**(2), 283–299 (2020). https://doi.org/10.1002/int.22211
20. Schmied, S., et al: Vertical integration via dynamic aggregation of information in OPC UA. In: Sitek P., Pietranik M., Krótkiewicz M., Srinilta C. (eds.) In Asian Conference on Intelligent Information and Database Systems. Communications in Computer and Information Science, vol 1178. Springer, Singapore (2020). https://doi.org/10.1007/978-981-15-3380-8_18
21. Schultz, P.T., Sartini, R.A., Mckee, M.W.: Aggregation and use of information relating to a users context for personalized advertisements. Google Patents (2019)
22. Gospodinov, N., Maasoumi, E.: Generalized aggregation of misspecified models: With an application to asset pricing. J. Econometrics (2019). https://doi.org/10.1016/j.jeconom.2020.07.010
23. He, X.: Typhoon disaster assessment based on Dombi hesitant fuzzy information aggregation operators. Nat. Hazards **90**(3), 1153–1175 (2018). https://doi.org/10.1007/s11069-017-3091-0
24. Cornelio, O.M., et al.: Competency assessment model for a virtual laboratory system at distance using fuzzy cognitive map. Inv. Oper. **38**(2), 169–177 (2018)
25. Liu, P., Xu, H., Geng, Y.: Normal wiggly hesitant fuzzy linguistic power Hamy mean aggregation operators and their application to multi-attribute decision-making. Comput. Ind. Eng. **140**, 106224 (2020). https://doi.org/10.1016/j.cie.2019.106224
26. Yager, R.R., Filev, D.P.: Induced ordered weighted averaging operators. IEEE Trans. Syst. Man, Cybern. Part B (Cybern.) **29**(2), 141–150 (1999). https://doi.org/10.1109/3477.752789
27. Sampson, T.R., et al.: A gut bacterial amyloid promotes α-synuclein aggregation and motor impairment in mice. Elife **9**, e53111 (2020). https://doi.org/10.7554/eLife.53111
28. Mar, O., Ching, I., González, J.: Operador por selección para la agregación de infor-mación en Mapa Cognitivo Difuso. Rev. Cubana de Cien. Informáticas **14**(1), 20–39 (2020)
29. Jin, L., Mesiar, R., Yager, R.: Ordered weighted averaging aggregation on convex poset. IEEE Trans. Fuzzy Syst. **27**(3), 612–617 (2019). https://doi.org/10.1109/TFUZZ.2019.2893371
30. Sha, X., Xu, Z., Yin, C.: Elliptical distribution-based weight-determining method for ordered weighted averaging operators. Int. J. Intell. Syst. **34**(5), 858–877 (2019)
31. Garg, H., Agarwal, N., Tripathi, A.: Choquet integral-based information aggregation operators under the interval-valued intuitionistic fuzzy set and its applications to decision-making process. Int. J. Uncertainty Quantification, **7**(3) (2017). https://doi.org/10.1615/Int.J.UncertaintyQuantification.2017020076

Multicriteria Analysis for LPWAN Selection for Industry 4.0 Based on TOPSIS and a Model of Proportionality

Alberto Ochoa-Zezzatti, Roberto Contreras-Masse, and José Mejía

Abstract By 2020, more than 50 billion devices will be connected through radio communications. In conjunction with the rapid growth of the Internet of Things (IoT) market, low-power wide-area networks (LPWAN) have become a popular low-rate long-range radio communication technology. Sigfox, LoRa, and NB-IoT are the three leading LPWAN technologies that compete for large-scale IoT deployment. This paper provides a comprehensive and comparative study of these technologies, which serve as efficient solutions to connect smart, autonomous, and heterogeneous devices. We show that Sigfox and LoRa are advantageous in terms of battery lifetime, capacity, and cost; other factors such as social and workplace are included and exemplified by Mexico's labor regulations. Meanwhile, NB-IoT offers benefits in terms of latency and quality of service. In addition, we analyze the IoT success factors of these LPWAN technologies, and we consider application scenarios and explain which technology is the best fit for each of these scenarios.

Keywords Internet of things · Industry 4.0 · Low power wide area network · Connectivity · TOPSIS · Proportionality

1 Introduction

The Internet of Things (IoT) refers to a network of physical devices, automobiles, home appliances, and all those items that are used in conjunction with actuators, electronics, sensors, software, and connectivity to enhance connection, collection, and data exchange. IoT involves the extension of internet connectivity beyond personal computers and mobile devices [1]. Nowadays, it is common that big enterprises generate a lot of data that is useful to describe the difference. The data could explicitly describe the process or give implicit insights that depend on how it is interpreted or handled. Nevertheless, the collected data could be useful to make decisions that may impact the daily performance of a company.

A. Ochoa-Zezzatti · R. Contreras-Masse (✉) · J. Mejía
Universidad Autónoma de Ciudad Juárez, Cd. Juárez, Chihuahua 32310, México
e-mail: rcontreras@itcj.edu.mx

W. Pedrycz et al. (eds.), *Computational Intelligence for Business Analytics*,
Studies in Computational Intelligence 953,
https://doi.org/10.1007/978-3-030-73819-8_2

There are different concerns that must be considered: data storage, if real time operations with the data are necessary, how often the data should be recollected, etc. Thus, an important part of the use of IoT targets to implement within this emerging technology is the way that the data is getting from one point to another. The last developments in information technologies have authorized small devices for carrying out smart services more efficiently. Although these devices are submitted to restrictions in terms of resources, they offer extraordinarily little energy consumption advantage. These devices can also gather diverse kinds of information through sensors or transducers, and then send this acquired data to other systems. All these devices can be linked to a communication network. There are several IoT services that are dedicated to collect information from devices connected to sensors. In addition, various smart services may be accomplished in this network by considering acquired data. In typical applications, the data is stored periodically by an IoT device, and then this data is transmitted to a data gathering gateway that makes possible the communication between sensor devices (also known as IoT devices) and gives the ability for cloud computing for big data applications [2]. The gateway devices could be either regular broadband WAN connected Edge Servers, Wi-Fi access points, cellular networks, or smartphones. The world of wireless IoT is categorized into several groups depending on their transmission range. NFC (Near Field Communication) and RFID (Radio Frequency Identification) are the technologies operating at the shortest range, that is limited to an average of ten to hundred centimeters at most for LF (Low Frequency) and HF (High Frequency) applications; for UHF (Ultra High Frequency) four to ten meters, being followed by home wireless technologies such as Bluetooth, Wi-Fi and ZigBee, and others, having a range up to 100 m depending on the size of the antenna and power of transmitter reader system and the environmental conditions. These ranges are somehow increased by other IEEE 802.11 × standard technologies reaching 1 km, known as Wireless Local Area Networks (WLAN). To achieve more transmission range, technologies such as cellular technologies, ZigBee NAN, and Wi-SUN can be implemented and used [3]. The most popular short-haul technologies are Wi-Fi, ZigBee, and Z-wave. Unlike ZigBee and Wi-Fi, Low Power Wide Area Network (LPWAN) technologies make possible large wireless connections by providing remarkable advantages such as extended coverage areas, extremely low power requirements, and no need for maintenance. LPWAN, also known as LPWA or LPN, in a more formal and accurate definition, is a wireless data transport protocol that is now one of the basic protocols for the implementation of IoT. Therefore, it is one of the numerous options to take advantage of the IoT potential inside an enterprise, as was previously mentioned. To have a better idea of the relevance of LPWAN we can consider the prediction made by statisca.com [2] of a steady increase in the number of LPWAN devices connected around the world, expecting this number to reach around 3.5 billion devices by 2021, as shown in Fig. 1 some years ago, it still was far to become a reality. Today, it is implementable.

Typically, the coverage area of LPWAN technologies can reach up to 15 km, and, therefore, these technologies are convenient for realizing IoT services in wide areas. In other words, small objects can convey information quite long distances, and other systems can gather them from a broad area. The smart objects linked to a LPWAN

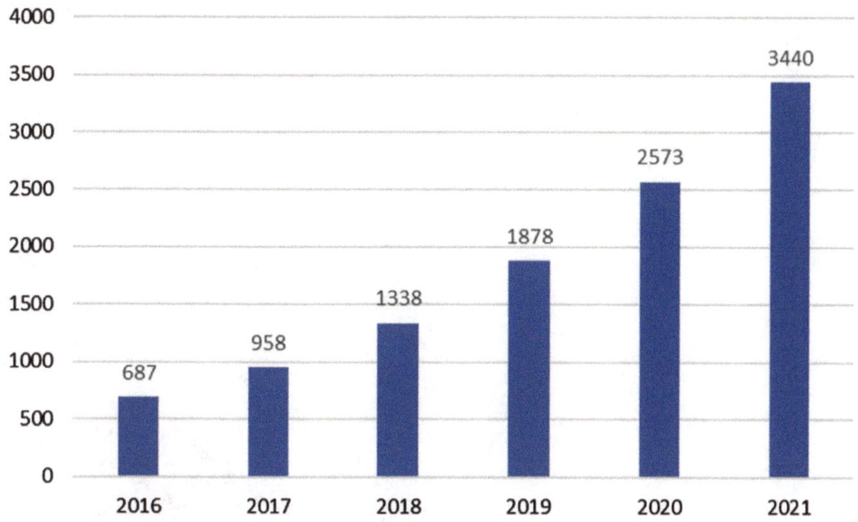

Fig. 1 Number of connected IoT devices (millions)

can convey measurement data acquired via sensors/transducers to cloud servers over a gateway [4]. As we can see, there are several wireless technologies allowing today to deploy wireless sensor network. They are suitable for different applications with regards to bandwidth and range, as it can be observed in Fig. 2. Most of IoT and M2M (Machine to Machine) solutions require long-range communication link with low bandwidth and are not well covered with traditional technologies. There is an appropriate time and place for LPWAN technology, which is quite good for these emerging sensor applications mentioned. Non-cellular wireless technologies such as Wi-Fi, Cellular Bluetooth, Zigbee, NRF, and ANT are not ideal for connecting low power devices distributed over large geographical areas. The transmission range of these technologies is limited to a few hundred meters. It can be enhanced by using of multi-hop networking principals and extending their geographical coverage, but there are significant challenges on reliability if used over large areas. WLANs typically have shorter coverage area and higher power consumption [5]. In previous years, cellular technologies like Narrowband-IoT and GPRS have been the only option designed to provide long-range services for various applications. This is a consequence of the mobile phone's success around the entire world. However, these options are being expensive and high energy consuming solutions and are not widely supported by different applications and vendors. The cellphone market covers a completely broad and general mission that is not optimal for every single IoT application. To overcome these problems, long range Low Power Wide Area Network (LPWAN) technologies like LoRa, Sigfox, NB-IoT, RPMA, NB-FI, etc., have developed and merged into successful solutions in different countries and circumstances [3].

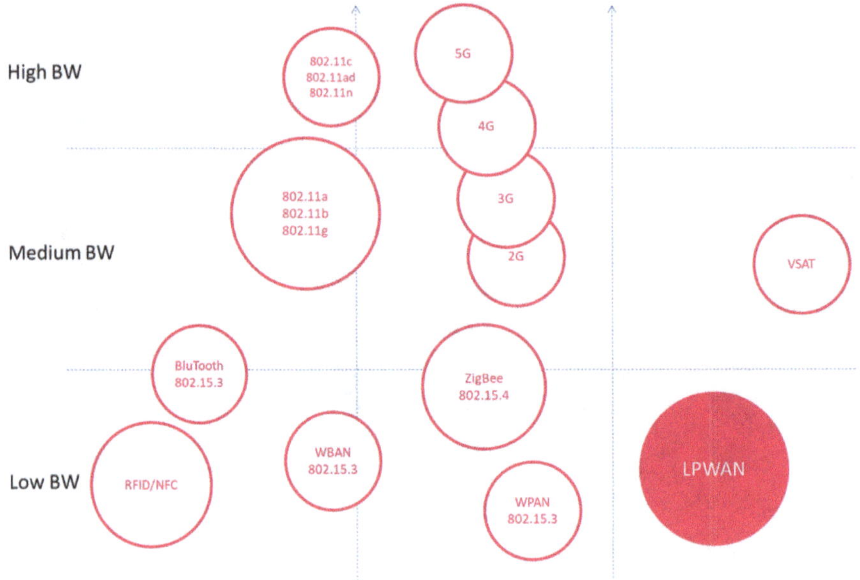

Fig. 2 Wireless technologies

The unique set of LPWAN characteristics in comparison with other wireless technologies are used here in the best way possible suited for IoT and M2M needs. The scheme in Fig. 3 shows the benefits of low-power network relative to traditional ones [6].

There are three fundamental technical features of Low Power Wide Area Networks that meet the requirements of IoT:

- Geographical range. LPWAN is designed for wireless data transport between devices separated by distances in the range of kilometers and not only meters.
- The amount of data transmitted. LPWAN's idea is to regulate the non-constant transport of lesser amounts of data due to the process of nature.
- Low power consumption. The protocol is based on the use of devices with batteries that last for years instead of weeks and months.

The potential of LPWAN is huge. There are several LPWAN technologies present on the market now. They differ from one another by frequency, bandwidth, RF modulation approach, spectrum utilization algorithms, and some different technical features that each technology concerns. And as a result, they have their key points to consider when choosing the right technology for IoT [6]. Now, there are several LPWANs technologies candidates to provide IoT-like connectivity for applications, such as wireless sensor/actuator networks previously mentioned, advanced infrastructure for smart metering, public lighting, smart cities, and more. The main ones are: LoRa, Wi-SUN, SIGFOX, RPMA, Weightless, DASH-7, INGENU and NBIoT. Each one of them has its pros and cons, regarding security, coverage, performance

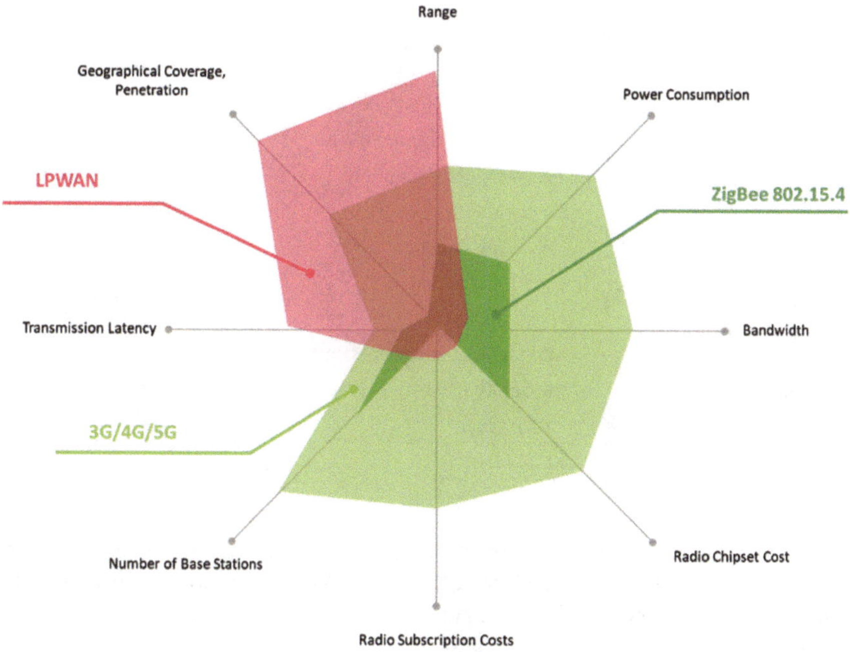

Fig. 3 Pros and cons from different wireless technologies

in non-line of sight conditions, network topology, business model, implementation/deployment/operation complexity, the data rate for up and downlink costs and other aspects. In spite of being of paramount importance when defining a connectivity technology for IoT, those aspects will not be explored extensively in this work due to lack of space and because they are out of the scope of this work, which is oriented to Physical Layer (PHY) evaluations and some extra evaluations that are highly important for any IoT implementation. Among those technologies, the most promising ones and which can be used in ISM (Industrial, Scientific, and Medical) and other free-use frequency bands are LoRa and Wi-SUN, as we will see later in the results [7].

It is not a trivial task to choose the better option for a specific application; thus, using a multicriteria analysis would help to give insights to the IoT project-leader. There are different attributes that could be considered: network characteristics as network topology; other features as if the LPWAN is suitable for private networks, suitable for public networks, also if it is an open standard or there are hardware considerations, nevertheless, all of them would be used for this analysis. The attributes from each technology that will be considered are shown in Table 1. A brief description of each one is shown next:

- **Battery life**. It turns out to be a factor of high importance, since, for the user, it is extremely comfortable or, in its counterpart, uncomfortable at the time of using a

Table 1 Attributes contemplated for the analysis

Attribute	Units
Battery life	(years)
Minimum cost (currency unit)	($)
RX current consumption	(mA)
TX current consumption	(mA)
Bandwidth	(Hz)
Maximum data rate	(bps)
Uplink Data Rate	(bps)
Downlink data rate	(bps)
Maximum messages per day	(mess)
Maximum payload length	(byte)
Urban range	(km)
Rural range	(km)

service of this type. Combining the cost that can be reduced to the final investment that a company must make to have a service or system functional, it is one of the most crucial factors. It is a benefit attribute: the more time the user can use every single device, the better for the company.

- **Minimum cost**. It is given in terms of installation and maintenance of a single unit inside the process. As it is expected, it will be considered as a cost attribute expecting the minimum possible for the cost.
- **RX current consumption**. It is the electrical current that the device that is working as a receiver demands to complete the communication process successfully (it affects the battery life). The less current the device demands, the best for the durability of operation without substituting the battery: it is a cost attribute.
- **TX current consumption**. It is the electric current that the device that is working as a transmitter demands to complete the communication process successfully (it affects the battery life). As in the RX current consumption case, the less current the device demands, the best for the durability of operation without substituting the battery: it is a cost attribute.
- **Bandwidth**. Describes the range in frequency that the LPWAN can operate and, therefore, how the devices must be tuned to properly work accordingly to each technology. It could be either licensed or unlicensed spectrums, depending on which benefits the network the most. It is a benefit attribute.
- **Maximum data rate**. The manufacturer usually provides a nominal value that could be understood as the maximum data rate (average) that can be implemented. For tasks where velocity is a fundamental factor, it must be considered as a benefit attribute.
- **Uplink data rate**. The uplink communication is often the most important part of this process due to the usual needs of a smart city network or a 4.0 industry process. It is the communication process where the field devices such as sensors send data. For the reasons previously described, it is a benefit attribute.

- **Downlink data rate**. Not always needed, but the option that will be selected has to be robust; thus, it allows the center gateway to send messages if needed to every single node whereas to acknowledge a previous message or as a signal that conditions the field device operability inside the process. As in the case of uplink data rate, it is a benefit attribute.
- **Urban range**. It is an estimated distance given in kilometers or miles to have a closer idea of how far the network devices could be inside an urban environment. It is a benefit attribute to guarantee the accurate communication process between devices.
- **Rural range**. It is an estimated distance given in kilometers or miles to have a closer idea of how far the network devices could be inside a rural environment. It is the same case as in urban range: it is a benefit attribute.
- **Receiver sensitivity**. This is the traditional measure of receiver performance. It is defined as the minimum received optical signal power at a specific Bit Error Rate in the back-to-back configuration. This parameter shows the quality of receiver design [3]. As it shows how good is the device parametrized as a receiver to get messages, it will be considered as a cost benefit.
- **Default transmitter power**. It could be interpreted in diverse ways for the LPWAN benefit; nevertheless, it will be considered as a cost attribute due to energy factors.
- **Maximum coupling loss**. Maximum Coupling Loss has been chosen by 3GPP as the metric to evaluate coverage of radio access technology. In theory, it can be defined as the maximum loss in the conducted power level that a system can tolerate and still be operational (defined by a minimum acceptable received power level). MCL can be calculated as the difference between the conducted power levels measured at the transmitting and receiving antenna ports as the reference point; the directional gain of the antenna is not considered when calculating MCL.
- **Nodes per gateway**. It not necessarily indicates a better performance of LPWAN technology. A considerable number does not mean that the performance with a few nodes will be the same as with the maximum. Nevertheless, there will be cases where an enormous number of devices need to be connected to the network; thus, it is a benefit attribute.

With increasing number of devices connected to the Internet, there is a renewed thrust on developing low-cost and low data-rate wireless technologies. SIGFOX built the first modern LPWAN. This became a reality when radio technology was becoming less expensive, and the tools for integrating applications were becoming easier to use. All these things have driven the emergence of LPWAN technologies. There are many commercially available LPWAN technologies that will be briefly described next [5].

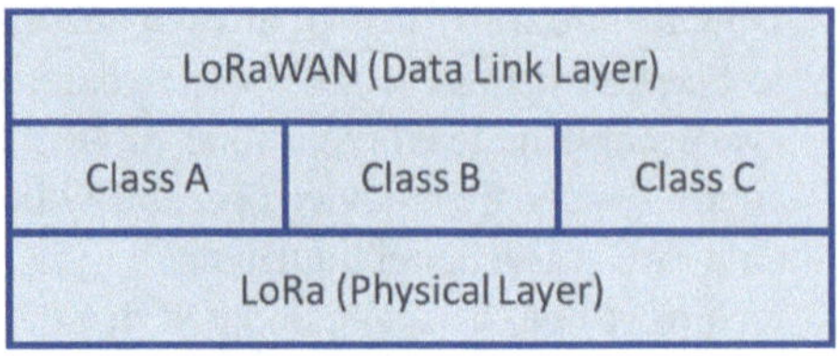

Fig. 4 LoRaWAN protocol architecture

1.1 LoRa

One of the emerging protocols in this scope is the Long- Range Wide-Area Network (LoRaWAN). LoRaWAN is one of the most popular and successful technologies in the LPWANs space. LoRaWAN consists of a protocol stack specified by LoRa Alliance that operates over the Long Range (LoRa) physical layer on unlicensed bands. The LoRaWAN features are low data rate, low complexity, different operating classes for various applications. It may exhibit an immense number of nodes per single gateway. In 2015, LoRaWAN v1.0 was declared by LoRa Alliance. In October 2017, LoRa Alliance announced LoRaWAN v1.1 [8].

LoRa is a physical layer technology that enables long range, low data rate and low power wireless communication. It is an unlicensed band technology that modulates the signals in the sub GHz ISM band using the spread spectrum technique. It was developed by Cycleo and commercialized by Semtech, Microchip, and others. LoRa can also be applied in P2P communications between nodes. LoRaWAN constitutes a data link layer protocol above the LoRa physical layer protocol, as shown in Fig. 4.

1.2 DASH7

Another well-defined standard is the DASH7 Alliance Protocol (D7AP). D7AP is an open-source Wireless Sensor and Actuator Network protocol (WSAN). It operates in the Sub-1 GHz bands based on the ISO/IEC 18,000–7 standard and specified by DASH7 Alliance. The ISO/IEC 18,000–7 standard defines the parameters of the active air interface communication at 433 MHz D7AP inherits the default parameters from ISO 18,000–7 and extends the standard by specifying a complete communication stack from the application layer to the physical layer. This stack contains an elevated level of functionality optimized for active RFID and WSAN. Also, it ensures interoperability among different operators. Conversely to legacy RFID systems, D7AP supports tag-to-tag communication. In 2013, the D7AP was announced by the DASH7 Alliance. In April 2016, the DASH7 Alliance published D7AP 1.1 [8].

1.3 NB-Fi (Narrowband Fidelity)

It is an open wireless protocol technology designed ground-up for machine-to-machine communication by WAVIoT. It covers a full stack of OSI model layers from the physical layer to the application layer. Like Sigfox, data received to the gateways is stored on a cloud and then accessed by end-users via the application.

NB-FI operates in 433, 500, 868 and 915 MHz bands, which makes it suitable for bypassing the limitations of regulations in different regions. Such for RFID applications in Europa can use 868 MHz, and USA can use 915 MHz, but you cannot use 915 MHz in Europa and vice versa in the USA. It can control up to 2 million devices with a single gateway and can achieve maximum coverage of 50 km in rural areas. Being bidirectional, its communication uses 256-bit encryption [8].

1.4 Sigfox

In today's unlicensed LPWAN market, Sigfox is the follower technology of LoRa. It uses Ultra Narrow Band modulation and operates at 200 kHz of a total band to transmit messages, where each message requires only 100 Hz bandwidth to be transmitted. Having a data rate between 100 and 600 bps, long distances can be achieved while being very robust against the noise. Sigfox, being a lightweight protocol, can handle its small messages (12 Bytes payload) very efficiently. As there is less data to be transmitted, less energy is needed, and hence battery life is enhanced. Due to less overhead, more space will be available for the user data to be transported, and hence this will increase the network capacity of the network to a greater extent. Sigfox limits its customers to transmit only 140 messages per day, and currently, it is available in 32 countries worldwide. Unlike LoRaWAN, multiple vendors supply it. Sigfox is suitable for remote metering and discrete data collection solutions for daily and hourly levels and values from various sensors [8].

1.5 Weightless

Weightless is a set of LPWAN technologies designed by the Weightless Special Interest Group. There are three several types proposed by the group: Weightless N, Weightless P, and Weightless W. Since Weightless-N and Weightless-W are focused on ultra-low cost and TV whitespace, respectively, it is Weightless-P which is more like LPWAN technologies. Adaptive data rates from 100 bps to 200 kbps, flexible channel assignment, and time-synchronized base stations enable optimization of power usage and efficient use of spectrum. It operates in 12.5 kHz bandwidth. FEC (Forward Error Correction), low latency, automatic retransmission request, adaptive channel coding, and other features make this technology extremely competitive to other similar technologies [8].

1.6 NB-IoT (Narrowband Internet of Things)

It is based on existing LTE functionalities. This standard is optimized to achieve low cost, ultra-low complexity, and indoor improvement coverage. It supports a substantial number of devices per cell-site sector, low-power consumption, low-data-rate, and latency less than 10 s. NB-IoT has been developed to operate in three modes: in-band, guard-band, and stand-alone. Whereas LoRaWAN and DASH7 use unlicensed frequencies that are globally available, NB-IoT uses the same frequencies as LTE, which is implemented worldwide. Those standards are developed to satisfy the needs of constrained IoT communication requirements. However, they almost consider static interconnected things and pay less attention to the mobility of things. There are three types of deployment of NB-IoT: In-Band, Guard-Band, and Standalone. NB-IoT has incredibly good coverage because it relies on 4G, so it would work well indoors and in dense urban areas. Also, having a faster response time compared to Lora increases its quality of service. Unlike Lora, it does not need a gateway to operate. It uses cells that would cover up to 50,000 end devices [8].

1.7 EC-GSM-IoT

It is another technology released by 3GPP. It is an optimized version of GSM to fulfill the IoT requirements. The current GSM network is re-used for deployment by a software upgrade without affecting existing GSM deployments. It is realized by mapping a new set of control and data channels over legacy GSM. The primary features are new logical channels designed for extended coverage. In EC-GSM-IoT, multiple access is accomplished as in 2G, i.e., by TDMA (Time-Division Multiple Access) and FDMA (Frequency-Division Multiple Access). TDMA divides the transmissions into different time slots, thereby allowing more than one user to utilize a single frequency band. Like GSM, the bandwidth per channel in EC-GSM-IoT is 200 kHz. FDMA divides communication over different frequency bands. Therefore, by multiplexing information in different time slots and bands, multiple simultaneous transfers can occur. It operates in a total system bandwidth of 2.4 MHz and is claimed to have the feature of roaming as it carries the architecture of GSM [8].

1.8 LTE Cat-M1

It is also a part of the same 3GPP Release standard. With uplink and downlink speeds of 375 kb/s in half duplex mode, Cat M1 specifically supports IoT applications with low to medium data rate needs. At these speeds, LTE Cat M1 can deliver remote firmware updates over-the-air (FOTA) within reasonable timeframes, making it well-suited for critical applications running on devices that may be deployed in the field for extended periods of time. The battery life of up to 10 years on a single charge or battery installation in some use cases also contributes to lower maintenance costs for deployed devices, even in locations where end devices may not be connected directly to the power grid. Compared to NB-IoT and EC-GSMIoT, LTE Cat M1 is ideal for mobile use cases because it handles hand-over between cell towers, much like high-speed LTE. Another benefit is the support of voice functionality via VoLTE (voice over LTE), which means it can be used for applications requiring a level of human interaction, such as for certain health and security applications.

1.9 RPMA (Random Phase Multiple Access)

It is a Low Power WAN technology developed by Ingenu and currently manufactured by a single company called u- blox. Like LoRa and Sigfox, RPMA is also a technology designed specifically to fulfill the IoT vision. But, unlike LoRa and Sigfox, it operates at 2.4 GHz ISM frequency band. Having a downlink data rate of 31 Kbps and an uplink data rate of 15.6 Kbps makes this technology compatible with IoT use. The most challenging property of RPMA is its link budget, which is 176 dB. Ingenu claims they provide IoT connectivity of RPMA to 29 countries worldwide.

2 Methodology

As we can see, there exists a lot of LPWAN technologies in the market. And probably, as the years come by, more will exist. The biggest issue with this tendency is the practical implementation. The different companies that develop these networks and give support to them obviously will sell their product as best as they can. And there is when the problem arrives: just the protocols that have been proved that works under the environments described can be contemplated for a huge task as a smart city or to provide a multi-purpose network on a factory. So, by now, we will consider just a few technologies. They will be used accordingly to the documentation that is available for the parameters previously described. And then, a multicriteria analysis will take place.

There are different techniques that could be used for choosing the best option; nevertheless, TOPSIS will be used. The TOPSIS technique is well known in textbooks, and it is a pragmatic method for dealing with real life multiple criteria decision-making problems. It helps decision-makers and engineers compare and rank a set of alternative decisions. In this method, the ranking of alternatives is based on the shortest distance from the ideal solution and the farthest from the negative ideal solution. The TOPSIS procedure includes the following steps [9]:

- **Step 1**: Determination of the weight of criteria and construction of the decision matrix. The primary step in the TOPSIS algorithm is creating a decision matrix and determining the weight of the criteria. Weighting the criteria is one of the most important and complicated steps of the MCDM methods. In this step, relative weights must be assigned not only to each criterion but to its quantitative and qualitative values based on their importance. Since weighting the criteria is the main step in the decision-making process, high precision is required to determine weights for each criterion and its values. There are a variety of methods that have been proposed to determine weights in MCDM methods. Usually, they are classified as weighting approaches into subjective or objective. The weights in the subjective methods are determined based on preference information of criteria, subjective opinions, and decision-maker's knowledge. In this paper, the Ratio Estimation procedure, which is a subjective method, is assigned to the most important criterion. Correspondingly, smaller weights are assigned to the remaining 8 criteria with lower order until a score is assigned to the least important criterion. The ratio is calculated by dividing each weight to the lowest weight. The ratio, therefore, is equal to w_j/w^*, where $w*$ is the lowest score assigned to the least important criterion and w_j is the score for the jth criterion. Finally, the weights are normalized by dividing each one by the total. In the next step, a decision matrix $\mathrm{X} = \left(x_{ij}\right)^{m\mathrm{xn}}$ is constructed using the data and the ratio estimation procedure.
- **Step 2**: Calculation of the normalized decision matrix. There are benefit attributes and cost attributes in a Multi Criteria Decision Making (MCDM) problem. To transform various attribute dimensions into non-dimensional units and facilitate inter-attribute comparisons, several known standardized equations are introduced to normalize each attribute value x_{ij} in decision matrix $\mathrm{X} = \left(x_{ij}\right)^{m\mathrm{xn}}$. The following equation is the most frequently used method of calculating the normalized value r_{ij}:

$$R = \left(r_{ij}\right)_{mxn} = \begin{pmatrix} r_{11} & \cdots & r_{1n} \\ \vdots & \ddots & \vdots \\ r_{m1} & \cdots & r_{mn} \end{pmatrix} \tag{1}$$

where

$$r_{ij} = \frac{x_{ij}}{\sqrt{\sum_{i=1}^{m}\left(x_{ij}\right)^2}} \tag{2}$$

For benefit attribute x_{ij}, i ∈ M, j ∈ N, and

$$r_{ij} = 1 - \frac{x_{ij}}{\sqrt{\sum_{i=1}^{m} \left(x_{ij}\right)^2}} \tag{3}$$

for cost attribute x_{ij}, i ∈ M, j ∈ N.

- **Step 3**: Calculation of the weighted normalized decision matrix. In the third step, the weighted normalized value vij is calculated by multiplying the normalized decision matrix by the normalized weights of criteria:

$$v_{ij} = w_j \ldots \times r_{ij} \tag{4}$$

where $i = 1, \ldots, m$; $j = 1, \ldots, n$. and m is the number of the attribute value in each criterion, n is the number of criteria, and w_j is the normalized weight of the jth criterion

$$w_j = \frac{W_j}{\sum_{j=1}^{n} W_j} \tag{5}$$

Subject to

$$\sum_{j=1}^{n} w_j = 1 \tag{6}$$

Note that W_j is the original weight assigned to each criterion.

- **Step 4**: Determination of the positive ideal solutions and negative ideal solutions. The positive ideal solution minimizes the cost criteria and maximizes the benefit criteria; on the contrary, the negative ideal solution maximizes the cost criteria and minimizes **the** benefit criteria. The equations are as follows:

$$A^+ = \left[v_1^+, \ldots, v_j^+, \ldots, v_n^+\right]$$
$$A^- = \left[v_1^-, \ldots, v_j^-, \ldots, v_n^-\right] \tag{7}$$

where A^+ denotes the positive ideal solution and A^- denotes the negative ideal solution and

$$\begin{cases} v_j^+ = \max\left(v_{ij}\right), i = 1, 2, \ldots, m \\ v_j^- = \min\left(v_{ij}\right), i = 1, 2, \ldots m \end{cases}$$
$$If j^{th} criterion is a benfit$$

$$\begin{cases} v_j^+ = \min(v_{ij}), i = 1, 2, \ldots, m \\ v_j^- = \max(v_{ij}), i = 1, 2, \ldots m \end{cases}$$
$$If j^{th} criterion is a \cos t \tag{8}$$

where v_{ij} denotes the attribute values of each cell for the *j*th layer.

- **Step 5**: Calculation of the separation of each alternative from the positive ideal solution and the negative ideal solution. In this step, the separation of each alternative from the positive ideal solution and the negative ideal solution is calculated and then two different layers S_i^+ and S_i^- are created. The equations are as follows. The separation from the positive ideal solution for each alternative is given as:

$$S_i^+ = \sum_{j=1}^{n} \left| v_{ij} - v_j^+ \right| = \sum_{j=1}^{n} D_{ij}^+$$
$$S_i^- = \sum_{j=1}^{n} \left| v_{ij} - v_j^- \right| = \sum_{j=1}^{n} D_{ij}^- \tag{9}$$

- **Step 6**: Calculation of the relative closeness to the positive ideal solution. The relative closeness of the *i-th* alternative A_j with respect to the positive ideal solution can be calculated as

$$C_i^+ = \frac{S_i^-}{S_i^+ + S_i^-} \tag{10}$$

where $0 \le c_i^+ \le 1, i = 1, 2, \ldots, m$.

- **Step 7**: Determination of the rank of the alternatives according to the relative closeness. In this step, the LPWAN sites can now be ranked by the descending order of the value of C_i^+. The best sites are those that have higher values of C_i^+ and since they are closer to the positive ideal solution, they are preferable and must be chosen.

Also, we will compare the provided method with the Multi-Objective Optimization based on Ratio Analysis (MOORA). It provides advantages; the biggest one is that there is a very minimal number of mathematical calculations used by this method; therefore, it is computationally faster than other multicriteria analysis that has been used in the last years. For MOORA, **steps 1, 2 and 3 are the same** as previously defined in the TOPSIS method. The differences start in step 4 for this methodology.

- **Step 8**. Estimate the Y_i value, which is the summation of beneficial attributes and non-beneficial attributes as follow:

$$Y_i = \sum_{j=1}^{g} v_{ij} - \sum_{j=g+1}^{n} v_{ij} (j = 1, 2, \ldots, n) \tag{11}$$

- **Step 9**. Determination of the rank of the alternatives according to the values of Y_i. As in TOPSIS, it should be done in a descending way.

2.1 Weights Calculation

The way that the weights will be calculated will be using the **Shannon Entropy Weight Method**. Entropy has been widely employed in social and physical sciences, economics, spectral analysis, and language modeling, as a few typical practical applications of entropy. More specifically, entropy evaluates the expected information content of a certain message. It can be considered as a criterion for the degree of uncertainty represented by a discrete probability distribution. Entropy can be employed in the process of decision making because it measures existent contrasts between sets of data and clarifies the average intrinsic information transferred to the decision maker. The method is performed in three steps.

- **Step 1**. Normalization of the arrays of decision matrix (performance indices) to obtain the project outcomes p_{ij}

$$p_{ij} = \frac{x_{ij}}{\sum_{i=1}^{m} x_{ij}} \tag{12}$$

- **Step 2**. Computation of the entropy measure of project outcomes using the following equation:

$$E_j = -k \sum_{i=1}^{m} p_{ij} ln(p_{ij}) \tag{13}$$

  In which $k = 1/ln(m)$.
- **Step 3**. Defining the objective weight based on the entropy concept

$$w_j = \frac{1 - E_j}{\sum_{j=1}^{n} (1 - E_j)} \tag{14}$$

  and finally, the ranking based on the score of each option can be done.

As the implementation of a LPWAN also needs a big monetary investment, there are several factors that are important for the project viability as the rate of return, usability and human factors that will affect each process involved with the LPWAN networks. There are diverse ways to understand the concept of rate of return on investment. Rate of return is the interest earned on the unpaid balance of an amortized loan. In other words, the rate of return is the break-even interest rate i^* at which the present worth of a project is zero or, in a mathematical expression

$$PW(i^*) = PW_{cif} - PW_{cof} \tag{15}$$

where *cif* stands for cash-in flows, and *cof* stands for cash-out flows. More specifically, it can be observed as

$$PW(i) = A_0 + \frac{A_i}{i + i^*} + \ldots + \frac{A_N}{(1+i)^N} = 0 \tag{16}$$

Thus, the internal rate of return is the interest rate charged on the unrecovered project balance of the investment such that, when the project ends, the unrecovered project balance is zero [10].

Additionally, other key factors to have a multicriteria analysis are the safety of the process. As an example, Mexico has The Mexican Official Standards issued by the Ministry of Labor and Social Security to determine the minimum safety necessary conditions, health, and the working environment, to prevent accidents and occupational diseases. NOM 035 aims to establish the elements to identify, analyze, and prevent psychosocial risk factors, as well as to promote a favorable organizational environment in the workplace.

According to the field of application, NOM 035 applies throughout the national territory and applies in all work centers. However, the provisions of this standard apply according to the number of workers who work in the workplace. There are three levels:

- Work centers where up to 15 workers work
- Work centers between 16 and 50 workers
- Work centers with more than 50 workers.

So, the workplace must determine what level it is at so that it follows the provisions that correspond to it according to the number of workers it employs. NOM 035 comes in two stages, which define the requirements with which the work centers will have to comply:

- Politics; prevention measures; the workers' identification exposed to severe traumatic events, and the dissemination of information.
- Identification and analysis of psychosocial risk factors; the evaluation of the organizational environment; control measures and actions; the practice of medical examinations, and records.

The norm considers the evaluation of the conditions in which the activities are carried out (environment and conditions of the organization), in no case is the stress on the worker, or his psychological profile evaluated [11]. Job rotation, i.e., a lateral transfer of an employee between different jobs in the same organization without a change in the hierarchical rank or salary grade, is often considered as a key instrument for management development, as lateral assignments typically coincide with a change in job content and necessary skills. But, lateral transfers also occur because underperforming employees are reallocated to different jobs to improve the quality of the person-job match or to motivate employees by giving them new tasks in the organization. In many firms, job transfers for underperforming employees are also a core element of their talent management strategies. Job rotation can be beneficial for several reasons:

- It speeds up skill acquisition and helps employees to learn on the job ('employee learning'),
- It motivates employees as it keeps work interesting ('employee motivation'),
- It helps firms to learn more about different dimensions of their employees' competencies ('employer learning') and improves the match quality of employees to jobs [12].

3 Results

In order to develop the math analysis described in the previous description, an Intelligence dashboard was designed and developed in JavaScript, using libraries for plotting as Plotly.js and also, with different resources to get visual content in HTML, CSS, with frameworks such as Bootstrap and AdminLTE templates for the front-end design with its own libraries.

The intelligence dashboard consists of nine modules. The first of them consists in a form where the user can put the attributes mentioned in the methodology section to be able to run the algorithm for TOPSIS in module 3, and the comparative with MOORA in module 4 as will be described further in this section. This form can be observed in Fig. 5a. Next, in the second module, we can visualize all the data that has been typed in module 1, as shown in Fig. 5b. The user can insert as many LPWANs as they want to compare priority for the dashboard development: give a wide use for the user to perform as many analyses as needed.

Module 3 shows three different sections that provide the user different information that can be observed in Fig. 5c. The first container, named "Results" allows performing the analysis with the desired number of LPWAN's. The number that must be used must be equal or less than the total of alternatives given by the user in module 1. If the alternatives number selected is less than the typed by the user, the algorithm will give priority to the options that were inserted first, i.e., the ones that are at the top of the table. The user can perform as many analyses as he wants. The results of the TOPSIS analysis will be shown in different containers, and the user can close the containers that will be no longer required as the data is interpreted, saved, or processed by another module. The list that is shown in each result container shows the number of LPWAN that corresponds with the number of the row that it occupies in the module's 2 table. Next, we can observe the percentage that the TOPSIS method provides for each option and thus, which is better according to the provided weights.

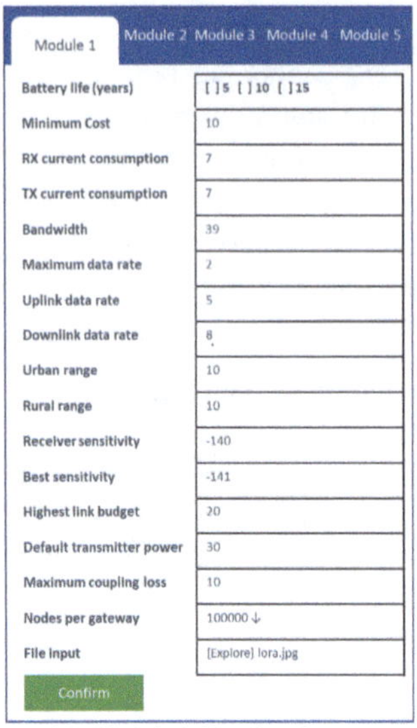

(a) Module 1. User interface to enter attributes for TOPSIS

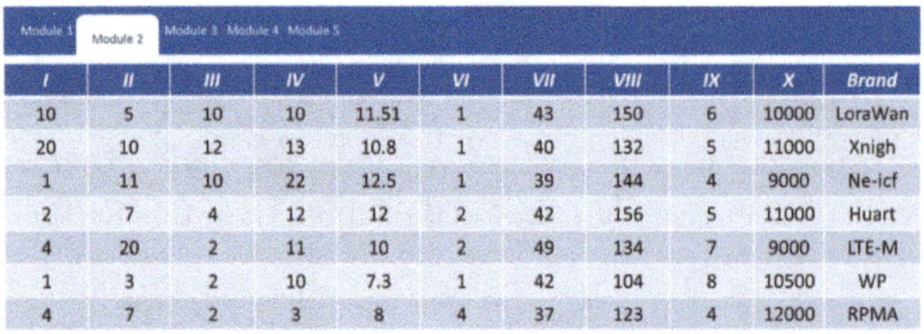

I	II	III	IV	V	VI	VII	VIII	IX	X	Brand
10	5	10	10	11.51	1	43	150	6	10000	LoraWan
20	10	12	13	10.8	1	40	132	5	11000	Xnigh
1	11	10	22	12.5	1	39	144	4	9000	Ne-icf
2	7	4	12	12	2	42	156	5	11000	Huart
4	20	2	11	10	2	49	134	7	9000	LTE-M
1	3	2	10	7.3	1	42	104	8	10500	WP
4	7	2	3	8	4	37	123	4	12000	RPMA

(b) Data entered by user

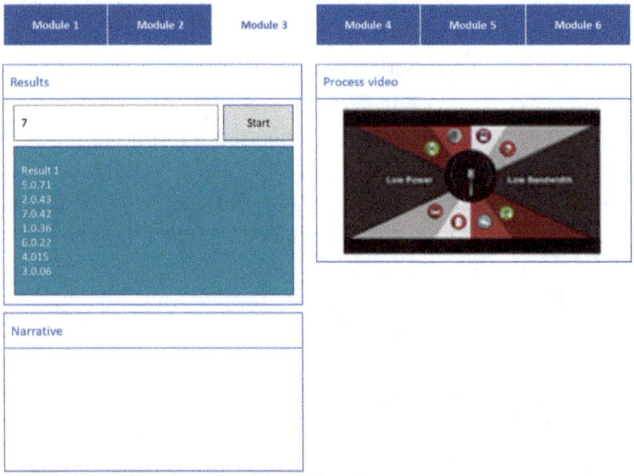

(c) Analysis performed with data entered by user

Fig. 5 First intelligence dashboard part for multicriteria analysis

The second container contains a narrative of the upper analysis result. In other words, it expresses the obtained results in words for the user.

Next, module 4 provides two different charts. The first graph shows a financial project supposed for each alternative, as it can be observed in Fig. 6a. It is a simple

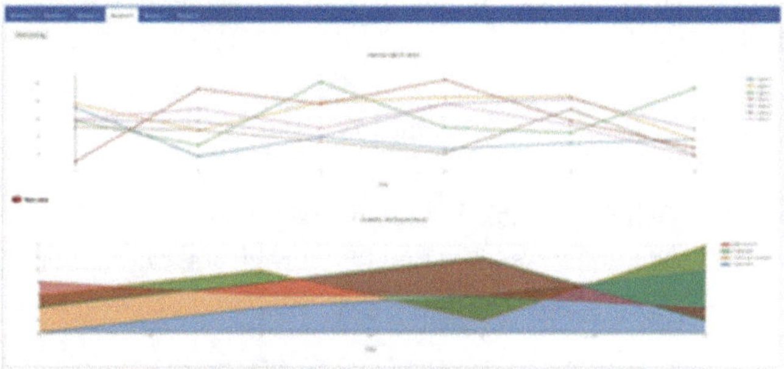

(a) Financial comparison

(b) Graphic representation of real world

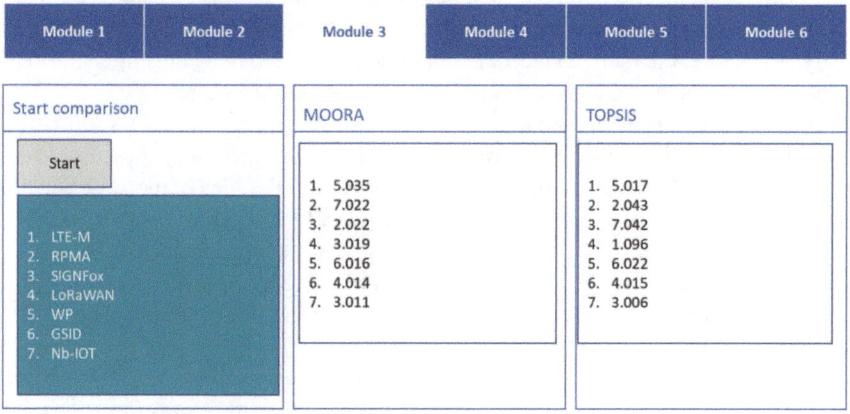

(c) Comparison against MOORA method

Fig. 6 Second intelligence dashboard part for multicriteria analysis for different LPWAN's model that gives the form of the plot, but it provides an idea of the processes that the financial investment in the implementation of one of the chosen technologies could have as the years pass and the company is holding up the industry 4.0 project. When the user presses the "Start plotting button", the first that the user can visualize is the starting investment, which is reflected as a negative number in the y-axis of the plot. This number can be interpreted as the basic initial inversion of the project, which is given by the following expression:

$$F_0 = \frac{Nodes}{4} * \min_{\cos t} \tag{17}$$

Then, as the "More data" switch is enabled, more data is generated and showed in the plot for each option. The simple model based on three attributes that impact positively the economics industry is the following expression:

$$F_{i+1} = F_i + 46000jb + 3500kr + 3500lt \tag{18}$$

where

F_{i+1}	Is the future balance
F_i	Is the actual balance.
b	Battery attribute
l	TX current consumption
r	RX current consumption
j, k, t	Random economic variables.

Under these conditions, it is possible to observe when the initial investment is fully recovered and, as more data is generated, the incomings that the company will get as the process continues. The subsequent plot (Fig. 6b) shows the usability predicted for each generation that will be part of the industry. As the years pass, the capability of older generations to adapt to LPWAN technology will decrease dramatically. Therefore, the maintenance and implementation will be younger generations' responsibility due to the learning curve that all those tasks require.

In module 5, is provided a visual simulation of the process. It allows to interact with a basic 3D industrial 4.0 scenario to understand the benefit of the process, as can be viewed in Fig. 6b. Its use is only illustrative to get a better idea of the huge benefits of a LPWAN. The user can navigate throw this world via its mouse and keyboard.

Module 6 shows the comparison of the last result that was obtained in module 3 with a MOORA analysis. First, the left container has the images provided in module 1 and is shown in an ascending way for the user comfortability. The number that is to the images left represents each technology. Thus, allowing a better experience as it can be observed in Fig. 6c. The middle and right containers have the TOPSIS and MOORA analyses results, respectively. The result is arranged in a list form. The first number corresponds to the LPWAN in the first container. The second is the score r that represents how good is the alternative, where

$$r \in \mathbb{R}, 0 < r < 1 \tag{19}$$

for both techniques. Now that the final dashboard has been discussed, the analyses with the seven technologies will be treated. The decision matrix build for the following results is shown in Table 2.

As were discussed in the previous section, first, the TOPSIS analyses take place in Module 3. In the following paragraphs, the results of each step of the method will be discussed. The normalized decision matrix, which is obtained from applying the normalization equation showed in the previous section, is shown in Table 3. Each row and column correspond to the same as in Tables 2 and 4.

Table 2 Decision matrix

Attribute	LoraWAN	Sigfox	NB-IoT	GSM	LTE-CAT	Weightless	RPMA
Battery life	10	10	10	5	5	10	10
Minimum cost	5	5	10	15	20	5	20
RX current consumption	10	20	46	50	50	13	85
TX current consumption	22	45	80	70	100	45	50
Bandwidth	125	200	180	200	1080	12.5	1000
Maximum data rate	50	0.1	200	240	1000	100	624
Uplink Data Rate	27	0.6	62.5	240	1000	100	634
Downlink data rate	27	0.6	27	240	1000	100	156
Maximum messages per day	3	6	0.6	2	2	1	2
Maximum payload length	18	30	10	15	20	20	20
Urban range	−137	−147	−137	−140	−142	−145	−143
Rural range	−137	−129	−142	−140	−123	−124	−137
Receiver sensitivity	157	146	164	154	156	147	176
Default transmitter power	20	15	23	23	20	20	20
Maximum coupling loss	157	162	160	160	159	160	161
Nodes Per Gateway	1,000,000	1,000,000	200,000	200,000	200,000	500,000	200,000

Table 3 Normalized decision matrix

0.43	**0.43**	**0.43**	**0.21**	**0.21**	**0.43**	**0.43**
0.86	0.86	0.71	0.57	0.42	0.86	0.42
0.92	0.84	0.62	0.59	0.59	0.89	0.31
0.87	0.73	0.52	0.58	0.41	0.73	0.70
0.08	**0.13**	**0.12**	**0.13**	**0.71**	**0.01**	**0.66**

Table 4 Weighted normalized decision matrix

0.03	**0.03**	**0.03**	**0.01**	**0.01**	**0.03**	**0.03**
0.04	0.04	0.03	0.02	0.02	0.04	0.02
0.05	0.04	0.03	0.03	0.03	0.05	0.02
0.04	0.04	0.03	0.03	0.02	0.04	0.04
0.01	**0.01**	**0.01**	**0.01**	**0.04**	**0.00**	**0.04**
0.00	**0.00**	**0.01**	**0.02**	**0.07**	**0.01**	**0.04**
0.00	**0.00**	**0.00**	**0.02**	**0.08**	**0.01**	**0.05**
0.00	**0.00**	**0.00**	**0.01**	**0.06**	**0.01**	**0.01**
0.03	**0.06**	**0.01**	**0.02**	**0.02**	**0.01**	**0.02**
0.01	**0.02**	**0.01**	**0.01**	**0.02**	**0.02**	**0.02**
0.03	0.03	0.03	0.03	0.03	0.03	0.03
0.04	0.04	0.04	0.04	0.04	0.04	0.04
0.02	**0.02**	**0.02**	**0.02**	**0.02**	**0.02**	**0.03**
0.02	**0.01**	**0.02**	**0.02**	**0.02**	**0.02**	**0.02**
0.03	**0.03**	**0.03**	**0.03**	**0.03**	**0.03**	**0.03**
0.05	**0.05**	**0.01**	**0.01**	**0.01**	**0.03**	**0.01**

After that, the weights provided by the experts' group is applied to get the weighted normalized decision matrix. We can observe that in some attributes, the resulting numbers for each LPWAN are very close due to the similarity that much of these technologies have in different technical aspects. Then, the determination of the positive ideal solution and negative ideal solution is performed by getting the A + and A- vectors, as we can see in Table 5. Each column is a different alternative, following the same numeration as in Table 2.

With the A+ and A− vectors, the distance that every LPWAN has from these ideal solutions is calculated, and it is shown in Table 6.

The results that were obtained are shown in Table 7, and the weights implemented for the algorithm can be visualized in Table 8 (already normalized).

We can observe that Weightless is the better option given the weights provided by the exerts group with a significant difference of 0.28 between this option an option number 2, RPMA. Next, the MOORA technique was implemented in Module 6. First,

Table 5 Positive ideal solution and negative ideal solution for the given weighted normalized decision matrix

A+	A−
0.0264	0.0132
0.0174	0.0353
0.0158	0.0473
0.0209	0.0448
0.0441	0.0005
0.0674	0.0000
0.0764	0.0000
0.0592	0.0000
0.0648	0.0065
0.0236	0.0079
0.0313	0.0327

Table 6 Separation of each alternative from the positive ideal solution and the negative ideal solution

S+	S−
7.5144	7.82229774
8.5975	8.905363616
8.4271	8.735003861
7.3771	7.684959386
9.4583	9.766228457
10.3969	10.70474564
7.0619	7.36983241

Table 7 TOPSIS results

Rank	LPWAN	Score
1	Weightless	0.71
2	RPMA	0.43
3	NB-IoT	0.42
4	LTE-M	0.36
5	GSM	0.22
6	LoraWAN	0.15
7	Sigfox	0.006

normalization of the arrays of decision matrix (performance indices) is obtained, and the results are shown in Table 9.

As it has been mentioned in the previous section, the weights for this technique will be obtained by applying the entropy method. The first step for computing this is multiplying the normalized decision matrix by the logarithm of the same element in the matrix. The result of the operation is in Table 10.

Table 8 TOPSIS weights

Attribute	Weight
Battery life	0.0619
Minimum cost	0.0412
RX current consumption	0.0515
TX current consumption	0.0515
Bandwidth	0.0619
Maximum data rate	0.0825
Uplink Data Rate	0.0928
Downlink data rate	0.0619
Maximum messages per day	0.0825
Maximum payload length	0.0412
Urban range	0.0515
Rural range	0.0619
Receiver sensitivity	0.0619
Default transmitter power	0.0412
Maximum coupling loss	0.0722
Nodes Per Gateway	0.0825

Table 9 Decision matrix normalization for MOORA

0.17	0.17	0.17	0.08	0.08	0.17	0.17
0.06	0.06	0.13	0.19	0.25	0.06	0.25
0.04	0.07	0.17	0.18	0.18	0.05	0.31
0.05	0.11	0.19	0.17	0.24	0.11	0.12
0.04	0.07	0.06	0.07	0.39	0.00	0.36
0.02	0.00	0.09	0.11	0.45	0.05	0.28
0.01	0.00	0.03	0.12	0.48	0.05	0.31
0.02	0.00	0.02	0.15	0.64	0.06	0.10
0.18	0.36	0.04	0.12	0.12	0.06	0.12
0.14	0.23	0.08	0.11	0.15	0.15	0.15
0.14	0.15	0.14	0.14	0.14	0.15	0.14
0.15	0.14	0.15	0.15	0.13	0.13	0.15
0.14	0.13	0.15	0.14	0.14	0.13	0.16
0.14	0.11	0.16	0.16	0.14	0.14	0.14
0.14	0.14	0.14	0.14	0.14	0.14	0.14
0.30	0.30	0.06	0.06	0.06	0.15	0.06

Table 10 Entropy weights method first part

−0.13	−0.13	−0.13	−0.09	−0.09	−0.13	−0.13
−0.08	−0.08	−0.11	−0.14	−0.15	−0.08	−0.15
−0.05	−0.08	−0.13	−0.13	−0.13	−0.06	−0.16
−0.07	−0.11	−0.14	−0.13	−0.15	−0.11	−0.11
−0.06	−0.08	−0.08	−0.08	−0.16	−0.01	−0.16
−0.04	0.00	−0.09	−0.10	−0.16	−0.06	−0.16
−0.02	0.00	−0.05	−0.11	−0.15	−0.06	−0.16
−0.03	0.00	−0.03	−0.13	−0.12	−0.08	−0.10
−0.13	−0.16	−0.05	−0.11	−0.11	−0.07	−0.11
−0.12	−0.15	−0.08	−0.11	−0.12	−0.12	−0.12
−0.12	−0.12	−0.12	−0.12	−0.12	−0.12	−0.12
−0.12	−0.12	−0.12	−0.12	−0.12	−0.12	−0.12
−0.12	−0.12	−0.12	−0.12	−0.12	−0.12	−0.13
−0.12	−0.10	−0.13	−0.13	−0.12	−0.12	−0.12
−0.12	−0.12	−0.12	−0.12	−0.12	−0.12	−0.12
−0.16	−0.16	−0.07	−0.07	−0.07	−0.12	−0.07

With the results in Table 10, it is possible to obtain the weights for each attribute with the equations discussed in the methodology section. They are shown in Table 11 with the results of each equation needed to accomplish the method.

After the weights were calculated, the final matrix can be obtained. The results of applying the weights to the normal decision matrix are observed in Table 12. It is worth mentioning that the blue rows are attributes that need to be maximized, whereas the red ones are minimal attributes.

Finally, the results of the method are shown in Table 13, with a descending list, where the first option is the best given the entropy weights and their respective obtained scores.

The scores are different between both methods, but the list is the same. Just two LPWAN's have a different position. Nevertheless, the first ranked option is the same. The weights obtained by the entropy method are shown in Table 14. This similarity could be understood as the weights, regardless of the method that was used, show a similar pattern. This can be observed in Fig. 7.

Last, but not least is fundamental to analyze the security and the energy consumption of the process. Anyway, this consideration is unique and different for each factory. In Module 8 different tools are given to do the Carbon footprint analyses, the Norm 035 rules to determine the risk level of the process, and a calculator to obtain the Job Rotation Index. All this data is different and must be entered by the operator.

For the carbon footprint, the aspects that are considered are:

- Aerial transport
  - Less than 4000 km

Table 11 Resulting weights for entropy method

Sum	Eij	1-Eij	W
−0.8283	0.6879	0.3121	0.0516
−0.7760	0.6445	0.3555	0.0588
−0.7557	0.6276	0.3724	0.0615
−0.8074	0.6706	0.3294	0.0544
−0.6306	0.5237	0.4763	0.0787
−0.6080	0.5049	0.4951	0.0818
−0.5539	0.4600	0.5400	0.0892
−0.4880	0.4053	0.5947	0.0983
−0.7518	0.6244	0.3756	0.0621
−0.8260	0.6860	0.3140	0.0519
−0.8450	0.7017	0.2983	0.0493
−0.8445	0.7013	0.2987	0.0494
−0.8443	0.7012	0.2988	0.0494
−0.8417	0.6990	0.3010	0.0497
−0.8451	0.7018	0.2982	0.0493
−0.7336	0.6092	0.3908	0.0646
		6.0510	1.0000

Table 12 Entropy weights applied to normal decision matrix for MOORA

0.02	**0.02**	**0.02**	**0.01**	**0.01**	**0.02**	**0.02**
0.01	0.01	0.02	0.03	0.03	0.01	0.03
0.01	0.01	0.02	0.03	0.03	0.01	0.04
0.01	0.01	0.03	0.02	0.03	0.01	0.02
0.01	**0.01**	**0.01**	**0.01**	**0.06**	**0.00**	**0.05**
0.00	**0.00**	**0.01**	**0.02**	**0.07**	**0.01**	**0.04**
0.00	**0.00**	**0.00**	**0.02**	**0.07**	**0.01**	**0.05**
0.00	**0.00**	**0.00**	**0.02**	**0.09**	**0.01**	**0.01**
0.02	**0.05**	**0.00**	**0.02**	**0.02**	**0.01**	**0.02**
0.02	**0.03**	**0.01**	**0.01**	**0.02**	**0.02**	**0.02**
−0.02	−0.02	−0.02	−0.02	−0.02	−0.02	−0.02
−0.02	−0.02	−0.02	−0.02	−0.02	−0.02	−0.02
0.02	**0.02**	**0.02**	**0.02**	**0.02**	**0.02**	**0.02**
0.02	**0.01**	**0.02**	**0.02**	**0.02**	**0.02**	**0.02**
0.02	**0.02**	**0.02**	**0.02**	**0.02**	**0.02**	**0.02**
0.04	**0.04**	**0.01**	**0.01**	**0.01**	**0.02**	**0.01**

Table 13 MOORA results

Rank	LPWAN	Score
1	Weightless	0.35
2	NB-IoT	0.22
3	RPMA	0.21
4	LTE-M	0.19
5	GSM	0.16
6	LoraWAN	0.14
7	Sigfox	0.11

Table 14 MOORA weights

Attribute	Weight
Battery life	0.0515
Minimum cost	0.0587
RX current consumption	0.0615
TX current consumption	0.0544
Bandwidth	0.0787
Maximum data rate	0.0818
Uplink data rate	0.0892
Downlink data rate	0.0982
Maximum messages per day	0.0620
Maximum payload length	0.0518
Urban range	0.0492
Rural range	0.0493
Receiver sensitivity	0.0493
Default transmitter power	0.0497
Maximum coupling loss	0.0492
Nodes per gateway	0.0645

- Between 40,001 and 10,0000 km
- Between 10,001 km and 20,000 km
- More than 20,001 km

The equation that gives de carbon emission is given by:

$$C_0 = \frac{1.126}{10000}A_0 + \frac{8.858}{100000}A_1 + \frac{7.632}{10000}A_2 + \frac{7.24}{100000}A_3 \tag{20}$$

where A represents the factors previously mentioned in the same order that the subscript of the letter. This notation will be used equally in the rest of the section.

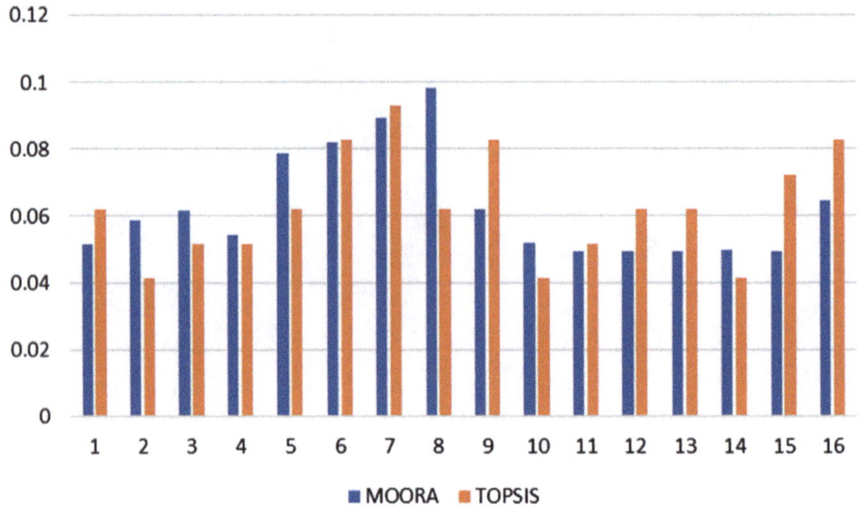

Fig. 7 Weights comparisons between methods

- Vehicle
  - Car—Gasoline
  - Car—Diesel
  - Car—Gas
  - Truck—Gasoline
  - Truck—Diesel
  - Truck—Gas
  - Truck with refrigeration—Diesel
  - Refrigerated Truck-diesel.

  With the representative equation:

$$C_1 = \frac{1.823}{10000}A_4 + \frac{1.913}{10000}A_5 + \frac{2.079}{10000}A_6 + \frac{25}{100000}A_7 + \frac{2.099}{10000}A_8 + \frac{2.631}{10000}A_9 + \frac{9.136}{10000}A_{10} + \frac{1.072}{1000}A_{11} \tag{21}$$

- Land transport
  - Taxi
  - Train
  - Bus
  - Foreign bus

And the carbon emission is calculated as follows:

$$C_2 = \frac{1.748}{10000}A_{12} + \frac{5.63}{100000}A_{13} + \frac{7.92}{100000}A_{14} + \frac{1.088}{10000}A_{15} \quad (22)$$

- Electrical consumption
  - Megawatts per hour

  Calculated with the equation:

$$C_3 = \frac{1}{2}A_{16} \quad (23)$$

- Use of the network

  Given in days that are converted to carbon emission with:

$$C_4 = \frac{3.3}{100}A_{18} \quad (24)$$

- Scrap
  - Normal
  - To recycle

  That, combined result the equation:

$$C_5 = \frac{4.59}{10000}A_{19} + \frac{2.1}{100000}A_{20} \quad (25)$$

- Water

$$C_6 = \frac{3.34}{10000}A_{21} \quad (26)$$

Finally, the result is the sum of every C, i.e., the total carbon emission given in tCO_{2e} is

$$C = \sum_{i=0}^{6} c_i \quad (27)$$

These equations are used and facilitated to the operator in the section of Module 8 shown in Fig. 8a. The second part of Module 8 presents a simple but efficient test of the NOM-035 norm. When the tests are fully performed, the results can be inserted in each cell, and the dashboard will show the result given the following equation

$$F = 6VH + 5H + 3M + L \quad (28)$$

(a) Cafbon footprint comparison

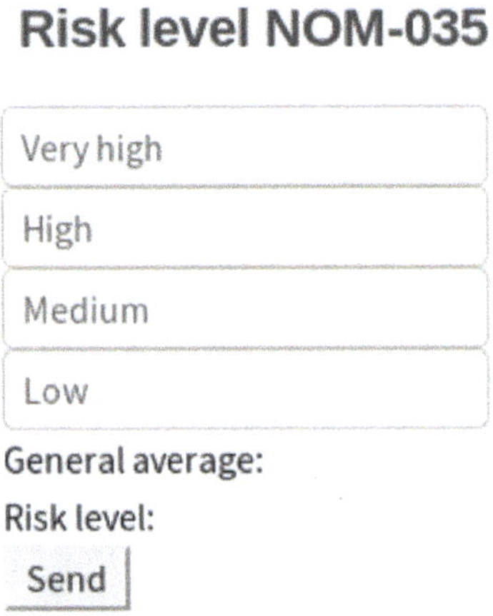

(b) Interface to enter NOM-35 risk level

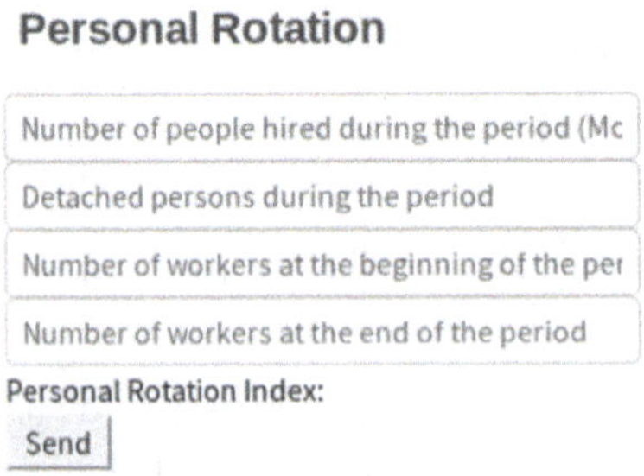

(c) Interface to enter personal rotation information

Fig. 8 Module 8

where

- VH is a very high counter
- H is a high counter
- M is medium counter
- L is a low counter.

Finally, the Job Rotation Index can be calculated in Sect. 3. The equation used for this is the following

$$JRI = \frac{2(A - D)}{F1 + F2} * 100 \tag{29}$$

where

- A is the number of persons hired during the period
- D is the number of persons released during the period
- F1 is initial workers in the period
- F2 is the final workers in the period.

The JRI, as was discussed in the methodology section, gives the factory the ability to perform basic environment analyses and, therefore, enough information about the workers' emotional status additionally to the NOM-035 safety analysis.

4 Conclusions and Future Work

We must, of course, accept that the final standards for IoT have yet to be defined; in fact, the protocols are in full development, and there are several projects that intend to define standards [2]. Today a lot of business sectors are facing several communication challenges. Some applications are using Internet of Thing with cellular modems just because it provides a positive Return on Investment (ROI) even when users are paying a few dollars per month per sensor for the connectivity. But there are many applications with lower ROI requirements, and it is still unconnected because of traditional communication technologies do not provide a positive return on investments with current monthly fees. Addressing the abovementioned problem, LPWA networks offer an effective solution. With this technology, the initial cost of a radio module and the amount of monthly service fee is much lower (cents/month as opposed to dollars/month). Whereas a cellular modem would cost over $30, and a WAVIoT IoT module can be built for a fraction of that.

References

1. Mallon, S.: IoT Is the Most Important Development of the 21st Century. SmartDataCollective. https://www.smartdatacollective.com/iot-most-important-development-of-21st-century/ (2018) Accessed on: Feb 2020
2. Fernandez, A.: What is LPWAN? An Introduction to the IoT Communications Protocol. Pandorafms. https://pandorafms.com/blog/what-is-lpwan/ (2019) Accessed on Jan 2020
3. Rama, Y., Özpmar, A.: A comparison of long-range licensed and unlicensed LPWAN technologies according to their geolocation services and commercial opportunities. In: 2018 18th Mediterranean Microwave Symposium, pp. 398–403. IEEE, Istanbul (2018). https://doi.org/10.1109/MMS.2018.8612009
4. Kabalcı, Y., Ali, M.: Emerging LPWAN technologies for smart environments: an outlook. In: 2019 1st Global Power, Energy and Communication Conference (GPECOM), pp. 24–29. IEEE, Nevsehir (2019). https://doi.org/10.1109/GPECOM.2019.8778626.
5. Aggarwal, S., Nasipuri, A.: Survey and performance study of emerging LPWAN technologies for IoT applications. In: 2019 IEEE 16th International Conference on Smart Cities: Improving Quality of Life Using ICT & IoT and AI (HONET-ICT), pp. 069–073. IEEE, Charlotte (2019). https://doi.org/10.1109/HONET.2019.8908117
6. (n.d.). What is LPWAN. Waviot https://waviot.com/technology/what-is-lpwan Accessed on Feb 2020
7. Prando, L., de Lima, E., de Moraes, L., Hamerschmidt, M., Fraindenraich, G.: Experimental performance comparison of emerging low power wide area networking (LPWAN) technologies for IoT. In: 2019 IEEE 5th World Forum on Internet of Things (WF-IoT), pp. 905–908. IEEE, Limerick (2019). https://doi.org/10.1109/WF-IoT.2019.8767343
8. Ayoub, W., Samhat, A., Nouvel, F., Mroue, M., Prévotet, J.: Internet of mobile things: Overview of Lorawan, Dash7, and NB-IoT in LPWANS standards and supported mobility. IEEE Commun. Surv. Tutorials **21**(2), 1561–1581 (2018). https://doi.org/10.1109/COMST.2018.2877382
9. Jozaghi, A., Alizadeh, B., Hatami, M., Flood, I., Khorrami, M., Khodaei, N., Ghasemi Tousi, E.: A comparative study of the ahp and topsis techniques for dam site selection using gis: a case study of sistan and baluchestan province, iran. Geosciences **8**(12), 494 (2018). https://doi.org/10.3390/geosciences8120494
10. Park, C., Kim, G., Choi, S.: Engineering economics. Pearson Prentice Hall, New Jersey (2007)
11. GOBMX. Factores de riesgo psicosocial. https://www.gob.mx/stps/articulos/norma-oficial-mexicana-nom-035-stps-2018-factores-de-riesgo-psicosocial-en-el-trabajo-identificacion-analisis-y-prevencion Accessed Mar 2020 (2020)
12. Kampkötter, P., Harbring, C., Sliwka, D.: Job rotation and employee perfor-mance—evidence from a longitudinal study in the financial services industry. Int. J. Hum. Resour. Manag. **29**(10), 1709–1735 (2018). https://doi.org/10.1080/09585192.2016.1209227

Comprehensive Minimum Cost Models Based on Consensus Measures

Álvaro Labella, Hongbin Liu, Rosa M. Rodríguez, and Luis Martínez

Abstract Nowadays, consensus is key in Group Decision Making (GDM). Many times, decision makers who participate in a GDM problem discuss and modify their initial opinions in order to reach a consensual solution; this process is known as Consensus Reaching Process (CRP). However, such a process can lead to endless negotiations in which the cost of achieving an agreement is too high. Several researchers have pointed out the importance of considering the cost of shifting the decision makers' opinions in CRPs. One of the most widespread proposals is the Minimum Cost Consensus (MCC) models. These models define consensus as the minimal distance between each decision maker's preference and the collective opinion and seek to minimize the overall cost of moving the experts' opinions by using different types of cost functions. However, small distances do not always guarantee an acceptable consensus level. Therefore, there is a need for defining new MCC models that not only consider the distances of each decision maker to the collective opinion, but also achieve a minimum agreement among decision makers to obtain better solutions. Furthermore, these novel MCC models have to be able to deal with preference structures commonly used in GDM problems such as fuzzy preference relations.

Keywords Consensus reaching process · Minimum cost · Fuzzy preference relation

Á. Labella (✉) · R. M. Rodríguez · L. Martínez
University of Jaén, 23071 Jaén, Andalucia, Spain
e-mail: alabella@ujaen.es

H. Liu
University of Henan, Zhengzhou 475001/475004, Henan, China

W. Pedrycz et al. (eds.), *Computational Intelligence for Business Analytics*,
Studies in Computational Intelligence 953,
https://doi.org/10.1007/978-3-030-73819-8_3

1 Introduction

Decision making (DM) is a common process in human beings' daily life characterized by a set of alternatives and the need for selecting one of them as the best solution to a DM problem. Nowadays, due to the increasing complexity of the DM problems, several decision makers with their own attitudes take part of the decision process and try to achieve a common solution, in this situation, we talk about *Group Decision Making* (GDM) [1]. The participation of several decision makers with different expectations, values and risk attitudes implies the inevitable appearance of disagreements among them, which might provoke unsatisfactory solutions or deadlocks. Despite the importance of this fact, GDM problems have been traditionally solved by applying a selection process that ignores disagreements among decision makers [2]. To oppose to this trend, *Consensus Reaching Process* (CRP) has been included as an additional phase in the resolution scheme of a GDM problem. In a CRP, decision makers discuss and change their initial preferences by trying to bring positions closer in order to increase the level of agreement within the group and achieve a consensual solution that satisfies most of decision makers who participate in the decision process [3].

Several researchers have pointed out the importance of considering the cost of shifting the decision makers' opinions in CRPs, and as a consequence, multiple consensus approaches have been presented in the specialized literature in order to decrease such cost. Ben-Arieh and Easton [3] defined the concept of *Minimum Cost Consensus* (MCC) and proposed a linear programming model to achieve the consensus by changing the decision makers' preferences as little as possible. Afterwards, Ben-Arieh et al. presented another proposal based on a quadratic cost function [4]. Other proposals based on the latter models have also been proposed in the literature [5, 6]. However, all these proposals consider only the distance of each decision maker to the collective opinion, ignoring the disagreement among them. It is essential to note that small distances among decision makers and collective opinions do not always guarantee an acceptable consensus level within the group. Therefore, it seems clear the need for new MCC models that take into account not only the distance of each decision maker to the collective, but also a minimum agreement between them to obtain better solutions.

This chapter aims to introduce new MCC models that achieve a solution from the consensus point of view. Furthermore, these models are also defined by supposing that decision makers provide their preferences by using *Fuzzy Preference Relations* (FPRs) [7], one of the most widely used preferences structures in DM.

This paper is structured as follows: Sect. 2 reviews several concepts related to GDM, CRP and MCC models. Section 3 introduces the novel MCC models that consider a minimum agreement among decision makers to achieve consensus. Section 4 shows a numerical example. Finally, in Sect. 5, some conclusions and future research are drawn.

2 Background

This section briefly reviews several concepts related to GDM, CRP and MCC models in order to facilitate the understanding of the proposal.

2.1 *Group Decision Making*

GDM is a task in which several decision makers try to find a collective solution to a decision problem by expressing their opinions over several alternatives. GDM has evolved enormously over the years, from the design of voting methods as those proposed by Borda or Condorcet [8] to the apparition of famous theories such as *social choice* [8] or *prospect theory* [9]. Nowadays, GDM is key in several fields of our society such as economic and management science [10] and, in addition, such core position has been reinforced even more due to the emergence of new technological trends such as social networks [11], e-democracy [12] or big data [13].

A GDM problem is formally characterized by a set of decision makers, $E = \{dm_1, \ldots, dm_m\}$ $(m \geq 2)$ who provide their preferences over a finite set of alternatives, $X = \{x_1, \ldots, x_n\}$ $(n \geq 2)$ with the aim of obtaining a common solution [1]. The latter can be reached as the decision makers express their own attitudes, opinions and assessments. The decision makers' preferences can be modeled by using different preference structures. In DM, preference relations have been used successfully in the decision makers' preferences elicitation. A preference relation P^k models the assessment p^k_{ij} of the decision maker dm_k that represents the preference degree of the alternative x_i over the alternative x_j, $i, j \in \{1, \ldots, n\}$. One of the most common preference structures used in DM are the FPRs. A FPR is defined as follows:

Definition 1 [7] A fuzzy preference relation $\boldsymbol{P}^k$, associated to a decision maker $\boldsymbol{dm}_k$ on a set of alternatives X, i s a fuzzy set on $\boldsymbol{X} \times \boldsymbol{X}$, characterized by the membership function $\boldsymbol{\mu}_{\boldsymbol{P}^k} : \boldsymbol{X} \times \boldsymbol{X} \to [0, 1]$. When the number of alternatives n is finite, $\boldsymbol{P}^k$, is represented by a $\boldsymbol{n} \times \boldsymbol{n}$ matrix of assessments $\boldsymbol{\mu}_{\boldsymbol{P}^k}(\boldsymbol{x}_i, \boldsymbol{x}_j) = \boldsymbol{p}^k_{ij}$ as follows:

$$P^k = \begin{pmatrix} p^k_{11} & \cdots & p^k_{1n} \\ \vdots & \ddots & \vdots \\ p^k_{n1} & \cdots & p^k_{nn} \end{pmatrix} \quad (1)$$

where each assessment p^k_{ij} represents the preference degree of the alternative x_i over the alternative x_j according to the decision maker dm_k. The FPR is usually assumed reciprocal $p^k_{ij} + p^k_{ji} = 1, \forall i, j = 1, 2, \ldots, n, k = 1, 2, \ldots, m$.

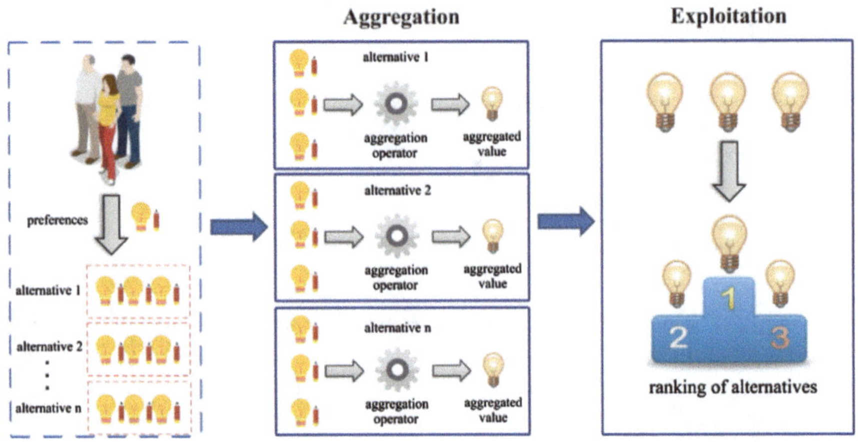

Fig. 1 Classical GDM resolution scheme

An example of FPR by considering three alternatives might be:

$$P^k = \begin{pmatrix} 0.5 & 0.7 & 0.4 \\ 0.3 & 0.5 & 0.1 \\ 0.6 & 0.9 & 0.5 \end{pmatrix}$$

The classical resolution scheme for GDM problems consists of a selection process of the best alternative divided into two phases (see Fig. 1):

1. *Aggregation*: the preferences relations provided by decision makers are aggregated by an aggregation operator obtaining a collective opinion.
2. *Exploitation*: from the collective opinion, the selection of the best alternative as the solution to the problem is carried out.

2.2 Consensus Reaching Processes

The classical resolution scheme for GDM problems described above does not guarantee that all the decision makers will agree with the final solution of the problem and some of them might feel that their opinions have not been taken into account [14]. However, in many real-life GDM problems, obtaining a solution that is widely accepted by the whole group is essential [15]. In these situations, an additional phase so-called CRP is included in the resolution process for GDM problems. A CRP is an iterative and dynamic process in which decision makers discuss and change their initial preferences bring them closer to each other in order to reach a consensual solution. Normally, a moderator supervises and guides the process by identifying those decision makers whose opinions are furthest from the collective one and makes

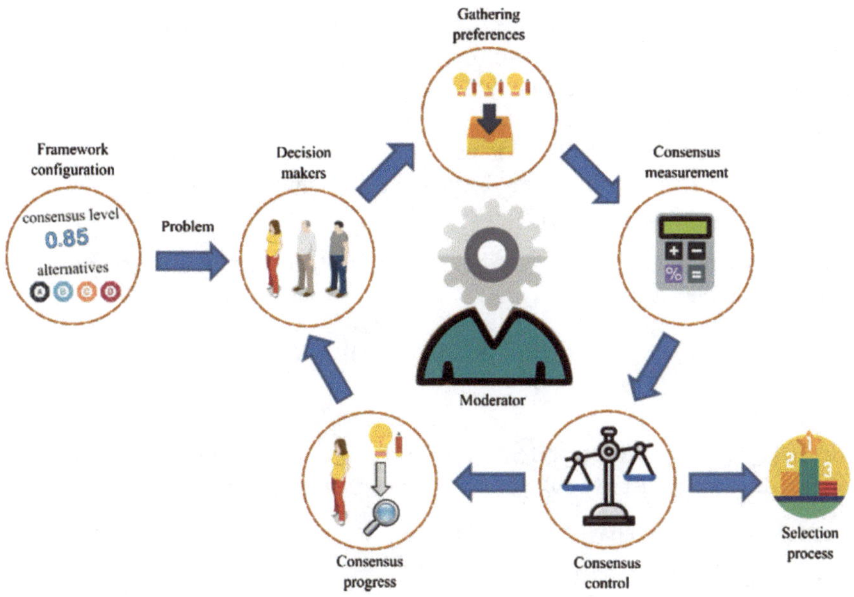

Fig. 2 CRP general scheme

suggestions to them in order to increase the level of agreement of the group. The general CRP scheme is composed by several phases described below (see Fig. 2):

1. *Framework configuration*: the set of alternatives, decision makers and the consensus level to reach in the group are defined.
2. *Gathering preferences*: decision makers provide their preferences by using preference relations.
3. *Consensus measurement*: the level of agreement of the group is computed by consensus measures based on distance measures and aggregation operators.
4. *Consensus control*: the current level of agreement $\mu \in [0, 1]$ is compared with a predefined consensus threshold $\alpha \in [0, 1]$ that represents the minimum level of consensus to achieve in the group. If the level of agreement is greater than α, a selection process starts; otherwise, another discussion round is necessary.
5. *Consensus progress*: moderator identifies decision makers whose opinions are furthest from the rest of decision makers and recommends them to modify their preferences to increase the level of agreement in the next discussion round.

The concept of consensus has been deeply discussed in the specialized literature. Some points of view consider consensus as a synonym of total agreement or unanimity, practically unreachable in real decision situations. However, other views provide a more flexible definition of consensus. Saint et al. [15] defined *consensus* as "a state of mutual agreement among members of a group, where all legitimate concerns of individuals have been addressed to the satisfaction of the group". Kacpryck et al. [1] introduced another interpretation of consensus so-called *soft consensus* based on the concept of fuzzy majority, which established that the consensus is reached when "most of the individuals agree as to almost all the relevant options". Therefore, flexible notions of consensus imply that different levels of agreement can be reached in a group and, consequently, the need of defining consensus measures that allow computing the current level of agreement in the group. According to the taxonomy proposed by Palomares et al. in [16], there are two different consensus measures:

- Consensus measure based on the distance among decision makers and collective opinion defined as:

$$Consensus_1(P^1, \ldots, P^m) = 1 - \frac{\sum_{i=1}^{m} d(P^i, \gamma(P^1, \ldots, P^m))}{m} \in [0, 1] \quad (2)$$

where $P^1, \ldots, P^m$ are the preference relations provided by the decision makers, $d(\cdot)$ is a distance measure ϵ [0.1] and γ is an aggregation operator.

- Consensus measure based on the distance among decision makers defined as:

$$Consensus_2(P^1, \ldots, P^m) = 1 - \frac{\sum_{i=1}^{m-1} \sum_{j=i+1}^{m-1} d(P^i, P^j)}{\frac{m(m-1)}{2}} \in [0, 1] \quad (3)$$

where $P^1, \ldots, P^m$ are the preference relations provided by the decision makers and $d(\cdot)$ is a distance measure $\in$ [0,1].

Remark 1 Note that, the distance measures used to compute the consensus depend on the type of preference relation that decision makers use to provide their preferences, but they will always return a value in [0, 1].

2.3 *Minimum Cost Models*

The cost of the modification of decision makers' preferences is a pivotal key in a CRP. An excessive cost might be the cause of laborious negotiations and too long processes. Ben-Arieh and Easton introduced in [3] the concept of *minimum cost consensus* and proposed an MCC model whose aim was to minimize the overall cost of changing decision makers' preferences by using a linear cost function.

Definition 2 Let $\{o_{1i}, o_{2i} \ldots, o_{mi}\}$, $\{o_{ki} \in \mathbb{R} | 0 \le o_{ki} \le 1\}$, be the assessments provided by a set of decision makers $E = \{dm_1, \ldots, dm_m\}$ over the alternative a_i and c_{ki} the cost of shifting the decision maker dm_k's opinion from zero to one or vice versa [3]. The MCC model based on linear cost function is given as follows:

$$\min \sum_{k=1}^{m} c_{ki} \left| \tilde{o}_{ki} - o_{ki} \right|$$
$$s.t. \left| \tilde{o}_{ki} - \overline{o}_i \right| \leq \varepsilon, k = 1, 2, \ldots, m, i = 1, 2, \ldots, n \tag{4}$$

where $\{\tilde{o}_{1i}, \ldots, \tilde{o}_{mi}\}$ are the adjusted opinions of the decision makers, $\overline{o}_i$ is the adjusted collective opinion for the alternative a_i and ε is the maximum acceptable distance of each decision maker to the collective opinion.

Therefore, according to the model, a decision maker's opinion does not need to be changed if it is in the interval $[\bar{\tilde{o}} - \varepsilon, \tilde{o} + \varepsilon]$.

Afterwards, Zhang et al. [17] studied the influence of the selected aggregation operator for computing the collective opinion on the level of agreement in the group and proposed a new MCC model:

$$\min \sum_{k=1}^{m} c_{ki} \left| \tilde{o}_{ki} - o_{ki} \right|$$
$$s.t. = \begin{cases} \overline{o}_i = F(\tilde{o}_{1i}, \ldots, \tilde{o}_{mi}) \\ \left| \tilde{o}_{ki} - \overline{o}_i \right| \leq \varepsilon, k = 1, 2, \ldots, m, i = 1, 2, \ldots n. \end{cases} \tag{5}$$

where F is an aggregation function.

Remark 2 Note that the main difference between Eqs. (4) and (5) is that, although both models compute the adjusted collective opinion iteratively, the latter takes into account the influence of the selected aggregation operator in the consensus computation since decision makers consensus level varies according to it.

Recently, several researchers have paid attention to the previous MCC models and multiple proposals have been presented in the specialized literature [5, 6]. However, all these proposals only consider the distance among decision makers and collective opinion ignoring a minimum agreement among decision makers to reach consensus. In this situation, the computed collective opinion cannot guarantee a required consensus degree for all the decision makers.

3 MCC Models Considering Distance Among Decision Makers and Consensus Degree

This section introduces novel MCC models that deal with the limitations of the existing MCC models. The proposed models deal with numerical values and afterwards, they are extended to FPRs.

Taking into account previous drawbacks of the existing MCC models, it is necessary to define a new MCC model that takes into account the agreement among decision makers in order to obtain better consensual solutions. Therefore, the model presented in Eq. (5) is modified by including the computation of consensus as follows:

$$\min \sum_{k=1}^{m} c_{ki} \left| \tilde{o}_{ki} - o_{ki} \right|$$
$$s.t. = \begin{cases} \overline{o}_i = F(\tilde{o}_{1i}, \ldots, \tilde{o}_{mi}) \\ \left| \tilde{o}_{ki} - \overline{o}_i \right| \leq \varepsilon, k = 1, 2, \ldots, m, i = 1, 2, \ldots n \\ consensus(\tilde{o}_{1i}, \ldots, \tilde{o}_{mi}) \geq \alpha \end{cases} \tag{6}$$

where $consensus(\cdot)$ represents the level of consensus achieved, $\alpha \in [0, 1]$ is a predefined consensus threshold, F is an aggregation function and ε is a parameter that measures the distance among the decision makers and the collective opinion.

Due to there are two different ways of computing consensus according to the two different consensus measures, one MCC model is proposed for each one as follows:

Consensus measure based on the distance among decision makers and collective opinion:

$$\min \sum_{k=1}^{m} c_{ki} |\tilde{o}_{ki} - o_{ki}|$$
$$s.t. = \begin{cases} \overline{o}_i = \sum_{k=1}^{m} w_k \tilde{o}_{ki} \\ |\tilde{o}_{ki} - \overline{o}_i| \leq \varepsilon, k = 1, 2, \ldots, m, i = 1, 2, \ldots n \\ \sum_{k=1}^{m} w_k \left| \tilde{o}_{ki} - \overline{o}_i \right| \leq 1 - \alpha \end{cases} \tag{7}$$

where w_k is the weight assigned to the decision maker dm_k.

Remark 3 Note that, without loss of generality, Weighted Average Mean has been used to aggregate the distances. Furthermore, the original decision makers' preferences, $\{o_{k1}, o_{k2}, \ldots, o_{kn}\}$, are considered in the interval [0, 1].

Consensus measure based on the distance among decision makers:

$$\min \sum_{k=1}^{m} c_{ki} \left| \tilde{o}_{ki} - o_{ki} \right|$$
$$s.t. = \begin{cases} \overline{o}_i = \sum_{k=1}^{m} w_k \tilde{o}_{ki} \\ \left| \tilde{o}_{ki} - \overline{o}_i \right| \leq \varepsilon, k = 1, 2, \ldots, m, i = 1, 2, \ldots n \\ \sum_{k=1}^{m-1} \sum_{j=k+1}^{m} \frac{w_k + w_j}{m-1} |\tilde{o}_{ki} - \tilde{o}_{ji}| \leq 1 - \alpha \end{cases} \tag{8}$$

where w_k is the weight assigned to the decision maker dm_k.

Remark 4 Note that, without loss of generality, Weighted Average Mean has been used to aggregate the distances as follows:

$$\frac{w_1}{m-1}\sum_{\substack{k=1\\k\neq 1}}^{m}|\tilde{o}_{ki}-\tilde{o}_{1i}|+\frac{w_2}{m-1}\sum_{\substack{k=1\\k\neq 2}}^{m}|\tilde{o}_{ki}-\tilde{o}_{2i}|+\ldots$$

$$+\frac{w_m}{m-1}\sum_{\substack{k=1\\k\neq m}}^{m}|\tilde{o}_{ki}-\tilde{o}_{mi}|=\sum_{k=1}^{m-1}\sum_{j=k+1}^{m}\frac{w_k+w_j}{m-1}\left|\tilde{o}_{ki}-\tilde{o}_{ji}\right| \tag{9}$$

Furthermore, the original decision makers' preferences, $\{o_{k1}, o_{k2}, \ldots, o_{kn}\}$, are considered in the interval [0, 1].

Once the new MCC models have been presented, they can be specified to using FPR (see Definition 1), which is one of the preferences structures most widely used in GDM.

Consensus measure based on the distance among decision makers and collective opinion:

$$\min\sum_{k=1}^{m}\sum_{i=1}^{n-1}\sum_{j=i+1}^{n}c_k\left|p_{ij}^k-\tilde{p}_{ij}^k\right|$$

$$s.t.=\begin{cases}\tilde{p}_{ij}=\sum_{k=1}^{m}w_k\,\tilde{p}_{ij}^k\\ \left|\tilde{p}_{ij}^k-\tilde{p}_{ij}\right|\leq\varepsilon, k=1,\ldots,m, i=1,\ldots,n-1, j=1,\ldots,n\\ \frac{2}{n(n-1)}\sum_{k=1}^{m}\sum_{i=1}^{n-1}\sum_{j=i+1}^{n}w_k\left|\tilde{p}_{ij}^k-\overline{p}_{ij}\right|\leq 1-\alpha\end{cases} \tag{10}$$

Consensus measure based on the distance among decision makers:

$$\min\sum_{k=1}^{m}\sum_{i=1}^{n-1}\sum_{j=i+1}^{n}c_k\left|p_{ij}^k-\tilde{p}_{ij}^k\right|$$

$$s.t.=\begin{cases}\overline{p}_{ij}=\sum_{k=1}^{m}w_k\,\tilde{p}_{ij}^k\\ \left|\tilde{p}_{ij}^k-\overline{p}_{ij}\right|\leq\varepsilon, k=1,\ldots,m, i=1,\ldots,n-1, j=1,\ldots,n\\ \frac{2}{n(n-1)}\sum_{i=1}^{n-1}\sum_{j=i+1}^{n}\sum_{k=1}^{m-1}\sum_{l=k+1}^{m}\frac{w_k+w_l}{m-1}\left|\tilde{p}_{ij}^k-\overline{p}_{ij}\right|\leq 1-\alpha\end{cases} \tag{11}$$

4 Numerical Examples

To show the advantages and usefulness of the proposed MCC models, a numerical example is shown in this section. First, a GDM problem is introduced and then, it is solved by considering that decision makers provide their preferences by using both numerical values and FPRs.

Let us suppose three decision makers $E = \{dm_1, dm_2, dm_3\}$ with weights $W = \{0.375, 0.250, 0.375\}$ and cost $C = \{2, 3, 4\}$ respectively and three alternatives $X = \{x_1, x_2, x_3\}$.

4.1 Numerical Values Assessments

In this example, decision makers' assessments are represented by real values as follows (for the sake of simplicity the assessments are provided just for the alternative x_1):

$$(o_1, o_2, o_3) = (0.1, 0.75, 0.33), o_k \in [0, 1]$$

Table 1 shows the decision makers' preferences evolution obtained by the classical model initially proposed by Ben-Arieh and Easton [3] and lately modified by Zhang et al. [17]. Therefore, the parameter ε is the one that determines the changes applied to the decision makers' preferences. The level of agreement in the group is computed from the modified decision makers' preferences and according to the two different consensus measures previously introduced (see Eqs. (2) and (3)).

On the one hand, according to the results shown in Table 1, when the value of ε is high (0.5), it is not necessary to change the decision makers' preferences, since the initial preferences satisfy the ε restriction. On the other hand, the consensus measure based on the distance among decision makers and collective opinion (Eq. 2) provides a value of 0.8, and the one based on the distance among decision makers (Eq. 3) provides a low value of 0.58, both far from a desired consensus situation. For the rest of the cases, decision makers' preferences are modified in order to satisfy the ε restriction. Regarding the level of agreement, the lower the value of ε the greater the

Table 1 Numerical values evolution and consensus achieved without considering consensus among decision makers

ε	Preferences evolution	μ (Eq. 2)	Preferences evolution	μ (Eq. 3)
0.50	(0.10, 0.75, 0.33)	0.8	(0.10, 0.75, 0.33)	0.58
0.15	(0.23, 0.48, 0.33)	0.93	(0.23, 0.48, 0.33)	0.84
0.05	(0.30, 0.38, 0.33)	0.98	(0.30, 0.38, 0.33)	0.94

Table 2 Numerical values evolution with different values of ε and α for the models of Eqs. (7) and (8)

ε	Equation (7)			Equation (8)		
	$\alpha = 0.65$	$\alpha = 0.85$	$\alpha = 0.95$	$\alpha = 0.65$	$\alpha = 0.85$	$\alpha = 0.95$
0.50	(0.10, 0.75, 0.33)	(0.13, 0.63. 0.33)	(0.26, 0.43, 0.33)	(0.33, 0.61, 0.33)	(0.33, 0.45, 0.33)	(0.35 0.37, 0.33)
0.15	(0.23, 0.48, 0.33)	(0.23, 0.48, 0.33)	(0.26, 0.43, 0.33)	(0.23, 0.48, 0.33)	(0.33, 0.45, 0.33)	(0.32 0.35, 0.33)
0.05	(0.30, 0.38, 0.33)	(0.30, 0.38, 0.33)	0.30, 0.38, 0.33)	(0.30, 0.38, 0.33)	(0.30, 0.38, 0.33)	(0.34, 0.37, 0.33)

agreement reached because the distances between preferences are reduced. Therefore, taking into account only the ε restriction, it is possible to increase the level of agreement in the group, but it might be, on certain occasions, unsatisfactory.

Once analyzed the results obtained from the classical MCC model, the next step consists of carrying out the resolution of the same problem but, on this occasion, by using the novel MCC models proposed in this contribution, in which a predefined consensus threshold α is considered. Table 2 shows the decision makers' preferences evolution according to different values of ε and consensus, α.

The comparison among the results shown in Tables 1 and 2 exposes clearly the influence of the parameter α in the proposed MCC models. Considering same values of ε in both tables, we can see that, when the level o f agreement reached μ is greater than the predefined consensus threshold α, the resulting decision makers' preferences are the same since they are modified according to the ε restriction and the parameter α has no influence. On the contrary, when the value μ is lower than the predefined consensus threshold α, the latter parameter guarantees to reach a desired consensus situation. A very clear example of this situation happens when $\varepsilon = 0.50$ and the consensus is computed based on the distance among decision makers (Eqs. 3 and 8). If α is not considered, the level of agreement obtained within the group is 0.58, which is not a high value of consensus. However, if parameter α is taken into account, the decision makers' preferences are modified in order to achieve such a level of consensus b y guaranteeing the agreement in the group and obtaining a consensual solution.

4.2 FPRs Assessments

In this example, decision makers' assessments are modeled by means of the following FPRs:

Table 3 FPR values evolution and consensus achieved without considering consensus among decision makers

ε	Equation (10) (α = 0)		Equation (11) (α = 0)	
	Preferences evolution	μ	Preferences evolution	μ
0.50	(0.99, 0.45, 0.12)	0.58	(0.99, 0.45, 0.12)	0.75
0.15	(0.59, 0.44, 0.32)	0.83	(0.60, 0.45, 0.32)	0.91
0.05	(0.49, 0.45, 0.41)	0.94	(0.43, 0.38, 0.34)	0.97

Table 4 FPR values evolution with different values of ε and α for the models of Eqs. (10) and (11)

ε	Equation (10)			Equation (11)		
	$\alpha = 0.65$	$\alpha = 0.85$	$\alpha = 0.95$	$\alpha = 0.65$	$\alpha = 0.85$	$\alpha = 0.95$
0.50	(0.89, 0.45, 0.12)	(0.52, 0.35, 0.18)	(0.23, 0.16, 0.20)	(0.99, 0.45, 0.12)	(0.82, 0.44, 0.12)	(0.60, 0.45, 0.31)
0.15	(0.58, 0.43, 0.30)	(0.58, 0.44, 0.30).	(0.22, 0.14, 0.15)	(0.99, 0.45, 0.12)	(0.60, 0.45, 0.33)	(0.51, 0.45, 0.40)
0.05	(0.49, 0.44, 0.40)	(0.49. 0.45, 0.41)	(0.48, 0.44, 0.41)	(0.99, 0.45, 0.12)	(0.50, 0.44, 0.41)	(0.43, 0.38, 0.34)

$$P^1 = \begin{pmatrix} 0.5 & 0.58 & \mathbf{0.99} \\ 0.42 & 0.5 & 0.99 \\ 0.01 & 0.01 & 0.5 \end{pmatrix}, P^2 = \begin{pmatrix} 0.5 & 0.73 & \mathbf{0.45} \\ 0.27 & 0.5 & 0.23 \\ 0.55 & 0.77 & 0.5 \end{pmatrix},$$

$$P^3 = \begin{pmatrix} 0.5 & 0.49 & \mathbf{0.12} \\ 0.51 & 0.5 & 0.34 \\ 0.88 & 0.66 & 0.5 \end{pmatrix}$$

Tables 3 and 4 show in detail the evolution of the FPRs by applying the new MCC models under different conditions: without considering a minimum level o f agreement among decision makers (Table 3) and considering different values of ε and α (Table 4). For the sake of simplicity, just one value for each FPR has been chosen, in this case, the one corresponding with the pairwise comparison among the alternatives a_1 and a_3 (in bold type).

From Table 3, we see in both models that, when consensus among decision makers is not considered, with a high value of ε, decision makers' preferences are not modified; thus, the consensus measures computation show a low level of agreement among them, 0.58 and 0.75 respectively. On the contrary, the more the value of ε decreases, the more change in preferences and the greater the level o f agreement achieved.

From Table 4 we see that, contrary to the previous case, the parameter α guarantees the desired agreement among decision makers when the value of ε is high since the minimum cost is determined by the parameter α, not ε. On the other hand, when the value of α is low and the value of ε is high, the minimum cost is determined by the latter parameter. Furthermore, the parameter α allows predefining the desired level of agreement to reach, which is impossible to determine just considering the ε parameter.

Therefore, from this numerical example, we can see that the parameter α also plays a pivotal role in the new models presented based on FPRs.

5 Conclusions and Future Work

Consensual decisions are increasingly important in several DM problems since they allow removing the disagreement among decision makers and obtaining better and more appreciated solutions by the group, giving rise to the CRPs.

The cost of shifting decision makers' preferences is a pivotal issue in a CRP. Recent MCC models try to reduce the cost by means of MCC models that only consider the distance among decision makers and collective opinion. However, it does not always guarantee a desired level of agreement within the group and achieve a consensual solution.

This paper has introduced new MCC models in which consensus among decision makers is taken into account. Consensus is computed by using two different consensus measures: based on the distance among decision makers and collective opinion and based on the distance among decision makers. In this way, it is possible to predefine a consensus level to reach within the group. Furthermore, these models have been presented for numerical values and extended for FPRs.

As future research work, new MCC models focused on large-scale GDM problems will be studied.

Acknowledgements This work is supported by the National Natural Science Foundation of China (grant numbers 11872175, 61803144) and partially supported by the Spanish Ministry of Economy and Competitiveness (grant number PGC2018-099402-B-I00) and Postdoctoral fellow Ramón y Cajal (RYC-2017-21978).

References

1. Kacprzyk, J.: Group decision making with a fuzzy linguistic majority. Fuzzy Sets Syst. **18**(2), 105–118 (1986). https://doi.org/10.1016/0165-0114(86)90014-X
2. Herrera, F., Herrera-Viedma, E., Verdegay, J.L.: A sequential selection process in group decision making with a linguistic assessment approach. Inf. Sci. **85**(4), 223–239 (1995). https://doi.org/10.1016/0020-0255(95)00025-K
3. Ben-Arieh, D., Easton, T.: Multi-criteria group consensus under linear cost opinion elas-ticity. Decis. Support. Syst. **43**(3), 713–721 (2007). https://doi.org/10.1016/j.dss.2006.11.009
4. Ben-Arieh, D., Easton, T., Evans, B.: Minimum cost consensus with quadratic cost functions. IEEE Trans. Syst. Man Cybern. Part A Syst. Hum. **39**(1), 210–217 (2009). https://doi.org/10.1109/TSMCA.2008.2006373

5. Gong, Z., et al.: Two consensus models based on the minimum cost and maximum return regarding either all individuals or one individual. Eur. J. Oper. Res. **240**(1), 183–192 (2015). https://doi.org/10.1016/j.ejor.2014.06.035
6. Li, Y., Zhang, H., Dong, Y.: The interactive consensus reaching process with the minimum and uncertain cost in group decision making. Appl. Soft Comput. **60**, 202–212 (2017). https://doi.org/10.1016/j.asoc.2017.06.056
7. Orlovsky, S.A.: Decision-making with a fuzzy preference relation. Fuzzy Sets Syst. **1**(3), 155–167 (1978). https://doi.org/10.1016/0165-0114(78)90001-5
8. Arrow, K.J.: Social Choice and Individual Values, vol. 12. Yale University Press (2012)
9. Tversky, A., Kahneman, D.: Prospect theory: an analysis of decision under risk. Econometrica **47**(2), 263–292 (1979). https://doi.org/10.1142/9789814417358_0006
10. Lu, J., et al.: Multi-Objective Group Decision Making, vol. 6. Imperial College Press (2007)
11. Squillante, M.: Decision making in social networks. Int. J. Intell. Syst. **25**(3 (Spec. Iss.)), 225–285 (2010). https://doi.org/10.1002/int.20397
12. Mateos, A., Jiménez-Martín, A., Ríos-Insua, S.: A group decision-making methodology with incomplete individual beliefs applied to e-democracy. Group Decis. Negot. (2014). https://doi.org/10.1007/s10726-014-9401-y
13. Mayer-Schönberger, V., Cukier, K.: Big data: a revolution that will transform how we live, work, and think. Houghton Mifflin Harcourt. **179**(9), 1143–1144 (2013). https://doi.org/10.1093/aje/kwu085
14. Butler, C.T., Rothstein, A.: On Conflict and Consensus: A Handbook on Formal Consensus Decision Making. Food Not Bombs Publishing (1991)
15. Saint, S., Lawson, J.R.: Rules for Reaching Consensus. A Modern Approach to Decision Making. Wiley (1994)
16. Palomares, I., et al.: Consensus under a fuzzy context: taxonomy, analysis framework AFRYCA and experimental case of study. Inf. Fusion. **20**, 252–271 (2014). https://doi.org/10.1016/j.inffus.2014.03.002
17. Zhang, G., et al.: Minimum-cost consensus models under aggregation operators. IEEE Trans. Syst. Man Cybern. -Part Syst. Hum. **41**(6), 1253–1261 (2011). https://doi.org/10.1109/TSMCA.2011.2113336

Strategies for the Development and Success of Women Entrepreneurs Through SWOT Analysis and Compensatory Fuzzy Logic

Magaly Oyervides Villarreal, Liliana Guerrero Ramos, Rafael Alejandro Espin-Andrade, and Israel Sánchez López

Abstract This article shows a formal system that establishes logical priorities useful for decision making in the implementation of planned actions, thus was selected the order of strategies to implement in the organization of women entrepreneurs; these strategies were proposed based on the results of the study conducted in the state of Coahuila, which analyzes in detail the current situation. This analysis was complemented with the use of compensatory fuzzy logic, through the Fuzzy Tree Studio program, which allows us to calculate with the geometric mean and other calculations integrated into the system, prioritizing strategies based on the veracity of their presence, the intensity of the impacts between the characteristics of the company (strengths and weaknesses) and the characteristics of the environment (weaknesses and threats). Offering innovative and attractive products and/or services benchmarking and constantly updating the first 3 of the 16 proposed strategies.

Keywords Fuzzy logic · Decision making · Strategic planning

1 Introduction

Nowadays, the participation of women in the business environment is more and more frequent, which has increased the presence of women entrepreneurs. In the state of Coahuila, Mexico, a study was conducted to determine the characteristics of women entrepreneurs in that state [1]. This article aims to determine the order of importance of the strategies proposed for implementation in businesses of Coahuila women through the application of fuzzy logic in the SWOT analysis, which was diagnosed

M. Oyervides Villarreal (✉)
Tecnológico Nacional de México, 27000 Torreón, Coahuila, México

L. Guerrero Ramos · R. A. Espin-Andrade
Universidad Autónoma de Coahuila, 27000 Torreón, Coahuila, México

I. Sánchez López
Universidad Juárez del Estado de Durango, 35010 Gómez Palacio, Durango, México

W. Pedrycz et al. (eds.), *Computational Intelligence for Business Analytics*,
Studies in Computational Intelligence 953,
https://doi.org/10.1007/978-3-030-73819-8_4

with the results of the research mentioned above, determining the Strengths, Weaknesses, Opportunities, and Threats of the woman entrepreneur and her organization, the present method is a mixed investigation, since the analysis begins by describing the characteristics of the woman and her company, and after the application of fuzzy logic through the Fuzzy Tree software, quantifies by assigning a numerical value to each characteristic of the SWOT analysis establishing an order of priority for the implementation of such strategies. It was determined that the strategy in the first order was to offer innovative products and/or services, as the main weakness that must be addressed with greater priority to resolve is the reconciliation between work and family life, since 68% of women are mothers, and the most important strength is that 67% of women entrepreneurs have a higher level, 57% in the business area. The opportunities to prioritize in its use are the existence of women's associations and the opening to women entrepreneurs for government financing through the Ministry of Economy, among other results that help women entrepreneurs in decision making.

2 Theoretical Framework

The fuzzy logic was formulated by mathematician and engineer Lotfi A. Zadeh, professor at the University of California at Berkeley. It is a discipline that emerged motivated by the study of vagueness, vague information, or difficult to specify, also allows to study and model decision-making processes with a high level of uncertainty, but vagueness and uncertainty are different concepts since uncertainty is associated with ignorance of the value of a variable while vagueness is related to knowledge of the value of a function (called degree of belonging) of a variable whose exact value is known. In other words, fuzzy logic is a computational intelligence technique that allows working with a high degree of imprecision, which tries to copy the way humans make decisions [2].

Compensatory fuzzy logic is a multivalued logical axiomatic approach different from the axiomatic norm and conorma. It is a transdisciplinary logical theory focused on a purpose defined as interpretability according to language, which gives it great strength as an optimal tool to make administrative models using knowledge engineering. Interpretability occurs according to logical theories and paradigms associated with many social practices in relation to natural and professional language, as well as in relation to classical Logic, theories and decision-making methods, mathematical statistics, and other disciplines and fields of knowledge [3]. Compensatory Fuzzy Arquimedian logic is compatible with the classical normative and conorma approach of Fuzzy Logic, as discussed in an article introducing this new contribution to fuzzy logic [4].

The compensatory fuzzy logic has been applied to create the Model called SWOT-OA Strengths, Weaknesses, Opportunities and Threats—Objectives, Actions, which has had several applications, among others, in Brazilian companies of the automotive industrial sector. This model makes it possible to determine priorities for an organization within the framework of a strategic alignment process [5–7].

This application, in conjunction with others developed or under development, using a software called Fuzzy Tree Studio, which will be explained in some detail later, or other software options, such as analytical tools, are allowing the integration of a Semantic Organizational Intelligence methodology for knowledge management and decision making.

The Fuzzy Tree Studio software has its precedent in the ICPro presented for the first time in 2008 by professors from the University of Mar del Plata, Argentina, directed by Eng. Gustavo Meschino, conceptualized as a framework of data analysis with computational intelligence techniques. This software facilitated the calculation of truth values associated with models based on Compensatory Fuzzy Logic (LDC). As a later development to overcome ICPro limitations, the same creators created the Fuzzy Tree Studio software, "which among other functionalities has a module to help the user to formalize and calculate the truth value of partial predicates and operate properly with them, generalizing the concepts of the traditional Predicate Logic" [8].

The SWOT-OA Model was implemented in Fuzzy Tree Studio, and an application was developed to determine the hierarchy of the characteristics of an ideal company run by the Coahuila businesswoman (formed with the results of the study conducted in the state of Coahuila, Mexico [1]) namely its strengths and weaknesses, the characteristics of its environment (Threats and Opportunities); and the strategies to be implemented to enhance the growth and business success of women in this context.

3 Methodology

It is a research that has a mixed approach, as it describes the characteristics of women entrepreneurs and their organization, and then quantifies the value of the importance of these characteristics, as well as the strategies proposed for the development, growth, and success of women and their businesses.

Objective: To determine the order of priority of the strategies proposed to potentiate the strengths by taking advantage of the opportunities, as well as to address the weaknesses and prepare for the threats of the environment of women entrepreneurs and their organizations in the state of Coahuila, through the application of the compensatory diffuse logic (LDC) to support decision making.

The procedure developed was as follows: Based on the study carried out in the state of Coahuila [1], the current situation is analyzed in detail, and the business woman and her company were diagnosed through the SWOT tool. Once this tool was implemented, the 4 most representative strengths were selected, as well as 4 weaknesses, 4 opportunities, and 4 threats, taking into account the greatest positive impact on the growth and success of businesses run by women. After selecting the characteristics mentioned above, strategies were designed for each one.

Chernov et al. [9] examines the implementation of the classical SWOT analysis method widely used for decision making in various economical problems. As it is known, it is a qualitative comparison of the multicriteria degree of Strength, Weakness, Opportunity, Threat for the different types of risks, foreseeing the evolution of markets, the situation, and development perspectives of companies, regions and economic sectors, territories, among others. Chernov et al. [10] contemplates the uncertainties, ambiguities, vagueness typical of business processes combined with the use of fuzzy logic and fuzzy set theory for an adequate representation and post-treatment in SWOT analysis. Feili et al. [11] proposes the use of the SWOT approach and fuzzy logic for the generation of sustainable tourism development strategies in the tourism sector. SWOT analysis is a tool commonly used to analyze the external and internal environments simultaneously in order to acquire a systematic approach and support for a decision situation; based on the analysis and calculation of each factor emanating from SWOT analysis, Feili et al. [11] present the appropriate strategies according to the weighting carried out through fuzzy logic analysis.

Taghavifard et al. [12] proposes the use of the fuzzy approach of Strengths, Weaknesses, Opportunities, and Threats (SWOT) with the aim of establishing strategic planning based on fuzzy logic and thus, solving the traditional strategic problem by planning key problems as internal and external factors in an imprecise and ambiguous environment.

In the case of the objective proposed in this work of establishing proposals for the development and success of women entrepreneurs, it is ideal to combine these two contributions mentioned above.

The SWOT-OA model was adapted in the evaluation diagrams of the Fuzzy Tree Studio software (logic trees), according to the number of characteristics of the company (8 Strengths and 4 Weaknesses) and the environment (8-4 Threats and 4 Opportunities) for the characteristics diagram; as well as the strategies (16) in the objectives diagram. The diagram was designed in the above-mentioned program (see the logic trees with their linguistic expressions in Appendix A), in which the diagram of the Linguistic expression of the characteristics in the diagram is shown in schematic form in order to determine the strategies based on the SWOT analysis. Six instruments were designed for the collection of information to be captured later in the Fuzzy Tree Studio (The instruments can be found in Appendix B).

A focal group was convened to gather information, made up of six women (successful businesswomen, feminist creators, and/or leaders of civil associations in the area of women's empowerment, and the representative of the Secretary of Women of the state of Coahuila in La Laguna (See Appendix C for the biographical records of each of them).

In order to process the information and obtain results for analysis, we began by emptying the information collected into an Excel file with the purpose of calculating a geometric mean that integrates the result into the consensus of all the participants in the focal group.

The Excel files were exported to the Fuzzy Tree Studio software.

The imported Excel matrices were evaluated doing the calculations established by the system (characteristics diagram to establish the hierarchy of the company's characteristics (Strengths and Weaknesses) and objectives diagram (to establish the priorities in the implementation of the strategies); obtaining the results that are synthesized in the body of the work.

4 Results

The SWOT analysis is shown in Tables 1, 2, 3 and 4, describing the Strengths, Weaknesses, Opportunities, and Threats to the left and the strategy designed for their potentiation, attention, use, or prevention respectively, to the right, which are presented below.

Table 1 shows the strengths that have been diagnosed through the instruments used to collect data and process this research.

Table 2 shows the weaknesses that have been diagnosed through the instruments used in the data collection and processing of this research, as well as the strategies to minimize them.

It shows on the left side the weakness or strength of the Coahuila woman's company, followed by her description, and on the right side, the value calculated using

Table 1 Strengths and strategies of the profile of women and their company of the state of Coahuila

Strengths	Strategies
Academic Level: 67% are women with a higher level, which represents that they have the skills to run a business, and of these, 57% are trained in business and/or economics, 12% in exact sciences, 12% in humanities, 7% in engineering and the rest of social or natural sciences, which reflects that they can assume various sectors	To be in constant update exploiting its capacities of learning and implementing within its organization the necessary changes for the growth of its company
Age: 76% are women under 55, women of this age already have maturity and orientation of the objectives and goals they have and want, and 37% under 40, which are women with many opportunities to grow and develop successfully because they still have enough life to continue to improve and grow entrepreneurially	Women of this age already have maturity and orientation of what they have and want; it is recommended to establish in writing the short, medium, and long term goals because they are women with many opportunities to grow and develop successfully because they still have enough life to continue to improve and grow entrepreneurially taking advantage of what is available to them
Changes and improvements: Women are open and active in making changes and improvements in their products and services (75%) and regularly introduce new products (57%) and acquire new equipment (62%) for their organization	To offer innovative and attractive products and/or services, improving them day by day, being part of an innovative process, updating the infrastructure, equipment, and material, as far as possible without risking profitability and growth looking for sources of financing. Eliminate factors of resistance to the change of individuals in their different roles, both the entrepreneur and the workers. Purchase of equipment according to the product and scale of production
Accounting aspects: 68% of women's businesses reinvest profits	Determine and ensure a fixed percentage of profits so that they are constantly reinvested

Table 2 Weaknesses and strategies of the profile of the woman and her company of the state of Coahuila

Importance of the company's characteristics		Calculated value
Weakness	*Reconciliation of work and family*: 68% of women are mothers and have to divide their time between home and work	0.8667
Strength	*Academic Level*: 67% are women with a higher level, which represents that they have the skills to run a business, and of these, 57% are trained in business and/or economics, 12% in exact sciences, 12% in humanities, 7% in engineering and the rest of social or natural sciences, which reflects that they can assume various sectors	0.8022
Strength	*Changes and improvements*: Women are open and active in making changes and improvements in their products and services (75%) and regularly introduce new products (57%) and acquire new equipment (62%) for their organization	0.7770
Weakness	*ICT*: Less than half of women's businesses have a website, do not buy over the Internet, just over half do, micro does not do marketing through the Internet, and small and medium only 48% do, women's MSMEs do not have corporate Internet	0.7681
Strength	*Accounting aspects*: 68% of women's businesses reinvest profits	0.7466
Strength	*Age*: 76% are women under 55, women of this age already have maturity and orientation of the goals and objectives they have and want, and 37% under 40, which are women with many opportunities to grow and develop successfully because they still have enough life to continue to improve and grow entrepreneurially	0.7410
Weakness	*Strategic Planning*: 47% of companies that perform strategic planning do so for a period of 1 year or less, 7% more than 1 year, and 53% do not perform strategic planning	0.7280
Weakness	*Formal control systems*: 66% of women's micro and SME's hardly use managerial control systems, do not carry out internal auement quality controls, fail to reaffirm economic and financial analyses, as well as budgetary control, and 47% of micro-enterprises do not implement accounting and costs	0.6834

fuzzy logic, showing the order of importance to be considered in decision-making. Own elaboration.

Table 3 shows the importance of each opportunity and threat to the company in descending order, showing the calculated value of each one, thus making known the priority that we must attend for decision making.

It shows on the left side the opportunity or threat presented in the Coahuila woman's company, followed by the description of the same, and on the right side, the value calculated using fuzzy logic, giving the order of importance to be considered in decision making. Own elaboration.

Whether to improve strengths, combat weaknesses, take advantage of opportunities, or to prepare for latent threats, a series of strategies were designed, which when

Table 3 Order of importance of environmental characteristics

Importance of the characteristics of the environment		Calculated values
Opportunity	*Associations*: The existence of INMUJERES, GEM, AMMJE, RME	0.8488
Opportunity	*Government loans*: Opening to women entrepreneurs for government financing through the Ministry of Economy	0.8129
Opportunity	*Presence*: The considerable increase in female participation in the business environment	0.8020
Threat	*Discrimination and inequity*: 39% of women have suffered labor discrimination. 80% of the women affirm the existing inequality, which causes a greater effort in comparison with the men to demonstrate their capacities	0.7858
Threat	*Business environment*: They perceive that new companies easily enter to compete (45%), a high competition (68%), 37% perceive the facility to create products substitutes to those manufactured ´for their sector	0.7805
Opportunity	*Laws*: Gender equality program in Coahuila	0.7751
Threat	*The working environment*: The working climate and environment are in favor of gender man	0.7172
Threat	*Funding*: Women find it difficult to access credit more often than men, as they are more easily denied than men	0.5398

subjected to fuzzy logic, throw the priority or importance for implementation in the organization, thus achieving not only combat or take advantage of the characteristics for which they were designed, Table 7 shows the importance of each strategy in descending order, showing the calculated value of each one, thus making known the strategy with the highest priority that we must take into account when making decisions for its implementation.

Table 4 shows on the left side the Strategies that were designed to be implemented in the Coahuila woman's enterprise and on the right side, the value calculated through fuzzy logic, showing the order of importance to be considered in decision making. Own elaboration.

In this way, with the help of fuzzy logic, the correct strategies can be implemented by knowing the priorities to be addressed and thus increasing the probability of success of the Coahuila women's business. This is the reason for this article, which shows that the application of this discipline gives added value to the SWOT tool thanks to the calculation of truthfulness, converting the analyses into controllable measurements and, at the same time, supporting the success of the women entrepreneurs.

Table 4 Order of importance of strategies to be implemented in the Coahuila women's enterprise

Importance of strategies	Calculated values
To offer innovative and attractive products and/or services, improving them day by day, being part of an innovative process, updating the infrastructure, equipment, and material, as far as possible without risking profitability and growth looking for sources of financing. Eliminate factors of resistance to the change of individuals in their different roles, both the entrepreneur and the workers. Purchase of equipment according to the product and scale of production	0.7900
Perform benchmarking knowing the market and scope of competitors to improve, become more competitive, and seek leadership, achieving customer satisfaction and preference. Offer a diversity of products with the same function and different prices covering different markets and socioeconomic levels	0.7837
To be in constant update exploiting its capacities of learning and implementing within its organization the necessary changes for the growth of its company	0.7780
It is necessary to know the sources of financing available and within reach, to have the personal will to obtain external financing, knowing beforehand that the degree of independence and control may diminish	0.7603
Subscribe to an association of women entrepreneurs, promoting the unity, support, knowledge, and motivation that these associations offer	0.7325
Make changes and/or improvements that promote the inclusion of women in the business environment	0.7203
Determine and ensure a fixed percentage of profits so that they are constantly reinvested	0.7147
Establish the Mission and Vision of the company to know the reason for the organization and where you want to go. Diagnose the current situation of the company identifying the problems to be solved. Establish possible solutions and study the resolution process. Make plans with their respective strategies to implement for the resolution of problems, in addition to establishing goals or objectives in the short, medium and long term aligned with the Mission and Vision already established	0.7020
Seek and participate in the support programs offered by the federal government through different entities such as the Secretariat of Economy in its multiple calls	0.6834
Make use of the laws that the state of Coahuila offers through the equity and gender program	0.6648
Use available policies in favor of gender equity and promote new ones that benefit women in the workplace	0.6603

(continued)

Table 4 (continued)

Importance of strategies	Calculated values
To develop articulated controls or systematic registers in the areas that compose the company. Operate with objective and measurable criteria of profitability, with the purpose of knowing the real utilities or in what measure reasonable levels of profitability are reached. Implement an inventory control system. Cross information of suppliers and lists of critical inputs. Have an extensive amount of information. Make a Balanced Scorecard of the company. Carry out audits every 6 months. Implement the mandatory use of budget control, cost analysis. Point of equilibrium. Financial Balances, graphs, etc.	0.6412
Establish and comply with specific schedules for work and family. Establish specific spaces for work where areas in common with the home do not intervene	0.6329
Acquire a domain on the web. Offer and sell their products and/or services through the web. To be supplied by means of the Internet if the price is more convenient. To install a corporative network in the company maintaining better communication and to increase the efficiency of the times of information and productivity	0.6266
Take advantage of the active presence of women who are joining the business world and join them by encouraging the creation of networks or associations of women entrepreneurs in the region	0.5981
Establish in writing the short, medium, and long term goals since they are women with many opportunities to grow and develop successfully because they still have enough life to continue to improve and grow entrepreneurially taking advantage of what is at their disposal	0.5760

5 Conclusions

According to the results of the SWOT-OA analysis based on LDC, the weakness that must be addressed with the highest priority to resolve is the reconciliation between work and family life, since 68% of women are mothers, and the most important strength is that 67% of women entrepreneurs have a higher level, 57% in the business area. The opportunities to prioritize in their use are the existence of women's associations and the opening to women entrepreneurs for government financing through the Ministry of Economy. The first order strategy was to offer innovative products and/or services, followed by benchmarking and constant updating, these being the first 3 of the 16 proposed strategies.

Appendix A

Diagrams for the evaluation of the characteristics of the company and the environment and of the strategies.

Diagram of the Characteristics

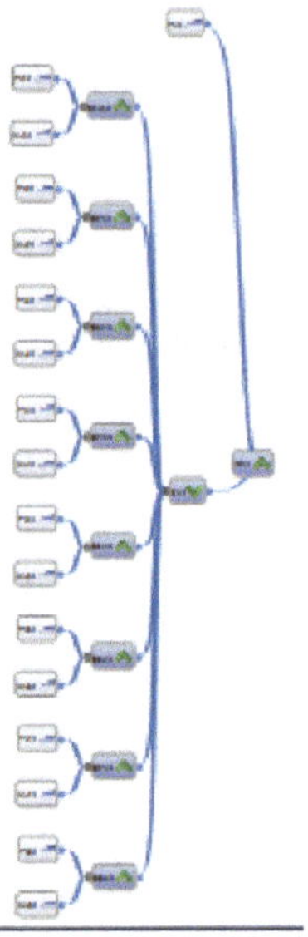

Linguistic expression of the characteristics diagram:

{The characteristic i of the organization (the environment) is present and {{The characteristic 1 of the environment (of the company) is present and i and 1 constitute a possible impact}. The characteristic 2 o f the intonation (of the organization) is present and i and 2 constitute a possible impact} The characteristic 3 of the environment is present and i and 3 constitute a possible in impact} The characteristic 4 of the environment is present and i and 4 constitute an impact}. {Characteristic 5 is present and i and 5 are a possible impact} {Characteristic 6 is present and i and 6 constitute a possible impact} {Feature 7 is present and (I(i,7))} or {Feature 8 of the is present and i constitutes a possible impact}}}.

Objectives diagram (used to evaluate and prioritize strategies).

Linguistic expression of the target diagram:

{{Characteristic 1 is important and characteristic 1 recommends the approach of objective i}, {(I2) and (R(i,2))}. (I3) and (R(i,3))} (I4) and (R(i,4))} {(I5) and (R(i,5))} (I6) and (R(i,6))} {(I7) and (R(i,7))} (I8) and (R(i,8))} {(I9) and (R(i,9))} (I10) and (R(i,10))} (I11) and (R(i,11))} (I12) and (R(i,12))} (I13) and (R(i,13))} {(I14) and (R(i,14))} (I15) and (R(i,15))} (I16) and (R(i,16))}.

Appendix B

Information-Gathering Tools (Swot-Oa)

Presence of general characteristics.

SWOT	Presence of characteristic
Academic level:	
Age:	
Changes and improvements:	
Accounting aspects:	
Reconciliation of work and family:	
Strategic planning:	
ICT:	
Formal control systems:	
Associations:	
Laws:	
Government credits:	

(continued)

(continued)

SWOT	Presence of characteristic
Presence:	
Discrimination and inequity:	
Funding:	
The working environment:	
Business environment:	

Intensity of the Impacts of the company's characteristics on the characteristics of the environment.

How true is it that each of the strengths allows us to take advantage of the opportunities?		Opportunities			
		Existence of women's associations	New laws that promote gender equality	Government credits and support for women	Increase in women's participation in the business environment
Strengths	High academic level				
	Productive age between 35 and 55, women with vision, maturity, and experience				
	Active and willing to implement changes and improvements in their organizations				
	Reinvest their profits				

How true is it that these strengths can counteract these threats?		Threats			
		Discrimination and inequity still exist	It is not easy to access bank financing	The working climate and environment are in favor of men	The business environment is very competitive
Strengths	High academic level				

(continued)

(continued)

How true is it that these strengths can counteract these threats?		Threats			
		Discrimination and inequity still exist	It is not easy to access bank financing	The working climate and environment are in favor of men	The business environment is very competitive
	Productive age between 35 and 55, women with vision, maturity, and experience				
	Active and willing to implement changes and improvements in their organizations				
	Reinvesting their profits				

How true is it that each of these weaknesses prevents us from taking advantage of these opportunities?		Opportunities			
		Existence of women's associations	New laws that promote gender equality	Government credits and support for women	Increase in women's participation in the business environment
Weaknesses	Difficulty reconciling work and family life				
	Do not carry out strategic planning				
	They have no control over the efficient management of the TIC				
	Lack of internal control systems				

How true is it that these weaknesses make the organization more vulnerable to threats?		Threats			
		Discrimination and inequity still exist	It is not easy to access bank financing	The working climate and environment are in favor of men	The business environment is very competitive
Weaknesses	Difficulty reconciling work and family life				
	Do not carry out strategic planning				
	Do not have mastery over the efficient management of ICTs				
	Lack of internal control systems				

Determination of the Veracity That the Company's Characteristics Recommend-Strategies.

ESTRATEGIAS	FORTALEZAS				DEBILIDADES				OPORTUNIDADES				AMENAZAS			
	Nivel Académico	Edad	Cambios y mejoras	Aspectos contables	Conciliación laboral y familiar	Planeación Estratégica	TIC	Sistemas formales de control	Asociaciones	Leyes	Créditos gubernamentales	Presencia	Discriminación e inequidad	Financiamiento	El entorno laboral	Entorno empresarial
Mantenerse en constante actualización explotando sus capacidades de aprendizaje e implementando dentro de su organización los cambios necesarios para el crecimiento de su empresa.																
Establecer por escrito las metas a corto, mediano y largo plazo ya que son mujeres con muchas oportunidades para crecer y desarrollarse con éxito debido a que aún les queda suficiente vida para seguir perfeccionándose y crecer empresarialmente aprovechando lo que hay a su disposición																
Ofrecer productos y/o servicios innovadores y atractivos, mejorándolos día a día, siendo parte de proceso innovador, actualizando la infraestructura, equipo y material, dentro de lo posible sin arriesgar la rentabilidad y crecimiento buscando fuentes de financiamiento. Eliminar factores de resistencia al cambio de los individuos en sus distintos roles, tanto de la empresaria como de las y los trabajadores. Compra de equipo de acuerdo con el producto y escala de la producción.																
Determinar y asegurar un porcentaje fijo de las utilidades para que se reinviertan constantemente.																
Establecer y cumplir horarios específicos para el trabajo y para la familia. Establecer espacios específicos para el trabajo donde no intervengan áreas en común con el hogar.																
Establecer la Misión y Visión de la empresa para conocer la razón de la organización y hacia donde se quiere llegar. Diagnosticar la situación actual de la empresa identificando los problemas a resolver. Establecer posibles soluciones y estudiar el proceso de resolución. Realizar planeaciones con sus respectivas estrategias a implementar para la resolución de los problemas, además de establecer metas u objetivos a corto, mediano y largo plazo alineadas con la Misión y Visión ya establecidas.																
Adquirir un dominio en la web. Ofrecer y vender sus productos y/o servicios a través de la web. Abastecerse por medio de internet si el precio es más conveniente. Instalar red corporativa en la empresa manteniendo una mejor comunicación y eficientando tiempos de información y productividad.																
Desarrollar controles articulados o registros sistemáticos en las áreas que componen la empresa. Operar con criterios objetivos y medibles de rentabilidad, con la finalidad de conocer las utilidades reales o en qué medida se alcanzan niveles razonables de rentabilidad. Implementar sistema de control de inventarios. Cruzar la información de las y los proveedores y los listados de insumos críticos. Disponer de una extensa cantidad de información. Realizar un Cuadro de mando integral de la empresa. Realizar auditorías cada 6 meses. Implementar el uso obligatorio de control presupuestario, Análisis de costos. Punto de equilibrio. Balances Financieros, gráficos, etc																
Suscribirse en alguna asociación de mujeres empresarias, fomentando la unidad, apoyo, conocimientos y motivación que dichas asociaciones ofrecen, siendo parte del																
Hacer uso de las leyes que el estado de Coahuila ofrece a través del programa de equidad y género.																
Es necesario conocer las fuentes de financiamiento disponibles y al alcance, tener la voluntad personal para obtener financiamiento externo, conociendo de antemano que puede disminuir el grado de independencia y control.																
Aprovechar la presencia activa de las mujeres que se van sumando al ámbito empresarial y unirse a ellas fomentando la creación de redes o asociaciones de mujeres empresarias en la región.																
Recurrir a las políticas disponibles a favor de la equidad de género e impulsar nuevas que beneficien a la mujer en el ámbito laboral																
Buscar y participar en los programas de apoyo que ofrece el gobierno federal a través de las distintas entidades como la Secretaría de economía en sus múltiples convocatorias																
Realizar Cambios y/o mejoras que fomenten la inclusión de las mujeres en el ámbito empresarial																
Realizar benchmarking conociendo el mercado y ámbito de los competidores para mejorar, ser más competitivos y buscar el liderazgo logrando la satisfacción y preferencia de los clientes. Ofrecer diversidad de productos con la misma función y diferentes precios abarcando diferentes mercados y niveles socioeconómicos.																

Appendix C

Biographical data sheets of the Focus Group were created for the application of the SWOT-OA system in Fuzzy Tree Studio based on compensatory fuzzy logic

Name	Biographical record
Lorena Torres Zamora	Responsible for regional coordination (La Laguna) of the secretariat of women in Coahuila
Erika García Graciano	Businesswoman of the metal-mechanic business with 15 years of experience in LUANER workshops
Magaly Villarreal Garza	Businesswoman of the decorative industry with 31 years of experience in the same business
Roxana Chávez Bermúdez	Businesswoman of the commercial business, jewelry, and accessories for women with 15 years of experience in the same business
Luz Elena Martínez García	Feminist. Founder of Civil Associations for women's rights. She currently directs the DIVERSA Foundation
María Evangelina Velázquez Reyes	Feminist. Director and Founder of MUSAS (Mujeres Solidarias en Acción Social de La Laguna)

References

1. Oyervides, M., Ramos, L.G., Chavarría, S.L.: La Mujer Empresaria en Coahuila: Sus Motivaciones Para Emprender y la Conciliación Entre la Vida Familiar y Laboral. Revista Internacional Administración y Finanzas **8**(7), 105–122 (2015)
2. D´Negri, C.E., De Vito, E.L.: Introducción al razonamiento aproximado: lógica difusa. Revista Argentina de Medicina respiratoria (4), 126–136 (2006).
3. Espín Andrade, R.A., González, E., Pedrycz, W., Fernández, E.: An interpretable logical theory: the case of compensatory fuzzy logic. Int. J. Comput. Intell. Syst. **9**(4), 612–626 (2016). https://doi.org/10.1080/18756891.2016.1204111
4. Espín-Andrade, R.A., González Caballero, E., Pedrycz, W., Fernández González, E.R.: Archimedian compensatory fuzzy logic system. Int. J. Comput. Intell. Syst. **8**, 54–62 (2015). https://doi.org/10.1080/18756891.2015.1129591
5. Knors, A.M., Vanti, A.A., Espín Andrade, R.A., Johann, S.L.: Aligning information security with the image of the organization and priorization based on fuzzy logic for the industrial automation sector. J. Inf. Syst. Technol. Manag. **8**(3), 555–580 (2011)
6. Espín Andrade, R.A., Vanti, A.A.: Administración Lógica: un estudio de ca-so en empresa de comercio exterior. Revista BASE **1**(3), 4–22 (2005)
7. Vanti, A. A., Espín Andrade, R. A., Goyer, D., Schripsema, D.: The importance of objectives and strategic lagging and leading indicators definition in the chain import and export process in the light of strategic planning through the use of fuzzy logic system. In: ACM SIGMIS Conference Personnel Resource (CPR), New York (2006). https://doi.org/10.1145/1125170.1125220
8. Chao-Ballester, A., Espín-Andrade, R. A.: Metodología para la gestión del conocimiento y la toma de decisiones basado en lógica difusa compensatoria. In: Más-Basnuevo, A., Pomín

Valentín, M.L. (eds.) Inteligencia Organizacional, pp. 163–194. Cultura Académica, Sao Paulo (2015)
9. Chernov, V., Dorokhov, O., Dorokhova, L.: Fuzzy approach to innovative programs development in conditions of partial and full uncertainty. Bull. Transilvania Univ. Braşov **9**(58), 20 (2016)
10. Chernov, V., Dorokhov, O., & Dorokhova, L.: Fuzzy logic approach to SWOT analysis for economics tasks and example of its computer realization. Bull. Transilvania Univ. Braşov **9**(58), 10 (2016) (2017)
11. Feili, H., Qomi, M., Sheibani, S., Azmoun, G.: SWOT analysis for sustainable tourism development strategies using fuzzy logic. Presented at the 3rd International Conference of Science & Engineering in the Technology Era, Copenhagen, Denmark, 11 (2017)
12. Taghavifard, M.T., Amoozad Mahdiraji, H., Alibakhshi, A.M., Zavadskas, E.K., Bausys, R.: An extension of fuzzy SWOT analysis: an application to information technology. Information **9**(3), 46 (2018). https://doi.org/10.3390/info9030046

Fuzzy Logic-Based Approaches in Supply Chain Risk Management: A Review

Alina Díaz-Curbelo, Ángel Manuel Gento Municio, and Rafael Alejandro Espin-Andrade

Abstract Uncertainty is inherent in the supply chains nature. In the context of various uncertainties, risk management plays a crucial role in effective supply chain management. The uncertainty involved in the risk assessment process can be divided into two types: random uncertainty and epistemic uncertainty. The fuzzy theory has been applied to address uncertainties in this context. The purpose of this paper is to develop a literature review of the major contributions of fuzzy logic in addressing uncertainty in supply chain risk management approaches. The results revealed that integration with disruptive analysis tools and multi-criteria decision-making methods are the most common types adopted, with the increasing trend of Petri nets and Bayesian approaches. The reviewed literature highlights some limitations related to the holistic complexity of risks in supply chains, the dynamic nature of the environment, and the reliability of the knowledge base in the assessment. In that sense, these observations reveal interesting future lines of research.

Keywords Fuzzy sets · Risk management · Uncertainty · Subjectivity · Supply chain · Decision making

1 Introduction

Supply chain risk management (SCRM) is considered an area that has gained increasing attention in recent years [1]. An effective risk management procedure can mitigate critical effects on Supply Chains (SCs).

The nature of SCs presents a variety of issues related to uncertainties, such as supply deliveries, unexpected changes in the flow of materials due to delays or interruptions, changes in demand, and the long logistics cycle that influences the

A. Díaz-Curbelo (✉) · Á. M. Gento Municio
University of Valladolid, 47011 Valladolid, Spain
e-mail: alina.diaz@uva.es

R. A. Espin-Andrade
Autonomous University of Cohauila, 25280 Saltillo, Coahuila, Mexico

W. Pedrycz et al. (eds.), *Computational Intelligence for Business Analytics*,
Studies in Computational Intelligence 953,
https://doi.org/10.1007/978-3-030-73819-8_5

availability of materials and stock alternatives [2]. This leads to imprecision and bias in the decision-making process. This inaccuracy and ambiguity, caused by unmeasured, incomplete, and unattainable information, is one of the largest disadvantages of the SCRM techniques.

The ability of organizations to understand and manage the increasingly interconnected and uncertain nature of risks enables better risk-based decision making. The effectiveness and efficiency of the organization are increased when the strategy to reduce uncertainty takes into account the context and realities of the environment. Context can be interpreted as a reference to the sources of risk, the magnitude of the risk, its relationship to business objectives, and the threat of disruption to SCs. The realities of the environment can be interpreted in terms of the degree of exposure to adverse events, the extent of extended SCs, supplier management practices, etc.

The uncertainty involved in the risk assessment process can be divided into two types: random uncertainty and epistemic uncertainty [3]. Random uncertainty, more related to probability theory, is irreducible. It refers to the inherent randomness that comes from natural variability. However, epistemic uncertainty is controllable and results from limited or inaccurate data, lack of information, and approximations in the mathematical model [1].

In most cases, due to the lack of sufficient data for probabilistic analysis, SC risks are often managed on the basis of expert judgment and experience. The word "uncertainty" has many different nuances, ranging from the randomness of events to the lack of knowledge of a system. This is due to the logical gap between "being an expert" and "being able to estimate probabilities accurately". In this sense, much of the type of data processed for risk studies are mainly qualitative rather than quantitative [4]. It is these qualitative data that are often found as linguistic variables such as "likely, very important or low", etc., rather than numerical values. These linguistic variables express imprecise and vague information rather than acute numerical values.

The linguistic variable has values such as words or sentences in a natural or artificial language. Fuzzy Logic (FL) is a powerful tool for modeling linguistic data [5]. Linguistic terms are converted into numerical terms using fuzzy operators [6, 7]. In this way, FL can reflect the human thought system, and it is a hot topic for understanding and assessing SC risks.

Hence, fuzzy set theory is an effective tool for quantifying or capturing the vagueness of language variables. Several researchers [8–12] argue that classical risk assessment tools could not accurately reflect human thinking and information needs. In this sense, linguistic expressions applied in fuzzy numbers have become recurrent tools to describe the linguistic variables in risk assessment.

In this sense, the aim of this paper is to develop a literature review of the main existing methods to SCRM that have integrated fuzzy theory in their approaches to articulate the epistemic uncertainty with the random and interdependent nature of risk events in the SC, in search of risk assessments more reliable.

The rest of this document is organized as follows. Section 2 presents the methodology carried out to develop the literature review. Section 3 shows the results of the reviewed approaches. Then, Sects. 4 and 5 show a brief discussion, identification of future lines, and concluding remarks with an integrated perspective of the SCRM.

2 Methodology

According to Denyer and Tranfield [13], the Systematic Literature Review (SLR) is a proven method that examines the literature sources of a specific topic with the aim of arriving at an organized result based on the current accumulated knowledge of the topic in question. An SLR seeks to classify and analyze contributions to the literature in a specific research area. The most important advantage of this method is that it consists of a series of commonly accepted steps so that it can be easily verified or replicated by other researchers.

The SLR is considered to be an aid to researchers, as it allows the synthesis of available studies on a particular topic and provides scientific knowledge to support practice. In this research, SLR has been used to integrate the information obtained from a set of individual studies of SCRM-FL. The results of the identified studies have been combined to provide a synthesis of the topic in this document and thus point to lines of future research.

This paper follows the five steps for an SLR proposed by Denyer and Tranfield [13] methodology (see Fig. 1).

These five steps are briefly described below and how they have been carried out in the specific framework of this research.

1. *Formulation of the research questions*

The first step in any SLR is to define the main research question or the specific questions to which the study is directed. The following research question (RQ) has been determined for the purpose of this study:

RQ: What are the main research contributions and directions in relation to the adoption of FL for SCRM?

2. *Identification of studies*

This step involves finding and locating relevant studies to answer the research question. A search of the Web of Sciences and Scopus bibliographic databases was carried out.

Three sets of keywords have been considered in the search chain. First, the keyword "supply chain" has been selected to frame the study and the set of related processes (e.g., supply, production, distribution, return). Secondly, the keyword "risk management" has been selected as an integrative discipline. RM involves four main stages: identification, assessment, mitigation, and monitoring. Thirdly, the integration set is represented by a series of words that reflect the uncertainty in the interdependent nature of risk events.

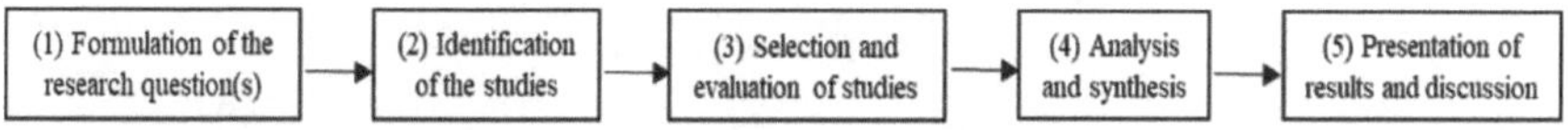

Fig. 1 Research methodology

It was determined that the terms should appear in the titles of articles, abstracts, or keywords. Literature was obtained from relevant journals in the areas of operations management, operations research, SC management, and a time period 2003–2018. According to Sodhi et al. [14], since 2003, there is an increase in the number of publications related to SCRM subject. This process identified a total of 376 documents.

3. *Selection and evaluation of studies*

After the first search phase, the summaries, methods, and tools used, the main contributions and the conclusions of the documents identified were carefully examined to determine whether they were relevant to the research question.

The following exclusion criteria were applied to select papers consistent with this research: (a) Papers that do not relate to the topic of this research; (b) Duplicate papers; (c) Publications non-refereed professional publications, such as textbooks, doctoral dissertations and conference proceedings, are excluded from our examination; and (d) Papers that were not indexed in journals with indicators of scientific quality, such as Journal Citation Reports (JCR) (Social Sciences Citation Index (SSCI) and Social Citation Index (SCI)), and Scimago Journal & Country Rank (SJR). In the final phase, the articles have been read in-depth. This process resulted in 70 papers for further analysis. In fact, we only include articles that report on an algorithm or model based on FL to represent inaccurate/vagueness knowledge and approximate reasoning integrated into SCRM methods, or studies that apply existing fuzzy models to address practical problems. This implies that articles that report fuzzy models that are not used to support decision-making in this discipline were excluded.

4. *Analysis and synthesis*

In this stage, the studies selected and evaluated in the previous stage were analyzed and synthesized. Each study was analyzed and grouped according to its thematic content. The studies were grouped by groups of methods and the RM stage. The main areas of application of the identified studies were also evaluated.

5. *Presentation of results and discussion*

In the following section, first, a descriptive analysis of the identified studies is made; then, the identified contributions are shown and grouped into five groups. Trends in identified methods and integration to RM processes are discussed. Then, the main areas that have been explored in practical applications are analyzed. Finally, research needs are identified as potential future lines of exploration.

3 Results and Discussion

FL emerged in the context of the theory of fuzzy sets [6]. A fuzzy set assigns a degree of belonging, typically a real number from the interval [0, 1] to the elements of a universe. A fuzzy number is a quantity with an imprecise value, rather than an exact value, as is the case with "ordinary" (single-value) numbers. A fuzzy number is a fuzzy subset of X. There are several types of fuzzy numbers, such as triangular and trapezoidal [15]. From many perspectives, fuzzy numbers represent the physical world more realistically than single valued numbers.

According to Feryal and Toktas [5] by citing Zadeh [6], FL not only consists of 0 and 1 as extreme cases of truth (or "the state of things" or "fact") but also includes the various intermediate states of truth [7]. Therefore, FL works closer to the way human brains work. In reality, FL can be seen as an approach to calculating with words rather than numbers. Although words are naturally less precise than numbers, their use comes closer to human intuition. The risk is more understandable as a fuzzy number than a crisp value.

Due to its adequacy for handling quantitative and qualitative data, FL has been increasingly used in the SC risk assessment process in recent years.

In recent years the use of fuzzy approaches integrated with SCRM methods in the search for a more effective and robust risk assessment has been highlighted. In the following section, we present a synthesis of the results of the literature review in this regard.

3.1 Fuzzy SCRM Approaches

Existing approaches to risk analysis can be grouped into three main categories: quantitative approaches, qualitative approaches, and the combination of quantitative and qualitative approaches. Its integration with multi-criteria decision-making tools (e.g. [10, 16–18]) has been notable, as well as with tools for disruption and dependency analysis in risk identification/modelling (e.g. [19–22]) problems of optimizing risk mitigation strategies (e.g. [23–25]), and combining them with other Artificial Intelligence tools (e.g. [19, 26]).

In practical problems, although it is very difficult to obtain sufficient statistical data, it may be possible to assess risk using these insufficient data in combination with subjective failure data based mostly on expert judgments. While some researchers have used a unique method, other researchers have focused on the integration and the combination of two or more methods depending on the aims. Table 1 shows a summary of the approaches identified in the literature review for this purpose.

Figure 2 shows the distribution of these publications in the studied period with a growing trend and Fig. 3 presents a list of 10 top journals in which the identified papers were published.

Table 1 Fuzzy theory-based approaches for SCRM

Methods	References
FL; AHP	[10, 17, 20, 27–34]
FL; AHP; SCOR model	[35]
FL; AHP, TOPSIS	[1, 8, 16]
FL; ANP	[36]
FL; ANP; goal programming; analysis of five forces; value at risk	[37]
FL; BN	[19, 20, 38–41]
FL; BN; AHP	[29, 42]
FL; BN; FMEA	[43]
FL; FMEA; geometric mean	[44]
FL; Bow-Tie analysis	[11, 45, 46]
FL; Bow-Tie analysis; FMEA; Lean Manufacturing	[47]
FL; DEA; DEA with restrictions; Monte Carlo simulation	[48]
FL; DEA; Monte Carlo simulation	[49]
FL, DEMATEL	[5, 50, 51]
FL; deterministic mathematical model; simulation	[25]
FL; ET	[52, 53]
FL; FTA	[21]
FL; BN; FTA	[54]
FL; FMEA	[22, 54–61]
FL; house of risk	[12]
FL; inoperability input–output model	[62]
FL; inoperability input–output model; global production network	[63]
FL; MCDM approaches	[60, 64–68]
FL; mean-risk optimization method	[24]
FL; mixed-integer non-linear mathematical model	[69]
FL; multi-objective mathematical programming	[23, 49, 70, 71]
FL; multi-objective stochastic programming	[9]
FL; neural networks; genetic algorithm	[15]
FL; PN	[72]
FL; PN; AHP; entropy method; cloud model	[26]
FL; QFD	[73]
FL; radial base function neural network	[19]

(continued)

Table 1 (continued)

Methods	References
FL; TOPSIS	[18, 74]

Abbreviations *AHP* analytic hierarchy process; *ANP* analytic network process; *BN* Bayesian network; *DEA* data envelopment analysis; *DEMATEL* decision making trial and evaluation laboratory; *ET* event tree; *FMEA* failure mode and effects analysis; *FTA* fault tree analysis; *MCDM* multi-criteria decision making; *PN* petri nets; *QFD* quality function deployment; *SCOR* supply chain operations reference model; *TOPSIS* technique of order preference for similarity with the ideal solution

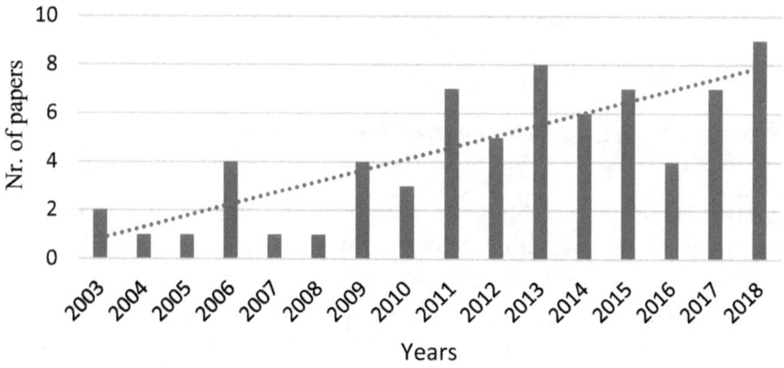

Fig. 2 Frequency of the reviewed papers

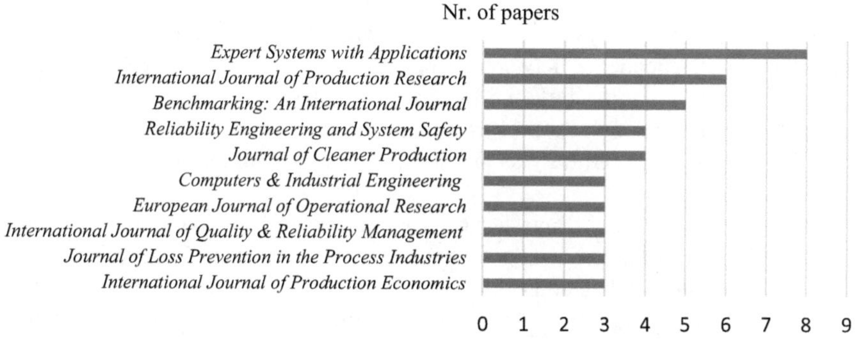

Fig. 3 Top 10 journals publishing articles exploring fuzzy SCRM-related methods

Considering the 70 reviewed papers, 19% adopt the integration of fuzzy theory with two or more methods. Disruption analysis tools (44.3%) and MCDM (43.01%) are the most common type adopted. In the stage of risk identification and modeling, ambiguous and inaccurate information is very frequent, based on the experience of possible interrelations between failure events. Among the disruption analysis artifacts, FMEA has been the most used (52.94%). However, the trend towards other methods (BN and PN) is observed, which are highlighted by their robustness in mitigating many of the limitations of the classical methods (ET, FTA, FMEA). In particular, AHP is the most frequent one of the MCDM methods (58.8%), and the trend in recent years of TOPSIS and DEMATEL methods is an interesting joined observation (25.53%).

Integration with Mathematical Programming methods is another remarkable combination (16.5%). Optimization of mitigation strategies is a resource frequently combined with simulation techniques. Considering all the 33 different techniques identified in the reviewed studies, many of them are used only once. This is the case of the techniques grouped under the label "Others", which includes common techniques in business analytics and SCM.

3.2 *Epistemic, Randomness and Dependency*

In this section, we want to highlight the triangulation of the fuzzy theory through the treatment of epistemic uncertainty with the nature of the other methods. In order to provide better visibility of the integrative analysis, we have grouped them in Fig. 4.

We have classified the methods into two dimensions: those based mainly on the interdependent analysis of failure events (Dependency) and those based mainly on probabilistic analysis and optimization of mitigation strategies (Randomness).

In the triggering analysis of risk events, the most frequent methods have been event tree, fault tree, and FMEA. These traditional models, despite their extensive use, present a number of limitations. The main shortcoming is that these approaches generally perform the analysis under unrealistic assumptions, e.g., consider statistical and stochastic independence between events; have a limited focus on capturing data on the causes of common failures; binary states of system components; and do not consider temporal behavior. However, in real-life systems, events have more conditioned dynamics, and this assumption could lead to an inadequate estimation of the reliability of the SC.

In this regard, Petri and Bayesian networks have been treated to address many of the limitations of the above approaches. They are two different approaches that are used either as individual approaches or in association with other methods. These approaches share potentialities such as allowing predictive analysis of system failure behavior, taking into account statistical, stochastic, and temporal dependencies of events. They, therefore, allow the analysis of practical systems with more realistic assumptions.

However, the main problem in most of these models is their exclusive focus on a specific problem without adapting the network to the SCRM domain and capturing the interdependent nature of risks in SC. In this sense, there is an added complexity in the synergistic performance of the network structure and processes that needs to be adapted to the SCRM context.

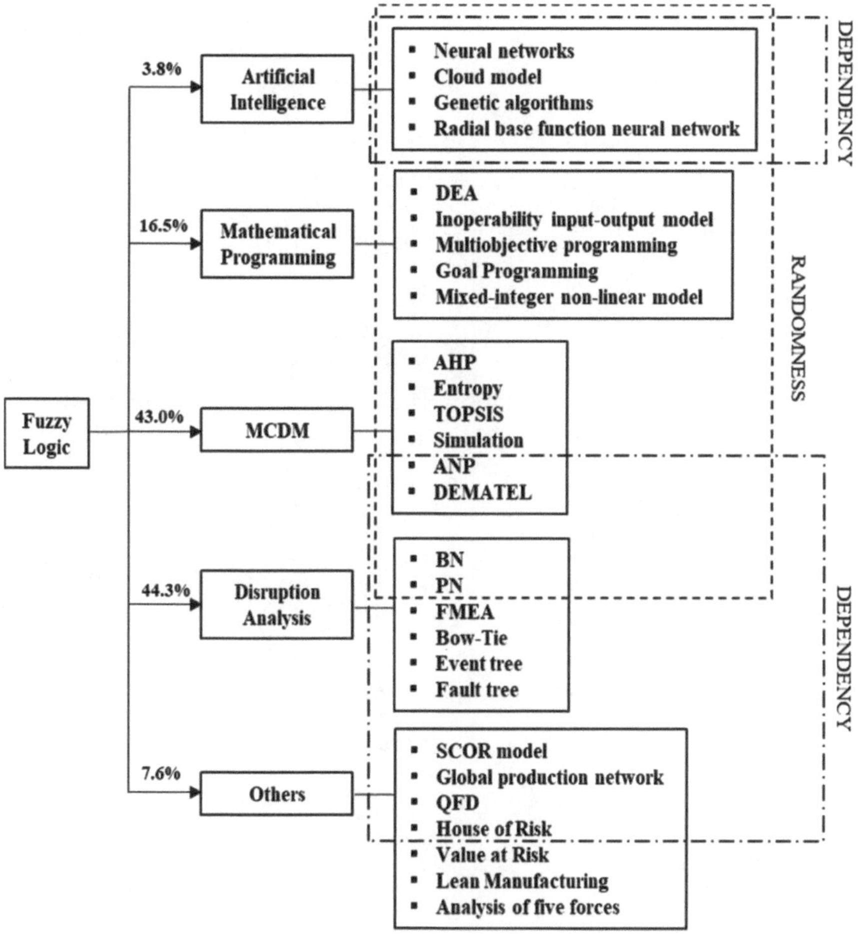

Fig. 4 Groups of integrated SCRM methods based on FL

3.3 RM Process

Another analysis considered in this research is related to the RM processes that have been identified in the reviewed publications. Table 2 groups the references identified in each RM process.

It is notable that the most studied stages are identification and assessment. These stages are mainly focused on MCDM and disruption analysis methods with aspects of modeling and disruptive analysis. In this analysis, some authors [5, 11, 45, 47, 53, 60] investigate interdependency analyses in risk assessment.

In the treatment process, optimization approaches are mostly found in the selection of risk mitigation strategies or alternatives. Monitoring is the least explored stage, and few authors consider all RM processes [11, 35, 47, 60, 64].

Table 2 RM processes and related studies

RM process	References
Identification	[1, 5, 10–12, 16–18, 20, 22, 25–31, 33, 35, 37, 42, 45–47, 49, 53, 55, 56, 60, 62–64, 67, 69, 72]
Assessment	[1, 5, 9–12, 16–18, 20, 22, 23, 25–31, 33, 35, 37, 42, 45–47, 49, 53, 55, 56, 60, 62–64, 67, 69, 72]
Treatment	[1, 9, 11, 12, 16, 22–24, 35, 36, 47, 48, 60, 64, 69]
Monitoring	[11, 35, 47, 60, 64]

Table 3 Application areas

Application areas	References
Aerospacial	[60]
Agrifood	[55, 56]
Automotive	[1, 27, 33]
Energy	[18, 26, 45, 46, 53, 62, 64]
Manufacturing	[5, 11, 12, 17, 22, 28, 31, 37, 64, 67]
Metallurgy	[16]
Textile	[16]
Naval, Maritime	[20, 35]
Chemical	[42, 47, 72]
Cases studies/simulations	[9, 10, 23–25, 29, 30, 36, 48, 49, 63, 69]

3.4 Application Areas

Table 3 shows the most important application areas obtained from the reviewed papers. While most of the research focused on a particular sector, a few focused on two or more sectors [16, 64].

It can be seen from Table 3 that SCRM methods have been mostly applied to manufacturing SCs, while service SCs are still under-explored. Also noteworthy is the energy sector, fundamentally in renewable energies, which emphasizes the world's concern for sustainable solutions. However, 17% of the papers reviewed used designed case studies or simulations as an alternative to real applications. Furthermore, applications are mainly limited to the private sector in relation to the public sector, so the literature could be extended in this sense.

3.5 Future Outlook

From the analysis of the reviewed studies, we have synthesized some research needs that may constitute future lines of research.

Many of the optimization studies focus on the solution of local optimal. This partial analysis underestimates the interdependent nature of the supply chain system. In this sense, holistic approaches challenge synergistic thinking and possible integrated optimization of mitigation strategies.

In turn, the probability values presented in the risk analysis are, in most cases, based on some background knowledge, but the strength of this knowledge is not reflected in the method. Future research may contribute to the reliability of risk level assessments by assessing the probability of human error or, instead, confidence levels of the estimation.

On the other hand, the use of probabilities to express uncertainty based on historical data when this information is available underestimates important aspects of the dynamics of the environment. How to learn from new information and integrate warning and decision support systems efficiently constitutes an ongoing challenge.

Time parameters are also an interesting topic for further discussion. Petri nets are one of the most widespread tools in this regard in risk assessment. However, in the representation of knowledge, it is an interesting question how to consider the duration of the failure or risk events in the severity of the consequences and in the system learning.

4 Concluding Remarks

In fact, FL is considered a useful tool with a greater tendency towards integrated approaches instead of individual methods, to facilitate informed reasoning in SC environments under uncertainty. These integrated perspectives allow for a more reliable risk assessment by combining the epistemic aspect with the interdependent and random nature of risk events. FL can also merge different kinds of parameters (e.g., quantitative and qualitative process). It is also useful when making decisions, since communicating the results of risk assessment in linguistic terms leads to a comprehensible approach for decision makers and the public.

This paper presented a literature review of 70 studies that propose integrated models to support SCRM based on FL. The results show a high concentration of studies published in 2013 and a growing trend over the evaluated period. Approximately 19% of the studies integrated methods of two or more methods. More than 40% of studies integrate FL to support the identification and evaluation of risk event interdependencies, and more than 70% is combined with random uncertainty. Disruption analysis tools and MCDM are the most explored types of methods. FMEA and AHP are the most common ones combined with Others, but growing trends towards Bayesian approaches are observed. Most of the studies do not include validation in real cases. In this case, some studies perform sensitivity analysis and simulation as validation tools. Once again, elements of integrative thinking can be appreciated, using the combination of different perspectives to assess and express uncertainty more reliably and accuracy in SCRM decision-making. Finally, we highlighted some challenges in order to provide motivation for future research related to the effectiveness of SCRM, where fuzzy theories can continue to play a relevant role. Therefore, investigating how to deal with the imprecise, uncertain, and vague nature of SCR knowledge information remains a path of research.

References

1. Salehi, S., Khanbabaei, M., Sabzehparvar, M.: A model for supply chain risk management in the automotive industry using fuzzy analytic hierarchy process and fuzzy TOPSIS. Benchmark. Int. J. **25**(9), 3831–3857. https://doi.org/10.1108/BIJ-11-2016-0167
2. Ho, W., Zheng, T., Yildiz, H., Talluri, S.: Supply chain risk management: a literature review. Int. J. Product. Res. **53**(16), 5031–5069 (2015). https://doi.org/10.1080/00207543.2015.1030467
3. Aven, T., Ylönen, M.: Safety regulations: implications of the new risk perspectives. Reliab. Eng. Syst. Saf. **149**, 164–217 (2016). https://doi.org/10.1016/j.ress.2016.01.007
4. Islam, M., Nepal, M.: A Fuzzy-Bayesian model for risk assessment in power plant projects. Proc. Comput. Sci. **100**, 963–970 (2016). https://doi.org/10.1016/j.procs.2016.09.259
5. Feryal, G., Toktas, P.: A novel fuzzy risk matrix based risk assessment approach. Kybernetes **47**(9), 1721–1751 (2018). https://doi.org/10.1108/K-12-2017-0497
6. Zadeh, L.: Fuzzy sets. Inf. Control **8**(3), 338–353 (1965). https://doi.org/10.2307/2272014
7. Ross, T.: Fuzzy logic with engineering applications, 3rd edn. Wiley, Chichester (2010). https://doi.org/10.1002/9781119994374
8. Berenji, H., Anantharaman, R., Karegar, M.: A new two-stage fuzzy decision making model in supply chain risk management. Int. Conf. Innov. Manage. Serv. **14**, 44–49 (2011)
9. Wu, D., Wu, D., Zhang, Y., Olson, D.: Supply chain outsourcing risk using an integrated stochastic-fuzzy optimization approach. Inf. Sci. **235**, 242–258 (2013). https://doi.org/10.1016/j.ins.2013.02.002
10. Radivojević, G., Gajović, V.: Supply chain risk modeling by AHP and fuzzy AHP methods. J. Risk Res. **17**(3), 337–352 (2014). https://doi.org/10.1080/13669877.2013.808689
11. Aqlan, F., Lam, S.: A fuzzy-based integrated framework for supply chain risk assessment. Int. J. Product. Econ. **161**, 54–63 (2015). https://doi.org/10.1016/j.ijpe.2014.11.013
12. Hoi-Lam, M., Wai-Hung, C.: A fuzzy-based house of risk assessment method for manufacturers in global supply chains. Indus. Manage. Data Syst. **118**(7), 1463–1476 (2018). https://doi.org/10.1108/IMDS-10-2017-0467
13. Denyer, D., Tranfield, D.: Producing a systematic review. In: The SAGE Handbook of Organizational Research Methods. Sage Publications Los, Ángeles (2009)
14. Sodhi, M., Son, B., Tang, C.: Researchers' perspectives on supply chain risk management. Product. Oper. Manage. **21**(1), 1–13 (2011). https://doi.org/10.1111/j.1937-5956.2011.01251.x
15. Huang, H., Chou, Y., Chang, S.: A dynamic system model for proactive control of dynamic events in full-load states of manufacturing chains. Int. J. Product. Res. **47**(9), 2485–2506 (2009). https://doi.org/10.1080/00207540701484913
16. Samvedi, A., Jain, V., Chan, F.: Quantifying risks in a supply chain through integration of fuzzy AHP and fuzzy TOPSIS. Int. J. Product. Res. **51**(8), 2433–2442 (2013). https://doi.org/10.1080/00207543.2012.741330
17. Kumar, S., Kumar, P., Kumar, B.: Risk analysis in green supply chain using fuzzy AHP approach: a case study. Resour. Conserv. Recycl. **104**, 375–390 (2015). https://doi.org/10.1016/j.resconrec.2015.01.001
18. Rostamzadeh, R., Ghorabaee, M., Govindan, K., Esmaeili, A., Khajeh, H.: Evaluation of sustainable supply chain risk management using an integrated fuzzy TOPSIS-CRITIC approach. J. Cleaner Product. **175**, 651–669 (2018). https://doi.org/10.1016/j.jclepro.2017.12.071
19. Zhang, L., Wu, X., Skibniewski, M., Zhong, J., Lu, Y.: Bayesian-network-based safety risk analysis in construction projects. Reliab. Eng. Syst. Saf. **131**, 29–39 (2014). https://doi.org/10.1016/j.ress.2014.06.006
20. John, A., Paraskevadakis, D., Bury, A., Yang, Z., Riahi, R., Wang, J.: An integrated fuzzy risk assessment for seaport operations. Saf. Sci. **68**, 180–194 (2014). https://doi.org/10.1016/j.ssci.2014.04.001

21. Kabir, S., Walker, M., Papadopoulos, Y., Rüde, E., Securius, P.: Fuzzy temporal fault tree analysis of dynamic systems. Int. J. Approx. Reason. **77**, 20–37 (2016). https://doi.org/10.1016/j.ijar.2016.05.006
22. Mangla, S., Luthra, S., Jakhar, S.: Benchmarking the risk assessment in green supply chain using fuzzy approach to FMEA. Insights from an Indian case study. Benchmark. Int. J. **25**(8), 2660–2687 (2018). https://doi.org/10.1108/BIJ-04-2017-0074
23. Yu, M., Goh, M.: A multi-objective approach to supply chain visibility and risk. Eur. J. Oper. Res. **233**(1), 125–130 (2014). https://doi.org/10.1016/j.ejor.2013.08.037
24. Yang, G., Liu, Y.: Designing fuzzy supply chain network problem by mean-risk optimization method. J. Intell. Manuf. **26**(3), 447–458 (2015). https://doi.org/10.1007/s10845-013-0801-7
25. Mostafaeipour, A., Qolipour, M., Eslami, H.: Implementing fuzzy rank function model for a new supply chain risk management. J. Supercomput. **73**, 3586–3602 (2017). https://doi.org/10.1007/s11227-017-1960-7
26. Guo, Y., Meng, X., Wang, D., Meng, T., Liu, S., He, R.: Comprehensive risk evaluation of long-distance oil and gas transportation pipelines using a fuzzy Petri net model. J. Nat. Gas Sci. Eng. **33**, 18–29 (2016). https://doi.org/10.1016/j.jngse.2016.04.052
27. Kutlu, A., Ekmekçioglu, M.: Fuzzy failure modes and effects analysis by using fuzzy TOPSIS-based fuzzy AHP. Expert Syst. Appl. **39**(1), 61–67 (2012). https://doi.org/10.1016/j.eswa.2011.06.044
28. Chan, F., Kumar, N.: Global supplier development considering risk factors using fuzzy extended AHP-based approach. Omega **35**(4), 417–431 (2007). https://doi.org/10.1016/j.omega.2005.08.004
29. Wang, X., Chan, H., Yee, R., Diaz-Rainey, I.: A Two-stage fuzzy-AHP model for risk assessment of implementing green initiatives in the fashion supply chain. Int. J. Product. Econ. **135**(2), 595–606 (2012). https://doi.org/10.1016/j.ijpe.2011.03.021
30. Viswanadham, N., Samvedi, A.: Supplier selection based on supply chain ecosystem, performance and risk criteria. Int. J. Product. Res. **51**(21), 6484–6498 (2013). https://doi.org/10.1080/00207543.2013.825056
31. Ganguly, K., Guin, K.: A fuzzy AHP approach for inbound supply risk assess-ment, Benchmark. Int. J. **20**(1), 129–146 (2013). https://doi.org/10.1108/14635771311299524
32. Gold, S., Awasthi, A.: Sustainable global supplier selection extended towards sustainability risks from (1+n)th tier suppliers using fuzzy AHP based approach. IFAC-PapersOnLine **48**(3), 966–971 (2015). https://doi.org/10.1016/j.ifacol.2015.06.208
33. Zimmer, K., Fröhling, M., Breun, P., Schultmann, F.: Assessing social risks of global supply chains: a quantitative analytical approach and its application to supplier selection in the German automotive industry. J. Cleaner Product. **149**, 96–109 (2017). https://doi.org/10.1016/j.jclepro.2017.02.041
34. Ganguly, K., Kumar, G.: Supply chain risk assessment: a fuzzy AHP approach. Oper. Supply Chain Manage. Int. J. **12**(1), 1–13 (2019). https://doi.org/10.31387/oscm0360217
35. Jiang, B., Li, J., Shen, S.: Supply chain risk assessment and control of port enter-prises: Qingdao port as case study. Asian J. Shipp. Logistics **34**(3), 198–208 (2018). https://doi.org/10.1016/j.ajsl.2018.09.003
36. Xiao, Z., Chen, W., Li, L.: An integrated FCM and fuzzy soft set for supplier selection problem based on risk evaluation. Appl. Math. Modell. **36**(4), 1444–1454 (2012). https://doi.org/10.1016/j.apm.2011.09.038
37. Hung, S.: Activity-based divergent supply chain planning for competitive Ad-vantage in the risky global environment: a DEMATEL-ANP fuzzy goal programming approach. Expert Systs. Appl. **38**(8), 9053–9062 (2011). https://doi.org/10.1016/j.eswa.2010.09.024
38. Wu, H.: Fuzzy reliability estimation using Bayesian approach. Comput. Indus. Eng. **46**(3), 467–493 (2004). https://doi.org/10.1016/j.cie.2004.01.009
39. Wu, H.: Fuzzy bayesian system reliability assessment based on exponential distribution. Appl. Math. Modell. **30**(6), 509–530 (2006). https://doi.org/10.1016/j.apm.2005.05.014
40. Ren, J., Jenkinson, I., Wang, J., Xu, D., Yang, J.: An offshore risk analysis method us-ing fuzzy bayesian network. J. Offshore Mech. Arct. Eng. **131**(4), 041101 (2009). https://doi.org/10.1115/1.3124123

41. Görkemli, L., Ulusoy, S.: Fuzzy Bayesian reliability and availability analysis of production systems. Comput. Indus. Eng. **59**(4), 690–696 (2010). https://doi.org/10.1016/j.cie.2010.07.020
42. Yazdi, M., Kabir, S.: A fuzzy Bayesian network approach for risk analysis in process industries. Process Saf. Environ. Protect. **111**, 507–519 (2017). https://doi.org/10.1016/j.psep.2017.08.015
43. Yang, Z., Bonsall, S., Wang, J.: Fuzzy rule-based Bayesian reasoning approach for prioritization of failures in FMEA. IEEE Trans. Reliab. **57**(3), 517–528 (2008). https://doi.org/10.1109/TR.2008.928208
44. Wang, Y., Chin, K., Poon, G., Yang, J.: Risk evaluation in failure mode and effects analysis using fuzzy weighted geometric mean. Expert Syst. Appl. **36**(2), 1195–1207 (2009). https://doi.org/10.1016/j.eswa.2007.11.028
45. Shahiar, A., Sadiq, R., Tesfamariam, S.: Risk analysis for oil and gas pipelines: a sus-tainability assessment approach using fuzzy based bow-tie analysis. J. Loss Prev. Process Indus. **25**(3), 505–523 (2012). https://doi.org/10.1016/j.jlp.2011.12.007
46. Ferdous, R., Khan, F., Sadiq, R., Amyotte, P., Veitch, B.: Analyzing system safety and risks under uncertainty using a bow-tie diagram: an innovative approach. Process Saf. Environ. Protect. **91**(1–2), 1–18 (2013). https://doi.org/10.1016/j.psep.2011.08.010
47. Aqlan, F., Mustafa, E.: Integrating lean principles and fuzzy bow-tie analysis for risk assessment in chemical industry. J. Loss Prev. Process Indus. **29**(1), 39–48 (2014). https://doi.org/10.1016/j.jlp.2014.01.006
48. Azadeh, A., Alem, S.: A Flexible deterministic, stochastic and fuzzy data envelopment analysis approach for supply chain risk and vendor selection problem: simulation analysis. Expert Syst. Appl. **37**(12), 7438–7448 (2010). https://doi.org/10.1016/j.eswa.2010.04.022
49. Wu, D., Olson, D.: Enterprise risk management: a DEA VaR approach in vendor selection. Int. J. Product. Res. **48**(6), 4919–4932 (2010). https://doi.org/10.1080/00207540903051684
50. Yadav, D., Barve, A.: Segmenting critical success factors of humanitarian supply chains using fuzzy DEMATEL. Benchmark. Int. J. **25**(2), 400–425 (2018). https://doi.org/10.1108/BIJ-10-2016-0154
51. Lin, K., Tseng, M., Pai, P.: Sustainable supply chain management using approximate fuzzy DEMATEL method. Resour. Conserv. Recycl. **128**, 134–142 (2018). https://doi.org/10.1016/j.resconrec.2016.11.017
52. Bidder, O., Arandjelović, O., Almutairi, F., Shepard, E., Lambertucci, S., Qasem, L., Wilson, R: A risky business or a safe BET? A fuzzy set event tree for estimating hazard in biotelemetry studies. Anim Behav **93**, 143–150 (2014).https://doi.org/10.1016/j.anbehav.2014.04.025
53. Javidi, M., Abdolhamidzadeh, B., Reniers, G., Rashtchian, D.: A multivariable model for estimation of vapor cloud explosion occurrence possibility based on a fuzzy logic approach for flammable materials. J. Loss Prev. Process Indus. **33**, 140–150 (2015). https://doi.org/10.1016/j.jlp.2014.11.003
54. Wang, Y., Xie, M., Ng, K., Meng, Y.: Quantitative risk analysis model of inte-grating fuzzy fault tree with Bayesian network. In: International Conference on Intelli-gence and Security Informatics (ISI), pp. 267–271. IEEE, Beijing (2011). https://doi.org/10.1109/ISI.2011.5984095
55. Braglia, M., Frosolini, M., Montanari, R.: Fuzzy criticality assessment model for failure modes and effects analysis. Int. J. Qual. Reliab. Manage. **20**(4), 503–524 (2003). https://doi.org/10.1108/02656710310468687
56. Pillay, A., Wang, J.: Modified failure mode and effects analysis using approximate reasoning. Reliab. Eng. Syst. Saf. **79**(1), 69–85 (2003). https://doi.org/10.1016/S0951-8320(02)00179-5
57. Sharma, R., Kumar, D., Kumar, P.: Systematic failure mode effect analysis (FMEA) using fuzzy linguistic modelling. Int. J. Qual. Reliab. Manage. **22**(9), 986–1004 (2005). https://doi.org/10.1108/02656710510625248
58. Meng, K., Peng, C.: Fuzzy FMEA with a guided rules reduction system for prioritization of failures. Int. J. Qual. Reliab. Manage. **23**(8), 1047–1066 (2006). https://doi.org/10.1108/02656710610688202

59. Liu, H., Liu, L., Bian, Q., Lin, Q., Dong, N., Xu, P.: Failure mode and effects analysis using fuzzy evidential reasoning approach and grey theory. Expert Syst. Appl. **38**(4), 4403–4415 (2011). https://doi.org/10.1016/j.eswa.2010.09.110
60. Chaudhuri, A., Mohanty, B., Singh, K.: Supply chain risk assessment during new product development: a group decision making approach using numeric and linguistic data. Int. J. Product. Res. **51**(10), 2790–2804 (2012). https://doi.org/10.1080/00207543.2012.654922
61. Rohmah, D., Dania, W., Dewi, I.: Risk measurement of supply chain organic rice product using fuzzy failure mode effect analysis in MUTOS Seloliman Trawas Mojokerto. Agric. Agric. Sci. Proc. **3**, 108–113 (2015). https://doi.org/10.1016/j.aaspro.2015.01.022
62. Aviso, K., Amalin, D., Promentilla, Angelo, M., Santos, J., Yu, K., Tan, R.: Risk assessment of the economic impacts of climate change on the implementation of mandatory biodiesel blending programs: A fuzzy inoperability input-output modeling (IIM) approach. Biomass Bioenergy **83**, 436–447 (2015). https://doi.org/10.1016/j.biombioe.2015.10.011
63. Niknejad, A., Petrovic, D.: Analysis of impact of uncertainty in global production networks' parameters. Comput. Indus. Eng. **111**, 228–238 (2017). https://doi.org/10.1016/j.cie.2017.07.011
64. Moeinzadeh, P., Hajfathaliha, A.: A combined fuzzy decision making approach to supply chain risk assessment. World Acad. Sci. Eng. Technol. **60**, 519–535 (2009). https://doi.org/10.5281/zenodo.1060613
65. Haleh, H., Hamidi, A.: A fuzzy MCDM model for allocating orders to suppliers in a supply chain under uncertainty over a multi-period time horizon. Expert Syst. Appl. **38**(8), 9076–9083 (2011). https://doi.org/10.1016/j.eswa.2010.11.064
66. Xia, D., Chen, B.: A comprehensive decision-making model for risk management of supply chain. Expert Syst. Appl. **38**(5), 4957–4966 (2011). https://doi.org/10.1016/j.eswa.2010.09.156
67. Khemiri, R., Elbedoui-Maktouf, K., Grabot, B., Zouari, B.: A fuzzy multi-criteria decisión making approach for managing performance and risk in integrated procurement–production planning. Int. J. Product. Res. **55**(18), 5305–5329 (2017). https://doi.org/10.1080/00207543.2017.1308575
68. Wang, Z., Ren, J., Goodsite, M., Xu, G.: Waste-to-energy, municipal solid waste treatment, and best available technology: comprehensive evaluation by an interval-valued fuzzy multicriterio decision making method. J. Cleaner Product. **172**, 887–899 (2018). https://doi.org/10.1016/j.jclepro.2017.10.184
69. Tabrizi, B., Razmi, J.: Introducing a mixed-integer non-linear fuzzy model for risk management in designing supply chain networks. J. Manuf. Syst. **32**(2), 295–307 (2013). https://doi.org/10.1016/j.jmsy.2012.12.001
70. Kumar, M., Vrat, P., Shankar, R.: A fuzzy programming approach for vendor selection problem in a supply chain. Int. J. Product. Econ. **101**(2), 273–285 (2006). https://doi.org/10.1016/j.ijpe.2005.01.005
71. Ji, G., Zhu, C.: A study on emergency supply chain and risk based on urgent relief service in disasters. Syst. Eng. Proc. **5**, 313–325 (2012). https://doi.org/10.1016/j.sepro.2012.04.049
72. Zhou, J., Reniers, G., Zhang, L.: A weighted fuzzy Petri-net based approach for security risk assessment in the chemical industry. Chem. Eng. Sci. **174**, 136–145 (2017). https://doi.org/10.1016/j.ces.2017.09.002
73. Liu, H.: The extension of fuzzy QFD: from product planning to part deployment. Expert Syst. Appl. **36**(8), 11131–11144 (2009). https://doi.org/10.1016/j.eswa.2009.02.070
74. Sahu, A., Sahu, N., Sahu, A.K.: Application of integrated TOPSIS in ASC index: partners benchmarking perspective. Benchmark. Int. J. **23**(3), 540–563 (2016). https://doi.org/10.1108/BIJ-03-2014-0021

Use of Fuzzy Logic in the Strategic Selection of Process Indicators

Israel Sánchez López, Rafael Alejandro Espin-Andrade, Liliana Guerrero Ramos, and Magaly Oyervides Villarreal

Abstract The need to process information for decision-making in companies and organizations is increasingly important in order to increase competitiveness, as well as the quality of products and services. The purpose of this research is to determine the importance of the processes for the achievement of the strategic objectives with their respective indicators used in the contemplated company; the perception of the organizational context by the managers of certain areas involved in the execution of the pertinent actions for the fulfillment of the strategic objectives was investigated, in order to see the conditions and the viability of their fulfillment. The analysis was carried out using the Fuzzy Tree Studio software, determining the processes to be prioritized for monitoring the fulfillment of strategic objectives. The processes most closely related to the objectives were: Use of the installed capacity in the Planning Area; Non-compliance in the Production Area, and Productivity in real units in the Production Area.

Keywords Strategic objectives · Strategic processes · Fuzzy logic · Decision-making

1 Introduction

The deficiency in industrial processes is a serious problem, especially for medium-sized companies in the metal-mechanical sector, so that decision makers in organizations take various measures to reduce errors and setbacks, as well as to optimize efforts to increase production and improve product or service delivery on time. Over time, organizations accumulate a large amount of information, a fundamental

I. Sánchez López (✉)
Universidad Juárez del Estado de Durango, 35010 Gómez Palacio, Durango, México

R. A. Espin-Andrade · L. Guerrero Ramos
Universidad Autónoma de Coahuila, 27000 Torreón, Coahuila, México

M. Oyervides Villarreal
Tecnológico Nacional de México, 27000 Torreón, Coahuila, México

W. Pedrycz et al. (eds.), *Computational Intelligence for Business Analytics*, Studies in Computational Intelligence 953,
https://doi.org/10.1007/978-3-030-73819-8_6

asset whose intelligent use can give a company a competitive advantage over other organizations [1].

The tendency is for companies in any field to pay special attention to the decision-making process they carry out throughout the management stages carried out at hierarchical levels [2]. The strategic planning process is oriented to key result areas and is considered functional when weaknesses are diminished, strengths are increased, the impact of threats is addressed in a timely manner, and opportunities are capitalized on in the achievement of the organization's strategic objectives [3]. The quantity and quality of information that companies possess, when faced with the decision-making process, plays a fundamental role in reducing the degree of uncertainty that generally characterizes the business environment.

Organizations make a large investment in hiring trained personnel to perform specific tasks that include decision-making in the management, operation, and evaluation of the function for which they are responsible. For this reason, companies have placed special value on the use of information systems; suppliers of the elements for decision making that correspond to the problem detected, as well as the achievement of the goal set. It can be recognized that the objective of this type of application is to support the personnel responsible for the administration of a function, area, or the entire organization in the best performance of their task, especially in decision making [1]. Knowledge, reasoning, and decision making have a profound relationship. The decision-making process for human agents involves a way of reasoning using a body of knowledge about decision alternatives, simultaneously incorporating their subjective reflection into the mind of the decision-maker [4].

The basis for decision making is the condition that arises when an individual is fully informed about a problem, knows solutions, alternatives, and knows what the results of each solution will be. Based on the above, when alternative solutions have been identified and the results expected from them, it is relatively easy to make the decision. It will, therefore, limit itself to choosing the solution that will produce the best result [5].

In addition, [5] it states that risk is the condition that prevails when individuals can define a problem, specify the probability that certain facts will occur, identify alternative solutions and establish the probability that each solution will lead to a result. In general, risk means that the problem and the alternative solutions are somewhere between the extreme of a certain fact and the extreme of an unusual and ambiguous one. The quality of the information that may be available about the relevant condition for decision making varies greatly, just as the risk entails such action. The type, quantity, and reliability of information influence the degree of risk of decision making.

Danielsson [6] and Díaz and Morillas [7] determines that there are two types of decision making in an organization, this in order to make or not an optimal decision for a given problem or strategy:

- Objective decision making: The objective probability refers to the possibility that a specific result is presented, based on figures and undeniable facts. Studying past records is useful in determining the likely outcome of a decision.

- Subjective decision making: Subjective probability refers to the possibility, based on a judgment of personal opinion that a specific outcome will be presented in the future. People have different opinion judgments, which depend on their intuition, previous experience in similar situations, technical ability, and personality traits.

In the book titled Organization and transformation of information systems in the company [5, 7] establishes that Business Information Systems have evolved from a low level of complexity and routine operations to levels that represent a high degree of complexity and integration, turning them into strategic instruments. Currently, what makes the difference in an organization, the main source of its competitive advantages, is its ability to convert data and information into knowledge applied to decisions, this based on what is determined by Danielsson [6].

For their part, [4] establish that organizational learning is the basis of knowledge management, and this, in turn, is the basis for the generation of intellectual capital in enduring organizations. In this sense, the generation and development of intellectual capital touches directly with the know-how of the company and its employees.

In the field of organizations, concepts such as knowledge society, information society, knowledge workers, information systems, tacit and explicit knowledge have arisen and evolved through the developments that have been made under the so-called Knowledge Management (KM).

In the KM, the main source of wealth and productivity is then tacit and explicit knowledge; the first is personal, difficult to communicate, and is rooted in the action and purpose of the action carried out by the employee within a given context, while the second is formal and systematic, and is available in information systems.

Danielsson [6] and DNegri and De Vito [8] determined knowledge CE is characterized by:

- Individual and organizational knowledge is increasingly becoming the main factor of organizational success.
- The term KE is the most appropriate to describe the current economic environment.
- The two main factors in the emergence of CE are globalization and the development of Information and Communication Technologies (ICT).

Decision-making modeling involves the simulation of this specific form of reasoning, and as such, is linked to the concept of Artificial Intelligence (AI). However, AI has concentrated its main focus on the diagnosis and representation of knowledge; most of its technologies do not model human preferences and maintain a narrow margin for subjectivity [8].

In this sense, it is important to prioritize modeling in a rational way, taking into account the importance of reflecting the subjectivity of decision makers, which is closely linked to decision analysis and Artificial Intelligence, which remains an important issue [9]. Mathematical methods of decision analysis continue to focus on the modeling of risk preferences and attitudes without emphasis on the representation of human experience and reasoning.

Knowledge of the preferences of decision makers (their decision-making policy) is represented by decisions to assign objects of the universe to certain pre-existing objects categories (classification). This knowledge is then exploited to produce decisions about new objects.

The fuzzy logic was formulated by mathematician and engineer Lotfi A. Zadeh, professor at the University of California at Berkely, in his work entitled "Fuzzy sets" in 1965 [10]. It is a discipline that arose motivated by the study of vagueness, vague or difficult to specify the information, it also allows the study and modeling of decision-making processes with a high level of uncertainty; but vagueness and uncertainty are different concepts since uncertainty is associated with the lack of knowledge of the value of a variable, while vagueness is related to the knowledge of the value of a function (called degree of belonging) of a variable whose exact value is known. In other words, fuzzy logic is a computational intelligence technique that allows working with a high degree of imprecision, which tries to copy the way humans make decisions [11].

Fuzzy logic has proven to be a useful approach to implementing decision systems when it comes to risk. When uncertainties and inaccuracies exist and are inevitable, the fuzzy strategy provides a robust methodology for dealing with that inaccuracy [10, 12]. It allows representing common knowledge, mostly of a qualitative linguistic type, in a quantitative mathematical language through two possible approaches: the theory of fuzzy sets and associated belonging functions or a generalization of the Logic of Predicates [13].

Expert Systems are pioneers in the idea of obtaining models from verbal expressions so that human agents can apply their essential experience to concrete problems [9]. The representation of knowledge based on logic is a central element recently called Soft Computing. Intelligence focused on decision-making has been addressed by Espin-Andrade et al. [12] based on Fuzzy Logic [14, 15]. Inaccuracy and uncertainty are modeled by Fuzzy Logic, which has allowed advances in knowledge modeling and decision-making based on verbal expressions [4, 16–18].

The main advantage of an approach to the representation of preferential knowledge based on Fuzzy Logic is the opportunity to use language as a communication and modeling element in the analysis of it, creating an explicit model of preferential knowledge; as well as using the inference capacity of the logical framework to propose decisions that better reflect the decision policy of the human agent [9].

That logic for decision-making would ultimately be a functional approach that is explicit in its predicates, but relationships of preference can be modeled as predicted logics as well. The axioms that form the basis of this logic must bring together the true characteristics of the decision-making processes and the way of reasoning of the people involved in them. Its affinity with approaches that adopt a descriptive approach to supporting decision-making must be natural. In this sense, it is possible to consider a logical approach to decision-making as a third position that combines normative and descriptive components. Multipurpose logics, with their ability to deal with imprecision and approximate reasoning, allow us to model properties that, although reasonable, lack general validity and, therefore, cannot be considered as axioms.

Artificial intelligence techniques are inspired by the functioning of the human brain, which has the ability to learn and use its knowledge for decision making. Fuzzy logic systems are capable of dealing with uncertainty and inaccurate information by using knowledge in the form of linguistic rules [19].

Fuzzy Logic systems can be applied to studies in strategic management, with the objective of being used as tools in decision making, due to their capacity to capture and model non-linearity in the relationship between variables [20]. Group Decision Making (GDM) is used to obtain the best solution or solutions to a given problem using the preferences or opinions expressed by a group of decision makers [21–23]. In such a situation, each decision-maker generally approaches the decision-making process from a different point of view. However, the decision of common interest obtains a consensus or agreement before making the decision. In particular, in a GDM situation, there is a set of different alternatives to solve the problem and a group of decision makers who are usually forced to express their views on the alternatives through a particular preference structure [23, 24].

One way of applying the "Principle of Gradualism" essential property of Fuzzy Logic is the definition of logic in which the predicates are functions of the universe within the range [0, 1], and the operations of conjunction, disjunction, negation, and implication are defined in such a way that their restriction to the domain [0, 1] results in Boolean logic. Different ways of defining operations and their properties determine different multivalent logics that are part of the Fuzzy Logic Paradigm. By using compensatory fuzzy logic, it allows elements to belong to groups with different degrees of belonging [20, 25]. In this case, a membership value equal to or close to one would identify "basic" points in a cluster, i.e., in the analysis, a result with a value close to one, the degree of membership would be high; otherwise the values close to zero; the above, using the geometric mean as a statistical tool.

2 Details Experimental

Compensatory Fuzzy Logic

Let n be a negation operator from [0, 1] to [0, 1], or a strictly decreasing operator fulfilling $n(n(x)) = x$, $n(0) = 1$ and $n(1) = 0$.

Let from now on $x = x1, x2, \ldots, xn)$, $y = (y1, y2, \ldots, y3)$, $z = (z1, z2, \ldots, zn)$ be any element of the Cartesian product $[0, 1]^n$.

A quartet of continuous operators (c, d, o, n), c and d from $[0, 1]^n$ to [0, 1], the operator o from $[0, 1]^2$ to [0, 1] and n a negation operator, constitute a Compensatory Fuzzy Logic (CFL) if the following group of axioms is satisfied:

- Compensation Axiom:
  $\min(x1, x2, \ldots, xn) \leq c(x1, x2, \ldots, xn) \leq \max(x1, x2, \ldots, xn)$.
- Commutativity or Symmetry Axiom:

  $c(x1, x2, \ldots, xi \ldots, xj \ldots, xn) = c(x1, x2, \ldots, xi, \ldots, \mathrm{xj}, \ldots, xn)$.

Strict Growth Axiom: if $xi = y1$, $y2 = y2$, $xi - 1 = Yi - 1$.

$xi + 1 = y1 + 1, \ldots, xn = yn$ are unequal to zero, and $xi > yi$ then $y\ (x1, x2, \ldots, xn) > c\ (y1, y2, \ldots, yn)$.

Veto axiom If $xi = 0$ for an i then $c(x) = 0$.

Fuzzy Reciprocity Axiom: $o\ (x, y) = n[o(x, y)]$.

Fuzzy Transitivity Axiom: if $o\ (x, y) \geq 0.5$ and $o\ (y, z) \geq 0.5$, then $o\ (y, z) \geq$ max $o\ (x, y), o\ (y, z)$.

For the case of delimited sets of $\mathcal{R}$nuniversal and existential quantifiers are defined naturally from conjunction and disjunction concepts, respectively, passing to the continuous case through integral calculus from Eqs. (1) and (2) [9]:

Definition 1 In the field of linguistics, X represents the fuzzy whole (%) depending on the parts of the function $\mu a\ \%(x)$ belonging to X. Actual numbers are in the range of [0–1]. Parts of the $\mu\%(x)$ function reflect the degree of association of x in $a\%$ [16]. The triangular fuzzy numbers used in the research are represented by the following expression (am, …, am… am).

Definition 2 $a\% = (a1, a2, a3)$ and $b\% = (b1, b2, b3)$, are a pair of triangular fuzzy numbers. According to [26]. the measurement of distance function ($a\%$, b%) is expressed as:

$$a\%, b\% = \sqrt{\left((1/3\left[(a1 - b1)^2 + (a2 - b2)^2 + (a3 - b3)^2\right]\right)} \tag{1}$$

Definition 3 $a\%$ is a triangular fuzzy number $a\% \propto$ is defined by:

$$a\% \propto = [(a2 - a1) * \propto + a1, a3 - (a3 - a2) \propto] \tag{2}$$

Definition 4 $a\% = (a, a, a)$ and $b\% = (b1, b2, b3)$, are a pair of triangular fuzzy numbers. The division of $= (a, a, a)$ by $b\% = (b1, b2, b3)$, is established as follows:

$$\begin{aligned}(a\%)/(b\%) = [&((a2 - a1)\alpha + a1)/(-(b3 - b1)\alpha + b3),\\ &(-(a3 - a2)\alpha + a3)/(-(b2 - b1)\alpha + b1)]\end{aligned} \tag{3}$$

$$\begin{aligned}When\, \alpha = 0, (a\%)/(b\%) = [a1/b3, a3/b1]\\ When\, \alpha = 1, (a\%)/(b\%) = [a2/b2, a2/b2]\end{aligned} \tag{4}$$

∴ the set of estimated values of $a\%/b\%$ is obtained as: $(a\%)/(b\%) = [a2/b3, a2/b2, a3/a1]$.

Definition 5 En the assumption that $a\% = (a1, a2, a3)$ y $b\% = (b1, b\ 2, b\ 3)$, don real number and the distance between them d($a\%$, b%) uses the Euclidean distance [6].

The estimated value by multiplication is: $a\% \otimes b\% = (a1 * b1, a2 * b2, a3 * b3)$.

The estimated value by addition is: $a\% \otimes b\% = (a1 + b1, a2 + b2, a3 + b3)$.

During the process of computation, a weight represents the subjective expert evaluation on an element through survey and research and reflects the level of importance for the element. Linguistic terminologies can be divided into several levels: very low (VL), Low (L), Intermediate (M), High (H), and Very High (VH). Supposing that all these terminologies can be displaced with triangular fuzzy numbers that fall in the interval of [0, 1]. According to some documents [24], every level corresponds to an evenly distributed membership function, an interval of 0.30 or 0.25.

Definition 6 A fuzzy predicate P i s a lnguistic expression (a proposition) with a degree of truth μp into [0, 1] interval. It applies the "principle of gradualism" which states that a proposition may be both true and false, having some degree of truth (or falsehood) assigned.

Definition 7 A simple fuzzy predicate ***Sp*** is a fuzzy predicate whose degree of truth μsp can be obtained by some of the next alternatives:

The application of a membership function associated with a fuzzy term, to a quantitative variable. For example, sp = "High intensity" is associated with the variable "intensity," which is measured in meters, and the concept "high by a membership function" is defined on the basis of the magnitude of the intensity.

The association of discrete values in the range [0, 1] to the language labels (usually adjectives) of a variable. For example: variable 'intensity', and its labels 'high': $\mu sp = 0.9$; 'medium': $\mu sp = 0.5$; 'low': $\mu sp = 0.1$.

Determination of the actual value in the range [0, 1] by an expert is usually in situations of a certain subjectivity that there is a variable that cannot be quantified using one of the two previous cases, for example, 'Infrastructure is adequate'.

Definition 8 Compound predicates can be represented as a tree structure, having their nodes associated with logical connectives (and, or, not, implication, double implication) and the successive branches related to lower hierarchical level predicates (simple or compound). Of course, the root of the tree corresponds to the main compound predicate, and the leaves will be a simple predicate.

2.1 The Decision-Making Process

The tendency is for companies in any industry to pay special attention to the decision-making process they carry out throughout the stages of administration exercised at the hierarchical levels [26]. An organization's performance can be improved through the use of the process-based approach. Processes are managed as a system by creating and understanding a network of processes and their interactions [27].

Authors such as Quinlan [28] and Vanegas et al. [29] agree that companies and organizations are as efficient as their processes; proposing that all work within the organization is carried out with the purpose of achieving some objective set by each area, as well as attending to the strategic plan established by the organization.

Quinlan [28] and Vanegas et al. [29] classifies the company's intangible assets (know how) into three categories, giving rise to a balance of intangible assets:

People's competencies: planning, producing, processing, or presenting products or solutions.

Internal structure: structured knowledge of the organization such as patents, processes, models, information systems, organizational culture, as well as the people in charge of maintaining such structure.

External structure: relationships with customers and suppliers, trademarks, and company image.

Wang and Lee [30] also make their knowledge classification but starting from the fact that they derive from operational knowledge (see the previous classification) oriented towards action and modification of the environment surrounding the agent [31]. In this regard, consider the following categories:

- Skills: these are non-formalized skills. The more skill is a more fuzzy knowledge is the specification of the problem it solves.
- Skills may contain well-defined subsets of knowledge, but the whole set has a very low level of logical and formal structure. The skill does not ensure that the kind of problem associated with it can be solved, rather than with a certain probability. Therefore, concrete problems may not be solved even if the appropriate knowledge is apparently possessed.
- Technologies: they are formalized action-oriented knowledge; they have a logical structure. They are operational knowledge, and their mission is not only to know but also to act. It must solve action problems in which the purpose of the decision-maker is to modify a specific attribute of the environment. It is a pragmatic knowledge.
- Pre-technological knowledge: it contains all the knowledge that is neither skills nor non-technological skills.

For its part [32] in his book, People focused knowledge management makes a more detailed classification. It defines three forms of knowledge: "public", "personal" and "shared experiences", and four types of knowledge: "factual", "conceptual", "explanatory" and "methodological".

In relation to the forms, it can be pointed out:

- Public: tacit, learned, and routine knowledge that is available in the public domain.
- Shared: knowledge communicated through languages and representations.
- Personal: this knowledge is more tacit than explicit since it is used unconsciously at work, in daily life, in daily tasks, etc.

In relation to the types of knowledge [33]:

- Factual: deals with data, events, measurements, and readings is regularly linked to the contents that are observable and verifiable. It corresponds to verifiable facts.
- Conceptual: deals with concepts, systems, and perspectives. Visualization and appreciation of reality, with much of the observer's abstraction.

- Expectations: it refers to the judgments, hypotheses, and expectations that connoisseurs have an introspective vision of reality.
- Methodological: it deals with reasoning, strategies, decision making, among other techniques. With verifiable and verifiable technical-theoretical characteristics.

2.2 *Fuzzy Tree Studio (FTS)*

Some other FTS General Specifications allow you to work with more than one project at a time. Project data can be stored in XML format for compatibility with other or future systems. During the design of the tree, the user may see errors or omissions that cause problems for the future evaluation stage. There are functions for undo, redo, copy, cut, and paste as usual. In addition, information on the tree is given at the time of design: weight, number of leaf knots, number of compound predicates, depth, and law. When a valid design is achieved, a linguistic expression is displayed for the main predicate. Taking the description of the nodes forms it. Simple predicates (sheet nodes) are intended to be evaluated by data. Its degree of truth can be defined by:

- Membership functions (triangular, trapezoidal, Gaussian, sigmoid, S-shaped, Z-shaped). In this case, a value is taken from the data set and evaluated using the membership function.
- User-defined labels, related to different degrees of truthfulness. In this case, the dataset must contain the description of the label (for example, "large", "sufficient", "small") and be associated with a previously defined true value.
- User values. In this case, the value is taken directly from the dataset. It is assumed that some experts gave the value according to his experience. For example, used in cases such as'Soil quality is good', which could not be quantified using a numerical variable.

Compound predicates are characterized by a logical operator (and, or not, implication, double implication) and associated with one or more simple predicates. Membership functions can be changed by changing their parameters; in addition, the form of the function can be displayed as the parameters are changed; the user can you can change the function interactively with the mouse by moving a few points.

Figure 1 is the proposed diagram made in the Fuzzy Tree Studio program. In this diagram it shows the relationship of the results of crossing the strategic objectives with the indicators, as well as with the considerations evidenced by internal experts in the company where the study was conducted.

The number and type of intervening variables were specified. Subsequently, the manipulation modality of the independent variable was chosen and translated into a quasi-experimental intervention (Analysis with Fuzzy Logic). The sample was selected; in this case, the operational manager of the company where the company was made. The management of the participants was planned. Collection instruments were designed. A focus group for the collection of information was convened, consisting

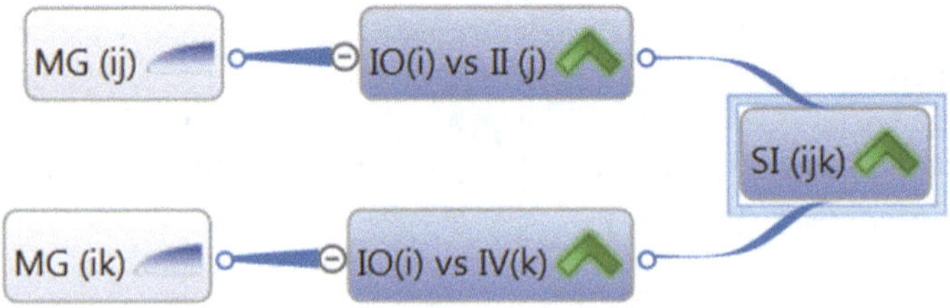

Fig. 1 Fuzzy tree studio diagram

of 3 area managers related to compliance and monitors the strategic objectives; Production, maintenance, and logistics. The analysis scheme was performed using the Fuzzy Tree Studio computer program. Excel files were exported to Fuzzy Tree Studio software.

Matrixes imported from Excel were evaluated by making the calculations established by the system (geometric mean), obtaining the results shown in the body of the work. The decision system was designed as a set of strategic objectives with their respective indicators of the same used in the contemplated company; Every linguistic arrangement connected by a Boolean operator consists of some predecessors and the corresponding consequence. The antecedents are linguistic variables that are described by fuzzy terms, and therefore it is the output variable [34].

The Technique of Order Preference by Similarity with the Ideal Solution is a Multi-Attribute Decision Making technique for the classification and selection of a series of externally determined alternatives over distances from the positive and negative ideal solution [35] that, in this sense, has similarities with compensatory fuzzy logic.

- The processes of each of the areas or departments of the organization were determined by means of a construct derived from the literary revision [33]. This choice of areas was chosen according to the organizational structure of the company, as well as according to the productive and structural conditions of the company where the study was carried out. Similarly, the number of indicators selected was determined by the literary review undertaken, as well as by the proposal emanating from the strategic managers of the organization.
- It was diagnosed by means of Strengths, Weaknesses, Opportunities, and Threats (SWOT) analysis to determine which processes are of an internal or external nature and thus propose strategies for improvement.
- The number and type of indicators involved in the total processes of a metal-mechanical company were specified. Therefore, it is considered a case study with the intention of expanding the sample of companies in the future.
- Subsequently, the mode of manipulation of the independent variable was chosen and translated into a quasi-experimental intervention (Fuzzy Logic Analysis).
- The sample was selected; in this case, managers and operational managers of the company where the company was carried out (37 participants).
- The management of the participants was planned.

- The analysis scheme was performed using the Fuzzy Tree Studio software.
- Excel files were exported to Fuzzy Tree Studio software.
- The matrixes imported from Excel were evaluated doing the calculations established by the system (geometric mean), obtaining the results that are shown synthesized in the body of the work.

3 Results and Discussion

The decision system was designed as a set of aspects to consider, in the importance of the processes for the achievement of the strategic objectives with their respective indicators of the same used in the contemplated company; any linguistic arrangement connected by a Boolean operator consists of some predecessors and the corresponding consequence. The antecedents are linguistic variables that are described by fuzzy terms, and therefore it is the output variable.

The analysis was performed with the use of Compensatory Fuzzy Logic (CFL), through the Fuzzy Tree Studio program, which allows calculating with the geometric mean and other calculations integrated into the system, based on the veracity of the presence of the above-mentioned characteristics and the intensity of the impacts between the areas of the organization involved in the achievement of the company's business objectives, with their respective operational indicators; providing a formal system that establishes logical priorities useful for decision making in the implementation of planned actions, thus obtaining the following results shown in Table 1.

Table 1 shows on the left hand side the area of the organization that meets various indicators presented in the second column; finally, the third column presents the result of the analysis of the weighting of the importance of the processes for the achievement of strategic objectives of the organization, through LGD, with the computer program Fuzzy Tree Studio. The processes most closely related to the objectives were: Utilization of installed capacity (0.996) in Area of Planning; Not compliance in production area (0.982), and Productivity in real units (0.954) in the production area.

One of the limitations of this study is the periodicity of monitoring of the above-mentioned evaluations due to the operative area of the organization and the singers' changes and adaptations to the production processes due to the different requirements of the client. In addition, when considering the evaluation of the achievement of strategic objectives based on the perception of the collaborators regarding the capacity to act for the improvement of the indicators contemplated, therefore, the importance of the evaluation of the indicators, since these are considered fundamental for the fulfillment of the objectives proposed by the organization. This, although it could not be thought of as something productive, the relevance of this that when weighing the indicators as a greater presence in the collaborators, can create the conditions of decision making on the part of the shareholders of the organization.

Table 1 Indicators

Area	Indicators	Calculated fuzzy value
Supplies	Inventory mobility	0.908
	Inventory turnover	0.890
	Rotation of passive credits	0.767
Human resources	Labor productivity	0.534
	Absenteeism	0.856
	Importance of wages	0.950
	Worker turnover indicator	0.467
	Sales indicator worker	0.612
Financial structure	Working capital indicator	0.534
	Break-even point indicator	0.734
	Break-even point	0.918
	Financial independence	0.343
Products services	Profitability by product	0.798
	Commerciality index	0.634
	Break-even point	0.834
	Quality level	0.512
Media production	Machinery productivity	0.665
	Maintenance/production indicator	0.698
Commercial area	Sales level	0.708
	Portfolio	0.583
	Missing per shipment	0.938
	Customer satisfaction (complaints and returns indicator)	0.796
	Supplier qualifications	0.844
Quality area	Compliance with audit program	0.886
	Compliance with and follow-up of corrective and preventive actions	0.693
	Compliance with calibration program for instruments and elements of control	0.913
	Training	0.756
Area of planning	Scheduled Production Compliance	0.852
	Utilization of installed capacity	0.996
	Overall efficiency	0.676
	Operational efficiency	0.849
	Quantity of raw material processed	0.407
	Factory production value	0.664
Area of production	Productivity in real units	0.954

(continued)

Table 1 (continued)

Area	Indicators	Calculated fuzzy value
	Productivity per employee	0.814
	Not in compliance	0.982
	% non-conforming product costs	0.636
	% non-compliant cost versus factory productivity	0.785
	Installation times	0.869
Area of maintenance	Fulfillment of requests	0.755
	Availability of machinery	0.587
	Plant maintainability	0.739
	Reliability of machinery	0.639

4 Conclusions

According to the results of the analysis based on Compensatory Fuzzy Logic, the opportunities to prioritize in its use are the existing priority of the importance of the processes involved in various areas of an organization. The processes that resulted in more ownership allow us to reassess the strategic objectives set and aim efforts to achieve them in a better way, as well as to diagnose the feasibility of compliance. It is recommended to operationalize what is proposed for the growth and business success of the companies, as well as to increase the sample of experts and to specify if there are more elements that can be contemplated for their analysis. It is advisable to follow others methodology in order to deepen and increase the elements present in the organizations for the purpose of implementing the Fuzzy Logic to analyze various items in terms of strategic management is also advised to identify which processes are internal and external to the organization.

References

1. Salgueiro-Plasencia, A.: Herramientas de Simulación para la Enseñanza de la Minería de Datos en la Robótica. In: V Latin American Congress on Biomedical Engineering CLAIB 2011 May 16–21, 2011, pp. 1272–1275. Springer, Heidelberg (2013). https://doi.org/10.1007/978-3-642-21198-0_323
2. Artieda, C.H.: Análisis de los sistemas de costos como herramientas estratégicas de gestión en las pequeñas y medianas empresas (PYMES). Revista Publicando **2**(3), 90–113 (2015)
3. Bindu, B., Jagdeep, K., Kumar, A.: A new fuzzy CCR data envelopment analysis model and its application to manufacturing enterprises. In: Collan, M., Kacprzyk, J. (eds.) Springer International Publishing AG, (Studies in Fuzziness and Soft Computing 357), vol. 357. Springer, Cham (2018). https://doi.org/10.1007/978-3-319-60207-3_21
4. Chen, S.J., Hwang, C.L.: Fuzzy multiple attribute decision making methods. In Fuzzy Multiple Attribute Decision Making, pp. 289–486. Springer, Heidelberg (1992)

5. Chiclana, F., Herrera-Viedma, E., Herrera, F.: Integrating Three Representation Models in Fuzzy Multipurpose Decision Making Based on Preference Relations. University of Granada, Granada (1998)
6. Danielsson, P.E.: Euclidean distance mapping. Comput. Graph. Image Process. **14**(3), 227–248 (1980). https://doi.org/10.1016/0146-664X(80)90054-4
7. Díaz, B., Morillas, A.: Some experiences applying fuzzy logic to economics. In: Seising, R., González, V.S. (eds.) Soft Computing in Humanities and Social Sciences, pp. 347–379. Springer, Heidelberg (2012). https://doi.org/10.1007/978-3-642-24672-2_19
8. DNegri, C.E., De Vito, E.L.: Introducción al razonamiento aproximado: lógica difusa. Revista Argentina de Medicina Respiratoria **4**, 126–136 (2006)
9. Dong, Y., Luo, N., Zhang, H.: Multiperson decision making with different preference representation structures: a selection process based on prospect theory. In: Fuzzy Systems (FUZZ-IEEE), 2014 IEEE International Conference on, pp. 18–24. IEEE, Beijing (2014). https://doi.org/10.1109/FUZZ-IEEE.2014.6891543
10. Dubois, D., Prade, H.: A review of fuzzy set aggregation connectives. Inf. Sci. **36**(1–2), 85–121 (1985). https://doi.org/10.1016/0020-0255(85)90027-1
11. Andrade R.A.E., González E., Fernández E., Gutiérrez S.M.: A fuzzy approach to prospect theory. In: Espin, R., Pérez, R., Cobo, A., Marx, J., Valdés, A. (eds.) In Soft Computing for Business Intelligence, pp. 45–66. Springer, Heidelberg (2014). https://doi.org/10.1007/978-3-642-53737-0_3
12. Espin-Andrade, R.A., Gonzalez, E., Pedrycz, W., Fernandez, E.: An interpretable logical theory: the case of compensatory fuzzy logic. Int. J. Comput. Intell. Syst. **9**(4), 612–626 (2016). https://doi.org/10.1080/18756891.2016.1204111
13. Andrade R.A.E., Fernández E., González E.: Compensatory fuzzy logic: a frame for reasoning and modeling preference knowledge in intelligent systems. In: Soft Computing for Business Intelligence, pp. 3–23. Springer, Heidelberg (2014). https://doi.org/10.1007/978-3-642-53737-0_1
14. Fodor, J.C., Roubens, M.R.: Fuzzy Preference Modelling and Multicriteria Decision Support, vol. 14. Springer Science & Business Media (2013)
15. Greco, S., Matarazzo, B., Slowinski, R.: Rough sets methodology for sorting problems in presence of multiple attributes and criteria. Eur. J. Oper. Res. **138**(2), 247–259 (2002). https://doi.org/10.1016/S0377-2217(01)00244-2
16. Hernández-Nariño, A., Medina-León, A., Nogueira-Rivera, D., Negrín-Sosa, E., Marqués-León, M.: La caracterización y clasificación de sistemas, un paso nece-sario en la gestión y mejora de procesos. Particularidades en organizaciones hospitala-rias. Dyna **81**(184), 193–200 (2014). https://doi.org/10.15446/dyna.v81n184.37309
17. Klir, G., Yuan, B.: Fuzzy Sets and Fuzzy Logic, vol. 4. Prentice hall, New Jersey (1995)
18. Llanes-Font, M., Isaac-Godínez, C.L., Moreno-Pino, M., García-Vidal, G.: De la gestión por procesos a la gestión integrada por procesos. Ingeniería Industrial **35**(3), 255–264 (2014)
19. Marakas, G.M.: Decision Support Systems in the 21st Century, vol. 134. Prentice Hall, Upper Saddle River (2003)
20. Martínez, A.M.: Gestión por procesos de negocio: Organización horizontal. Ecobook, Madrid (2014)
21. Mendel, J.M.: Uncertain Rule-Based Fuzzy Logic Systems: Introduction and New Directions. Prentice Hall, Upper Saddle River (2001)
22. Moreno Velo, F.J.: Un entorno de desarrollo para sistemas de inferencia complejos basados en lógica difusa. Universidad de Sevilla, Tesis doctoral (2013)
23. Morote, J.P., Serrano, G.L., Nuchera, A.H.: La gestión de la innovación y la tecnología en las organizaciones. Ediciones Pirámide (2014)
24. Morris Díaz, A., Rodriguez Monroy, C., Vizán Idoipe, A., Martínez Soto, M., Gil Araujo, M.: Sistema de gestión de la calidad y desempeño organizacional en la industria petrolera. Interciencia **38**(11), 793–802 (2013)
25. Olivos, P.C., Carrasco, F.O., Martínez Flores, J.L.M., Moreno, Y.M., Nava, G.L.: Modelo de gestión logística para pequeñas y medianas empresas en México. Contaduría y Administración **60**(1), 181–203 (2015). https://doi.org/10.1016/S0186-1042(15)72151-0

26. Pérez, I.J., Cabrerizo, F.J., Alonso, S., Herrera-Viedma, E.: A new consen-sus model for group decision making problems with non-homogeneous experts. IEEE Trans. Syst. Man Cybern. Syst. **44**(4), 494–498 (2014). https://doi.org/10.1109/TSMC.2013.2259155
27. Pesado, P.M., Bertone, R., Esponda, S., Pasini, A., Boracchia, M., Martorelli, S., Swaels, M.: Mejora de procesos en el desarrollo de sistemas de software y en procesos de gestión. Workshop de Investigadores en Ciencias de la Computación (WICC) **15**, 581–585 (2013)
28. Quinlan, J. R.: C4. 5: programs for machine learning. Elsevier (2014)
29. Vanegas, G., Botero, C., Restrepo, A.: Una aproximación mediante lógica di-fusa al análisis de la competitividad empresarial. Adm. Y Org. **17**(33) (2014). https://doi.org/10.24275/uam/xoc/dcsh/rayo/2019v22n43
30. Wang, T.-C., Lee, H.D.: Developing a fuzzy TOPSIS approach based on subjective weights and objective weights. Expert Syst. Appl. **36**(5), 8980–8985 (2009). https://doi.org/10.1016/j.eswa.2008.11.035
31. Wang, Y.-M., Chin, K.-S., Poon, G.K.K., Yang, J.-B.: Risk evaluation in failure mode and effects analysis using fuzzy weighted geometric mean. Expert Syst. Appl. **36**(2), 1195–1207 (2009). https://doi.org/10.1016/j.eswa.2007.11.028
32. Yang, Y., Haihua, P., Kejian, L., Haoran, S.: Fuzzy TOPSIS-based supply chain optimization of fresh agricultural products. AMSE J. **59**, 186–203 (2016)
33. Zadeh, L.A.: Fuzzy logic = computing with words. IEEE Trans. Fuzzy Syst. **4**(2), 103–111 (1996). https://doi.org/10.1109/91.493904
34. Zadeh, L.A.: Toward a theory of fuzzy information granulation and its centrality in human reasoning and fuzzy logic. Fuzzy Sets Syst. **90**(2), 111–127 (1997). https://doi.org/10.1016/S0165-0114(97)00077-8
35. Zadeh, L.A.: Fuzzy sets as a basis for a theory of possibility. Fuzzy Sets Syst. **1**(1), 3–28 (1978). https://doi.org/10.1016/0165-0114(78)90029-5

Evaluating Intelligent Methods for Decision Making Support in Dermoscopy Based on Information Gain and Ensemble

Newton Spolaôr, Rui Fonseca-Pinto, Ana I. Mendes, Leandro A. Ensina, Weber S. R. Takaki, Antonio R. S. Parmezan, Conceição V. Nogueira, Claudio S. R. Coy, Feng C. Wu, and Huei D. Lee

Abstract Melanoma, the most dangerous skin cancer, is sometimes associated with a nevus, a relatively common skin lesion. To find early melanoma, nevus, and other lesions, dermoscopy is often used. In this context, intelligent methods have been applied in dermoscopic images to support decision making. A typical computer-aided diagnosis method comprises three steps: (1) extraction of features that describe image properties, (2) selection of important features previously extracted, (3) classification of images based on the selected features. In this work, traditional data mining approaches underexploited in dermoscopy were applied: information gain for feature selection and an ensemble classification method based on gradient boosting. The former technique ranks image features according to data entropy, while the latter combines the outputs of single classifiers to predict the image class. After evaluating these approaches in a public dataset, we can observe that the results obtained are competitive with the state-of-the-art. Moreover, the presented approach allows a reduction of the total number of features and types of features to produce similar classification scores.

N. Spolaôr · L. A. Ensina · W. S. R. Takaki · A. R. S. Parmezan · F. C. Wu · H. D. Lee (✉)
Laboratory of Bioinformatics (LABI), Western Paraná State University (UNIOESTE), Foz do Iguaçu, Paraná 85867-900, Brazil
e-mail: huei.lee@unioeste.br

R. Fonseca-Pinto
ciTechCare—Center for Innovative Care and Health Technology, Polytechnic of Leiria. IT—Instituto de Telecomunicações—Leiria, 2411-901 Leiria, Portugal

R. Fonseca-Pinto · A. I. Mendes · C. V. Nogueira
Polytechnic of Leiria, 2411-901 Leiria, Portugal

C. S. R. Coy · F. C. Wu
University of Campinas (UNICAMP), Campinas, São Paulo 13083-887, Brazil

A. R. S. Parmezan
Laboratory of Computational Intelligence, Institute of Mathematics and Computer Science, University of São Paulo, São Carlos, São Paulo 13566-590, Brazil

C. V. Nogueira
Center of Mathematics (CMAT), University of Minho (Uminho), 4704-553 Braga, Portugal

W. Pedrycz et al. (eds.), *Computational Intelligence for Business Analytics*, Studies in Computational Intelligence 953,
https://doi.org/10.1007/978-3-030-73819-8_7

Keywords Machine learning · Computer-aided diagnosis · Image analysis

1 Introduction

Cancer and related diseases are of extreme importance to health agents and societies. In particular, skin cancer is one of the most deadly illnesses. Recent epidemiological studies have shown that the occurrence of new skin cancer cases each year is higher than the combined incidence of breast, prostate, lung, and colon cancers [1–3].

To understand and manage suspicious skin lesions, it is important to detect early morphological changes to the skin, in which digital imaging technologies play a central role. Due to the importance of early diagnosis in skin cancer, a detailed exam of the melanocytic skin lesions is crucial. A non-invasive imaging technique, known as epiluminescence microscopy or dermoscopy, has been used with great success to help health agents and researchers in this task [4].

It has established the advantage of combining dermoscopy with a detailed physical examination of the skin, mostly due to the human eye assessment's subjectivity, and to the proven increase in early detection of melanoma signs when this imaging technique is applied. For example, the most dangerous type of skin cancer (malignant melanoma) can often be associated with a form of nevus (a relatively common skin lesion). In this case, the use of digital imaging techniques is key to assessing specific clues to guide an accurate diagnosis. This clinical and digital information obtained by dermoscopy has been applied to support decision making using Computer-Aided Diagnosis (CAD) algorithms based on data mining and Machine Learning (ML) methodologies [5–7].

Besides the study reported by [5], recent papers presented reviews of ML techniques employed in Computer-Aided Diagnosis of melanoma and other skin lesions not only for dermoscopy but also for different modalities of medical imaging. These reports show that the most commonly used classification methods include support vector machines, logistic regression, artificial neural networks, K-nearest neighbor, and decision trees. They also agree that recent approaches tend to explore deep learning methods based on Convolutional Neural Network (CNN) for either image segmentation and lesion classification [7, 8].

In this context, CNN constitutes the base approach for many proposed methods [9–14]. As an example, in [15], the authors combined CNN with Gabor filtering capabilities for extracting features from images and proposed an approach called Gabor Convolutional Network (GCN). They state that Gabor filtering helps to reduce the burden on the convolutional network for feature extraction and selection. A melanoma classification and segmentation method based on a lightweight CNN model is proposed in [16], aiming to reduce the computational cost of the network compared with the one from the traditional CNN approach.

In [17], a multilevel feature selection framework focused on skin lesion identification is proposed. The authors present the proposed method in two phases. The first phase is dedicated to the extraction of features and the second one to feature selection and dimensionality reduction. For the feature extraction step, three pre-trained CNN models (Inception-V3, Inception-ResNet-V2, and DenseNet-201) were re-trained using pre-processed and segmented images of dermoscopy through transfer learning. After submitting the segmented images to the re-trained models, sets of features are obtained and then combined using feature fusing strategies. For the second phase, the authors employ a combination of techniques for feature selection called entropy controlled neighborhood component analysis (ECNCA) to achieve a feature selection and dimensionality reduction effect. The authors evaluate the resulting feature set by submitting it to 12 different classifiers, including three ensemble-based ones (AdaBoost Ensemble, Ensemble subset KNN, and Ensemble Random UnderSampling Boost).

Following a different path, in [18], a method for classification of wide-field images into suspicious and non-suspicious cases of skin lesions is proposed. After pre-processing the input images, features are extracted, and the most relevant ones are selected using a univariate feature selection strategy. Also, PCA dimensionality reduction is applied to the chosen features. Only then the features are input to a logistic regression-based classifier. Other works also utilize manual feature extraction and selection [13, 16] to reduce the computational burden and increase accuracy on the classification step.

CNN is considered the most successful neural network architecture for image analysis by different authors [7, 12, 19]. However, the high computational cost of the CNN training process may be regarded as a drawback of such an approach, in comparison with other ML techniques, particularly when compared to decision tree-based methods. The automatic feature extraction and selection aspect of CNNs is regarded as an advantage of this approach compared with other supervised learning techniques. Still, it does not entirely avoid problems like overfitting and bias [7, 11, 12, 19], and manual tuning is often required to enhance the model's performance.

Thus, using shallow learning methods with handcrafted features is still a competitive alternative to deep neural networks with automatic feature extraction for dermoscopic image analysis. One can also note that some combinations of traditional algorithms in computational intelligence have not been applied and evaluated in the problem of dermoscopy classification.

To the best of our knowledge, the well-known XGBoost ensemble approach and the Information Gain (IG) algorithm [20–24] have not been applied together in dermoscopy. In this context, this work's main contribution consists of introducing the use and evaluation of tree-based XGBoost, supported by IG, in the differentiation of melanoma from non-melanoma images. The results from this combination of algorithms are compared with other classical methodologies often applied in automatic classification in dermoscopy assisted diagnosis.

In the remainder of this work, Sect. 2 presents material and methods, and Sect. 3 shows the results. Finally, Sect. 4 addresses some conclusions and future work.

2 Material and Methods

ML methodologies have been applied in recent years to assist diagnosis in a broad range of medical areas [25–28]. In particular, due to (1) the absence of ionizing radiation in the acquisition, (2) the portable nature of the device, and (3) the huge commercial pressure and advances around photographic sensors, dermoscopy emerged as a natural imaging modality at the forefront in ML usage to assist clinician diagnosis in dermatology.

To test the proposed methodology, a dataset comprising 104 dermoscopic images (46 melanomas and 58 benign lesions) RGB true colored (24-bit color) and JPEG compressed with a minimum resolution of 300 dpi were used. The acquisition was conducted according to clinical protocols in dermoscopy and following all legal requirements. These images were already used in related works to validate image processing techniques in dermoscopic image processing, as presented in [5, 29].

Classical ML algorithms using several paradigms have been tested and published showing huge success, as can be found in [5, 25, 30–43]. In particular, for decision tree J48 [44], Nearest Neighbours [45] (NN) and Support Vector Machines [46] (SVM) ML algorithms, a comparative study regarding classification performance was conducted in [5]. The dataset, pre-processing methodology, and the results of the paper previously mentioned [5] will be considered to assess the results at the present work.

2.1 *Machine Learning Methodology*

In this work, regarding ML methodology, an approach of testing ensemble algorithms was followed by applying an eXtreme gradient boosting, the XGBoost. In fact, this is a distinct methodology when compared with the one considered in [5]. This ensemble method has been highlighted in scientific papers and ML challenges as a flexible, portable, and efficient implementation of the gradient boosting framework, providing a parallel tree boosting to solve scientific problems quickly and accurately [47, 48]. Unlike traditional classification methods, such as decision trees, the ensemble builds a series of N trees. Although XGBoost uses information gain internally as a feature selection measure (Sect. 2.2), we decided to perform feature selection before building the ensemble classifier to keep control of which features would be used as an input for classification.

The core of XGBoost is to optimize the value of the objective function, using the residual obtained at each iteration of the gradient boosting to correct the previous predictor, optimizing a specific loss function [49, 50]. Thereby, this process prevents overfitting and also improves classification on leaves.

Following the same approach as in [5], and for comparison purposes, 166 features grouped according to whether their nature in terms of Texture (T), Shape (S), or Local binary patterns (L) were extracted in each image. The allocation of the elicited

Table 1 Feature combinations and their properties

Features						
Group IDs	Shape and geometric features	NGTDM	Haralick's descriptors	Fractal dimension	Laws' energy measures	Local binary patterns
TSL	×	×	×	×	×	×
TS	×	×	×	×	×	
TL			×	×	×	×
T			×	×	×	

features was conducted to create an arrangement of feature groups, as shown in Table 1: Texture, Shape and Local binary patterns (TSL), Texture and Shape (TS), Texture, and Local Binary Pattern (TL) and texture.

2.2 Feature Selection

Regarding Feature Selection (FS), the approach presented in this work is distinct from the one followed in [5] where ReliefF [42] was applied. At the present work, IG FS was conducted. Information gain, also implemented in the Weka library [43], is a supervised technique that scores each feature by measuring its information gain (entropy-based) regarding the class [44]. We used IG to rank features, from higher to lower scores, providing us with an ordering of the most relevant attributes.

In particular, IG is a filter strategy for feature selection that measures the dependence between one feature and the class label based on the entropy concept [51, 52]. To do so, the IG measure evaluates each attribute A_j, $j = 1 \ldots n$, individually according to the dependence (relevance) between it and a class label C. Therefore, the difference between the entropy of the data set $D(A_1, A_2, \ldots, A_n, C)$ and the entropy weighted sum of each subset $D_v \subset D$, where D_v consists of the instances where A_j has the value v is calculated. So, if A_j has $v = 10$ different values in D, the weighted sum would be applied to 10 different D_v subsets. The entropy and IG are defined by Eqs. (1 and 2), respectively.

$$Entropy(D) = -\sum_{i=1}^{m} p_i * log_2(p_i) \quad (1)$$

$$IG(D, A_j) = Entropy(D) - \sum_{v \in A_j} \frac{|D_v| * Entropy(D_v)}{|D|} \quad (2)$$

Let m be the number of distinct class values and p_i the probability that an example in D belongs to a determined class.

By using IG to remove irrelevant features, the dimensionality of the training samples and the complexity of classification algorithms are effectively reduced. This step is central in ML performance, once it will generate a subset from the original feature set, obtained so that some specific evaluation criteria will be optimized.

2.3 Experimental Setting

The experimental setup followed the same pipeline as in [5] and was organized in four steps, as illustrated in Fig. 1.

After acquisition and cropping, each dermoscopic image in the dataset was initially converted to the corresponding grayscale version by selecting the channel with the highest entropy (Step 1—Medical Images Acquisition). Afterward, according to Table 1, features based on texture, shape, and local binary patterns were extracted from each of the preprocessed images. These features were discretized according to the equal width binning technique to ensure different values (bins) for IG calculation [43]. The FS was conducted according to the IG methodology for each feature group with discretized data (TSL, TS, TL, and T). By adopting a threshold value, four feature subsets were created. These 16 subsets (i.e., 4 feature groups $\times$ 4 threshold settings) are composed of 10, 20, 40 and 80% of the best-ranked features (Step 2—Data Preprocessing). Each subset was then submitted to the ensemble for classification (Step 3—Classification). We used the Caret[1] package from the R programming language to search for parameter values for the XGBoost ensemble method. Caret was configured with the following settings:

- Learning approach ("method" parameter): xgbtree;
- Number of XGBoost parameters values combinations to be tested ("tuneLength" parameter): five;
- Resampling approach to evaluating parameter tuning ("trControl" parameter): cross-validation with five folds.

Steps 2 and 3 were performed within the tenfold stratified cross-validation approach. In particular, we submitted the training folds for FS and classifier building, keeping the corresponding testing folds to evaluate the classifier in terms of Sensitivity (SE), Specificity (SP), and Accuracy (Acc) (Step 4—Evaluation of Results). We then averaged the results across the folds to estimate the performance of the proposed method.

[1] https://cran.r-project.org/web/packages/caret/index.html.

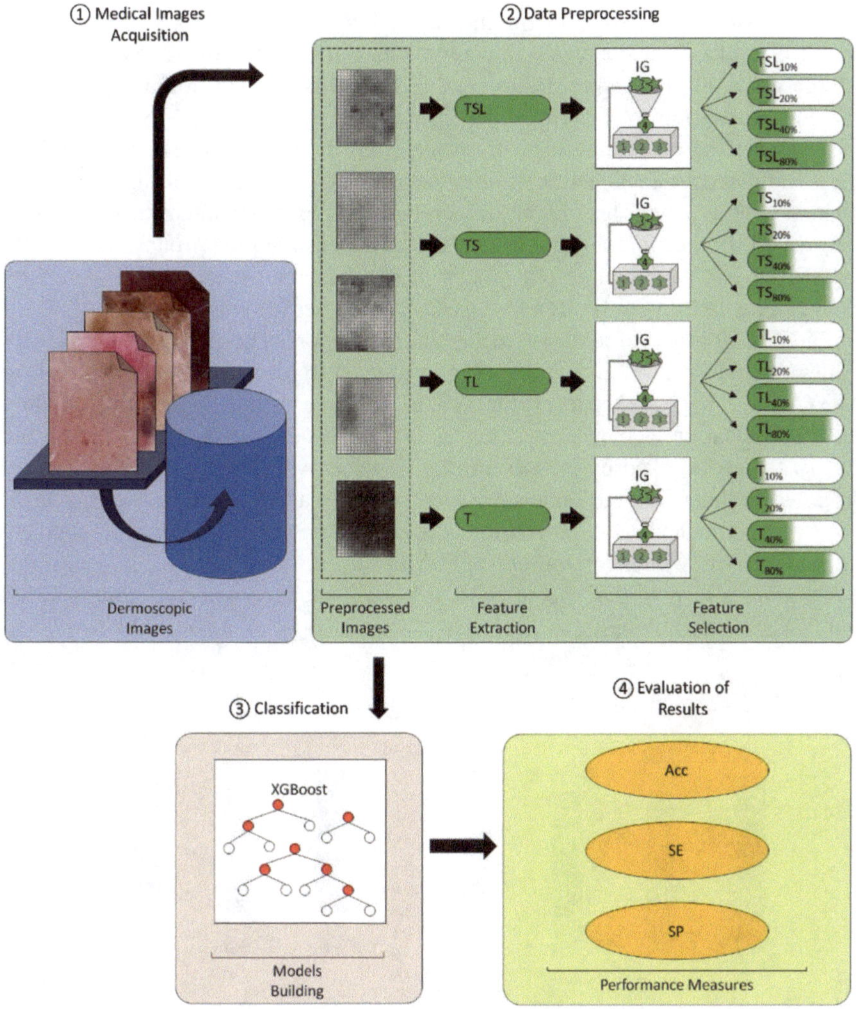

Fig. 1 Experimental setup. This figure is an adaptation of material published in [5]. Any citation to the material should consider that paper. Legend: TSL = Texture, shape and local binary patterns; TS = Texture and shape; TL = Texture and local binary patterns; T = Texture; IG = Information gain; Acc = Accuracy; SE = Sensitivity; SP = Specificity

3 Results and Discussion

One of the first aspects regarding classification results is the known difficulty in comparing methodologies due to the metrics used to assess classification outputs. In some cases, not all standard metrics (e.g., SE, SP and Acc) are shown. In others, the criteria used to select image subsets from the publicly available datasets are not explicitly stated (e.g., selecting a percentage of images instead of all the dataset regardless of the properties of the images). This scenario will induce bias in the obtained results when one needs to compare the performance of ML algorithms. In the present work, the same images used in [5] were enrolled in the experimental procedure, allowing a fair comparison of results.

To assess the performance of the results, the three metrics (SE, SP and Acc) were calculated for each feature group. As in the previous studies, a slight degree of dispersion was found. This finding was strengthened by an unpaired Kruskal–Wallis test, which did not find any statistical difference between the groups. The results for SE, SP and Acc for the four groups of features and the associated variability (standard deviation) can be seen in Figs. 2, 3 and 4, respectively.

The TSL10 subset of features reached in [5] the worst values for all metrics, with values ranging from 0.33 to 1.00 (SE), from 0.00 to 0.75 (SP), and from 0.20 to 0.73 (Acc). Table 2 shows the average SE, SP and Acc, as well as the corresponding standard deviation, achieved by the J48, SVM and NN classifiers built using the TSL10 features ranked by ReliefF in [5]. However, when using the methodology proposed in this work, the same subset led to higher average values in terms of SP and Acc, as indicated in the last row in Table 2. Also, this subset presented a relatively small variability with the proposed approach.

One can also note that the smallest subsets associated with the groups T, TS and TL achieved the best average performance in Figs. 2, 3 and 4. In particular,

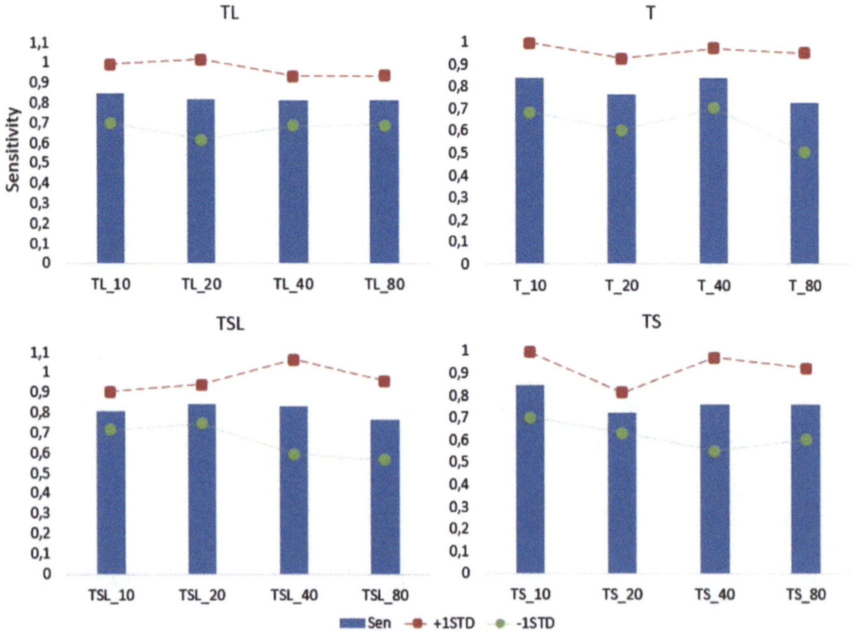

Fig. 2 Sensitivity (average and standard deviation) for the classifiers built from the four groups of features described in Table 1. Legend: TL = Texture and local binary patterns; T = Texture; TSL = Texture, shape and local binary patterns; TS = Texture and shape

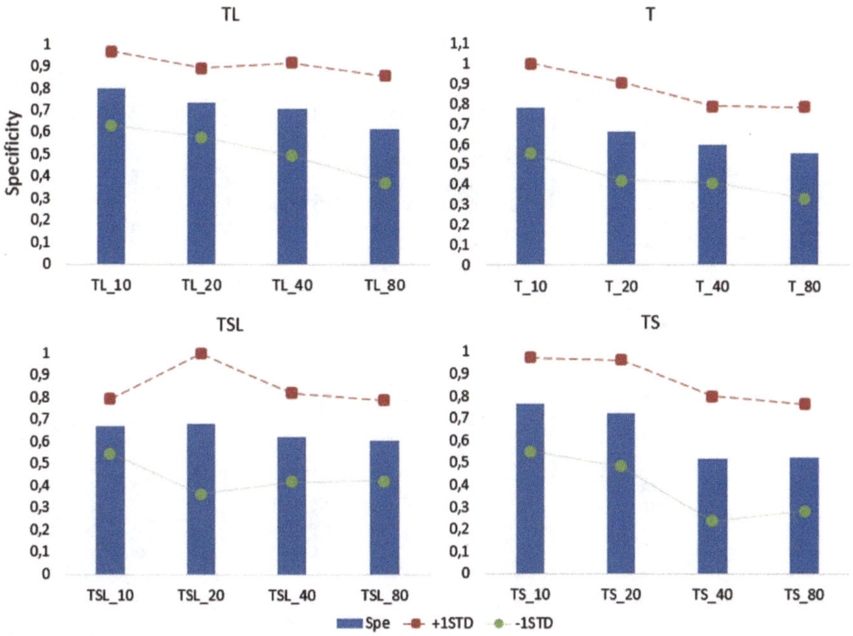

Fig. 3 Specificity (average and standard deviation) for the classifiers built from the four groups of features described in Table 1. Legend: TL = Texture and local binary patterns; T = Texture; TSL = Texture, shape and local binary patterns; TS = Texture and shape

T10, TL10 and TS10 yielded the highest average performance values, regardless of the evaluation measure. This finding is also relevant, as it agrees with FS papers suggesting that using a relatively small number of features can lead to better learning performance than using a large number of features [22].

To achieve a more concise perspective of the experimental results regarding different evaluation measures, we conducted the Multi-Criteria Decision Analysis (MCDA) method described in [5]. This work considered a set of measures different from previous work by replacing the area under the curve with SP, as the latter criterion is directly associated with confusion matrices. In what follows, the set components are listed:

1. Sensitivity;
2. Specificity;
3. Accuracy;
4. The complement of the Coefficient of Variation (CV) for SE;
5. The complement of the CV for SP;
6. The complement of the CV for Acc.

Note that we took into account the complement of the CV, i.e., 1-CV, to keep only measures that need to be maximized. The higher the CV's complement in a specific measure, the lower the ratio of the measure's standard deviation to its average.

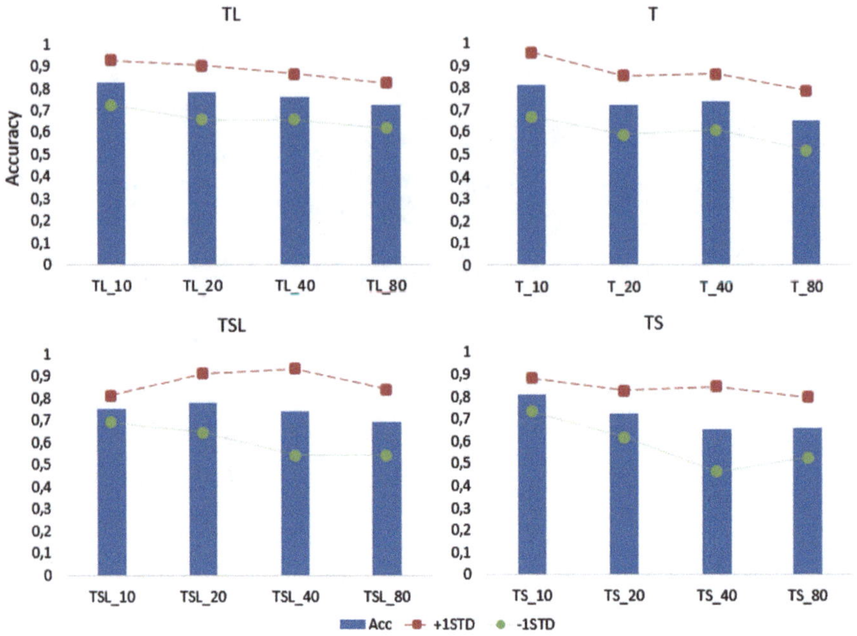

Fig. 4 Accuracy (average and standard deviation) for the classifiers built from the four groups of features described in Table 1. Legend: TL = Texture and local binary patterns; T = Texture; TSL = Texture, shape and local binary patterns; TS = Texture and shape

Table 2 Average performance (and standard deviation) of classifiers built from TSL10

Machine learning algorithm	Feature selection methods	SE	SP	Acc	Reference
J48	ReliefF	0.87 (0.19)	0.05 (0.10)	0.50 (0.09)	[5]
SVM		0.88 (0.14)	0.05 (0.10)	0.51 (0.10)	
NN		0.68 (0.20)	0.40 (0.25)	0.55 (0.16)	
Ensemble	IG	0.81 (0.09)	0.67 (0.12)	0.75 (0.06)	This work

Table 3 sorts, in descending order, the area of a six-dimension polygon built by MCDA for each subset of features. This polygon is depicted through six dimensions (axes), as each axis is associated with the value reached by one of the measures previously listed.

TL10 led the ranking, as it is associated with the highest value in three axes—average SE, SP and Acc. The corresponding CV's complement is also relatively high. TS10 and TSL10 also reached performance values equal or similar to the best ones in a few measures.

The multi-criteria analysis strengthened a finding already highlighted for isolated measures: the smallest subsets of features within each group (TL, T, TSL and TS)

Table 3 Area of the polygon summarizing the performance of the proposed method for each subset of features

Subset of features	Area value
TL10	4.45
TS10	4.30
TSL10	4.23
T10	4.13
TL20	4.01
TL40	3.96
TS20	3.76
TSL20	3.71
T40	3.69
TL80	3.58
T20	3.48
TSL40	3.35
TSL80	3.31
TS80	2.98
T80	2.91
TS40	2.63

reached the top-four performance. However, as the subsets' size increases, there was no direct relationship between the subset size and the area value.

In line with the conclusions in [5] and following the criteria previously mentioned for the selection of ML methodologies, the optimal score was obtained using a NN classifier and the TSL20 feature subset. Following a similar methodology as in the cited work, and using ensemble and IG as explained above, Table 4 shows the best combination of features.

Table 4 Comparison between the best result from [5] and the best results from this work in terms of the number of features. For each learning evaluation measure, the average (and standard deviation) is presented

Machine learning algorithm	Feature selection method	Subset of features	#features	SE	SP	Acc	Reference
NN	ReliefF	TSL2033		0.86 (0.16)	0.73 (0.18)	0.81 (0.10)	[5]
TL10			15	0.85 (0.15)	0.80 (0.17)	0.83 (0.10)	
Ensemble	IG	TS10	10	0.85 (0.15)	0.77 (0.21)	0.81 (0.08)	This work
		T10	9	0.84 (0.16)	0.78 (0.22)	0.82 (0.15)	

From the results in Table 4, it is possible to observe a remarkable decrease in the number of features necessary to produce similar classification scores when the ensemble ML and IG are applied. This finding can be explained, among other factors, by the use of the same measure (IG) during feature selection and the learning of trees in the ensemble method. In particular, in this work, IG FS was applied to all the 166 features extracted from images to yield 16 subsets of features. In turn, the tree learning procedure performed embedded feature selection based on IG on each subset. Although IG is a frequently used measure in the FS literature, it has been relatively unexplored in dermoscopy [5]. Finally, using IG for feature ranking allowed us to precisely define the number of selected features, as designed in Sect. 2.3. Such flexibility would be more complex to achieve in some feature subset selection methods, such as Correlation-based Feature Selection [43].

It should be emphasized that the ensemble method used by us builds a series of N trees (classifiers). In particular, the classification of instances incorrectly predicted by a tree is prioritized by the next tree in the series. As a result, this method can occasionally outperform algorithms that generate a single classifier, such as NN. This possibility was verified in the dataset evaluated in this work, after comparing the best ensemble models with the best classifier found in [5] in terms of the number of features. However, the NN model previously built is still useful due to its higher simplicity, as it does not require a chain of classifiers to label images.

Moreover, the best score using the work in [5] was obtained in a heterogeneous subset formed by all types of the enrolled features (T, S and L). In the results obtained using the presented methodology, texture features gained more relevance. By themselves, T10 achieved similar results using a reduced number and fewer types of features (e.g., nine texture features).

Another difference between the results of the proposed methodology and those reported in [5], is the trend observed in this work regarding the scores (for all metrics) when the percentage of features in each group increases. In fact, as previously mentioned, the best performance when the ensemble ML is applied is obtained mostly in the 10% group of features. Although a minor difference was observed, this trend was not noted for the TSL10 group.

It should be emphasized that the four steps of the competitive method proposed in this work, outlined in Sect. 2.3, are also used in the literature. Although these steps are not always explicitly represented, as was the case in [5], they are often used by CAD approaches based on ML to classify medical images properly. For example, a recent review [26] depicts a workflow for machine learning algorithms applied in medical image analysis. This workflow includes the feature extraction and feature selection tasks, which are inherent to Step 2—Data Preprocessing, and refers to Step 3—Classification. Implicitly, the remaining steps are also considered to feed Step 2 and assess Step 3 output. Another paper in medical image analysis uses procedures with different names but similar purposes to our Step 1—Medical Images Acquisition and Step 4—Evaluation of Results [53].

This work differentiates from related ones in dermoscopy, mainly due to the algorithms used in Steps 2 and 3. In particular, we used IG for feature selection and XGBoost for classification. This unexplored combination of algorithms performed well, as already discussed, suggesting that it is relevant in the studied domain. Moreover, when compared with the ReliefF algorithm used in [5], IG uses fewer parameters and has a smaller computational complexity. In turn, XGBoost benefits from the patterns learned from several trees working together to boost performance. On the other hand, many related papers (e.g., [5]) employ machine learning algorithms that do not cooperate. Although ensembles of shallow classifiers have recently been used in dermoscopy [17], this work differentiates from them by using XGBoost, an approach that led many challenges (e.g., Kaggle [54]) and experimental comparisons in papers dealing with different domains. Finally, the proposed methodology follows the shallow learning strategy, which can demand less computational resources during classifier training than deep learning alternatives.

4 Conclusion

In this work, a machine learning paradigm not widely used in dermoscopy was tested against classical ML methodologies. The Ensemble algorithm, together with IG for FS, was applied to a set of dermoscopic images already tested for traditional ML algorithms and Relief. Although the conclusions regarding data dispersion were broadly the same and, in both cases, no statistical differences were found among the tested group of features for both methodologies, the proposed approach discussed in this work revealed similar performance with fewer features. This result is associated with the proposed approach using IG for FS and to choose the features that compose the trees learned by the ensemble classification method.

These results need to be reinforced by using other datasets of dermoscopic images with distinct properties regarding acquisition, number, and type of images. The binary classification between malignant melanoma and non-melanoma can be expanded to different types of dermatological diagnosis (e.g., typical versus atypical nevus; inflammatory versus non-inflammatory lesion).

Acknowledgements We would like to acknowledge EurekaSD: Enhancing University Research and Education in Areas Useful for Sustainable Development—grants EK14AC0037 and EK15AC0264. We would like to thank Araucária Foundation for the Support of the Scientific and Technological Development of Paraná through a Research and Technological Productivity Scholarship for H. D. Lee (grant 028/2019). The Portuguese team was partially supported by Fundação para a Ciência e a Tecnologia (FCT). Rui Fonseca-Pinto under the projects UIDB/05704/2020, UIDP/05704/2020 and UID/EEA/50008/2020 and Conceição V. Nogueira was financed by the project Pest-C/MAT/UI0013/2014. We also would like to thank PGEEC/UNIOESTE through a postdoctoral scholarship for N. Spolaôr, the Brazilian National Council for Scientific and Technological Development (CNPq) through the grant 140159/2017-7 for A. R. S. Parmezan and the Coordenação de Aperfeiçoamento de Pessoal de Nível Superior—Brasil (CAPES)—Finance Code 001 through an MSc. scholarship for L. A. Ensina. These agencies did not have any further involvement in this paper. The authors thank J. G. Martins for his help in local binary patterns implementation.

References

1. Argenziano, G., Soyer, H.P., Chimenti, S., Talamini, R., Corona, R., Sera, F., Binder, M., Cerroni, L., Rosa, G.D., Ferrara, G., Hofmann-Wellenhof, R., Landthaler, M., Menzies, S.W., Pehamberger, H., Piccolo, D., Rabinovitz, H.S., Schiffner, R., Staibano, S., Stolz, W., Bartenjev, I., Blum, A., Braun, R., Cabo, H., Carli, P., Giorgi, V.D., Fleming, M.G., Grichnik, J.M., Grin, C.M., Halpern, A.C., Johr, R., Katz, B., Kenet, R.O., Kittler, H., Kreusch, J., Malvehy, J., Mazzocchetti, G., Oliviero, M., Ozdemir, F., Peris, K., Perotti, R., Perusquia, A., Pizzichetta, M.A., Puig, S., Rao, B., Rubegni, P., Saida, T., Scalvenzi, M., Seidenari, S., Stanganelli, I., Tanaka, M., Westerhoff, K., Wolf, I.H., Braun-Falco, O., Kerl, H., Nishikawa, T., Wolff, K., Kopf, A.W.: Dermoscopy of pigmented skin lesions: results of a consensus meeting via the Internet, J. Am. Acad. Dermatol. **48(5)**, 679–693 (2003). https://doi.org/10.1067/mjd.2003.281
2. ACS: In: Cancer facts and figures. American Cancer Society. https://www.cancer.org/research/cancer-facts-statistics/all-cancer-facts-figures/cancer-facts-figures-2016.html (2016). Accessed 12 Dec 2018
3. Robbins, S., Cotran, R., Kumar, V.: Pathologic basis of disease. Saunders Company, Heidelberg (1984). https://doi.org/10.1002/path.1711470413
4. Carli, P., De Giorgi, V., Crocetti, E., Mannone, F., Massi, D., Chiarugi, A., Giannotti, B.: Improvement of malignant/benign ratio in excised melanocytic lesions in the 'dermoscopy era': a retrospective study 1997–2001. Br. J. Dermatol. **150**(4), 687–692 (2004). https://doi.org/10.1111/j.0007-0963.2004.05860.x
5. Lee, H., Mendes, A., Spolaôr, N., Oliva, J., Parmezan, A., Chung, W., Fonseca-Pinto, R.: Dermoscopic assisted diagnosis in melanoma: reviewing results, optimizing methodologies and quantifying empirical guidelines. Knowl. Based Syst. **158**, 9–24 (2018). https://doi.org/10.1016/j.knosys.2018.05.016
6. Celebi, M., Codella, N., Halpern, A.: Dermoscopy image analysis: overview and future directions. IEEE J. Biomed. Health Informatics **23**(2), 474–478 (2019). https://doi.org/10.1109/JBHI.2019.2895803
7. Vocaturo, E., Perna, D., Zumpano, E.: Machine learning techniques for automated melanoma detection. In: 2019 IEEE International Conference on Bioinformatics and Biomedicine (BIBM), pp. 2310–2317. IEEE, San Diego (2019). https://doi.org/10.1109/BIBM47256.2019.8983165
8. Cui, X., Wei, R., Gong, L., Qi, R., Zhao, Z., Chen, H., Song, K., Abdulrahman, A.A.A., Wang, Y., Chen, J.Z.S., Chen, S., Zhao, Y., Gao, X.: Assessing the effectiveness of artificial intelligence methods for melanoma: a retrospective review. J. Am. Acad. Dermatol. **81**(5), 1176–1180 (2019). https://doi.org/10.1016/j.jaad.2019.06.042
9. Ninh, Q.C., Tran, T., Tran, T.T., Tran, T.A.X., Pham, V.: Skin lesion segmentation ba-sed on modification of segnet neural networks. In: 2019 6th NAFOSTED Conference on Information and Computer Science (NICS), pp. 575–578. IEEE, Hanoi (2019). https://doi.org/10.1109/NICS48868.2019.9023862
10. Rashid, F., Irtaza, A., Nida, N., Javed, A., Malik, H., Malik, K.M.: Segmenting melanoma lesion using single shot detector (SSD) and level set segmentation technique. In: 13th International Conference on Mathematics, Actuarial Science, Computer Science and Statistics (MACS), pp. 1–5. IEEE, Karachi (2019). https://doi.org/10.1109/MACS48846.2019.9024823
11. Al-masni, M.A., Kim, D-H., Kim, T-S.: Multiple skin lesions diagnostics via integrated deep convolutional networks for segmentation and classification. Comput. Methods Programs Biomed. **190**, (2020). https://doi.org/10.1016/j.cmpb.2020.105351
12. Gu, Y., Ge, Z., Bonnington, C.P., Zhou, J.: Progressive transfer learning and adversarial domain adaptation for cross-domain skin disease classification. IEEE J. Biomed. Health Informatics. **24**(5), 1379–1393 (2020) https://doi.org/10.1109/JBHI.2019.2942429
13. Namboodiri, T.S., Jayachandran, A.: Multi-class skin lesions classification system using probability map based region growing and DCNN. Int. J. Comput. Intell. Syst. **13**(1), 77–84 (2020). https://doi.org/10.2991/ijcis.d.200117.002

14. Sies, K., Winkler, J.K., Fink, C., Bardehle, F., Toberer, F., Buhl, T., Enk, A., Blum, A., Rosenberger, A., Haenssle, H.A.: Past and present of computer-assisted dermoscopic diagnosis: performance of a conventional image analyser versus a convolutional neural network in a prospective data set of 1,981 skin lesions. Eur. J. Cancer. **135**, 39–46 (2020). https://doi.org/10.1016/j.ejca.2020.04.043
15. Adjobo, E.C., Mahama, A.T.S., Gouton, P., Tossa, J.: Proposition of convolutional neural network based system for skin cancer detection. In: 15th International Conference on Signal-Image Technology & Internet-Based Systems (SITIS), pp. 35–39. IEEE, Sorrento (2019). https://doi.org/10.1109/SITIS.2019.00018
16. Wei, L., Ding, K., Hu, H.: Automatic skin cancer detection in dermoscopy images based on ensemble lightweight deep learning network. IEEE Access **8**, 99633–99647 (2020). https://doi.org/10.1109/ACCESS.2020.2997710
17. Akram, T., Lodhi, H.M.J., Naqvi, S.R., Naeem, S., Alhaisoni, M., Ali, M., Haider, S.A., Qadri, N.N.: A multilevel features selection framework for skin lesion classification. Hum. Centric Comput. Inf. Sci. **10**(12), 1–26 (2020). https://doi.org/10.1186/s13673-020-00216-y
18. Birkenfeld, J.S., Tucker-Schwartz, J.M., Soenksen, L.R., Avilés-Izquierdo, J.A., Marti-Fuster, B.: Computer-aided classification of suspicious pigmented lesions using wide-field images. Comput. Methods Programs Biomed. 195, (2020). https://doi.org/10.1016/j.cmpb.2020.105631
19. Goyal, M., Oakley, A., Bansal, P., Dancey, D., Yap, M.H.: Skin lesion segmentation in dermoscopic images with ensemble deep learning methods. IEEE Access **8**, 4171–4181 (2020). https://doi.org/10.1109/ACCESS.2019.2960504
20. Zhou, Z.: Ensemble methods: foundations and algorithms. Chapman & Hall/CRC, London (2012)
21. Esfahlani, F.Z., Visser, K., Strauss, G.P., Sayama, H.: A network-based classification framework for predicting treatment response of schizophrenia patients. Expert Syst. Appl. **109**, 152–161 (2018). https://doi.org/10.1016/j.eswa.2018.05.005
22. Liu, H., Motoda, H.: Computational methods of feature selection. Chapman & Hall/CRC, London (2007)
23. Bolón-Canedo, V., Alonso-Betanzos, A.: Recent advances in ensembles for feature selection. Springer, Cham (2018)
24. Brancati, N., Frucci, M., Gragnaniello, D., Riccio, D., Di Iorio, V., Di Perna, L., Simonelli, F.: Learning-based approach to segment pigment signs in fundus images for retinitis pigmentosa analysis. Neurocomputing **308**, 159–171 (2018). https://doi.org/10.1016/j.neucom.2018.04.065
25. Rastgoo, M., Garcia, R., Morel, O., Marzani, F.: Automatic differentiation of melanoma from dysplastic nevi. Comput. Med. Imaging Graph. **43**, 44–52 (2015). https://doi.org/10.1016/j.compmedimag.2015.02.011
26. Latif, J., Xiao, C., Imran, A., Tu, S.: Medical imaging using machine learning and deep learning algorithms: a review. In: 2nd International Conference on Computing, Mathematics and Engineering Technologies (iCoMET), pp. 1–5. IEEE, Sukkur (2019). https://doi.org/10.1109/ICOMET.2019.8673502
27. Fonseca-Pinto, R., Machado, M.: A textured scale-based approach to melanocytic skin lesions in dermoscopy. In: 40th International Convention on Information and Communi-cation Technology, Electronics and Microelectronics (MIPRO), pp. 279–282. IEEE, Opatija (2017). https://doi.org/10.23919/MIPRO.2017.7973434
28. Wernick, M.N., Yang, Y., Brankov, J.G., Yourganov, G., Strother, S.C.: Machine learning in medical imaging. IEEE Signal Process. Mag. **27**(4), 25–38 (2010). https://doi.org/10.1109/MSP.2010.936730
29. Machado, M., Pereira, J., Fonseca-Pinto, R.: Classification of reticular pattern and streaks in dermoscopic images based on texture analysis. J. Med. Imaging. **2**(4), 044503 (2015). https://doi.org/10.1117/1.JMI.2.4.044503
30. Barata, C., Marques, J.S., Rozeira, J.: The role of keypoint sampling on the classifica-tion of melanomas in dermoscopy images using bag-of-features. In: Pattern Recognition and Image

Analysis: 6th Iberian Conference, IbPRIA 2013. In: Sanches, J.M., Micó, L., Cardoso, J.S. (eds.), pp. 715–723. Springer, Heidelberg (2013). https://doi.org/10.1007/978-3-642-38628-2_85

31. Barata, C., Marques, J.S., Rozeira, J.: Evaluation of color based keypoints and features for the classification of melanomas using the bag-of-features model. In: Advances in Visual Computing: 9th International Symposium, ISVC 2013, Rethymnon, Crete, Greece, July 29–31, 2013. In: Part I. Bebis, G., Boyle, R., Parvin, B., et al. (eds.), pp. 40–49. Springer, Heidelberg (2013). https://doi.org/10.1007/978-3-642-41914-0_5
32. Barata, C., Marques, J.S. Mendonça, T.: Bag-of-features classification model for the diagnose of melanoma in dermoscopy images using color and texture descriptors. In: Kamel, M., Campilho, A. (eds.) Image Analysis and Recognition. ICIAR 2013. Lecture Notes in Computer Science, vol. 7950, pp. 547–555 (2013). https://doi.org/10.1007/978-3-642-39094-4_62
33. Barata, C., Marques, J.S., Celebi, M.E.: Improving dermoscopy image analysis using color constancy. In: IEEE International Conference on Image Processing, pp. 3527–3531. Paris (2014). https://doi.org/10.1109/ICIP.2014.7025716
34. Barata, C., Ruela, M., Francisco, M., Mendonça, T., Marques, J.S.: Two systems for the detection of melanomas in dermoscopy images using texture and color features. IEEE Syst. J. **8**(3), 965–979 (2014). https://doi.org/10.1109/JSYST.2013.2271540
35. Barata, C., Celebi, M.E., Marques, J.S.: Improving dermoscopy image classification using color constancy. IEEE J. Biomed. Health Informatics. **19**(3), 1146–1152 (2015). https://doi.org/10.1109/JBHI.2014.2336473
36. Kaur, R., Albano, P.P., Cole, J.G., Hagerty, J., Leander, R.W., Moss, R.H., Stoecker, W.V.: Real-time supervised detection of pink areas in dermoscopic images of melanoma: importance of color shades, texture and location. Skin Res. Technol. **21**(4), 466–473 (2015). https://doi.org/10.1111/srt.12216
37. Abuzaghleh, O., Faezipour, M., Barkana, B.D.: A comparison of feature sets for an automated skin lesion analysis system for melanoma early detection and prevention. In: 2015 Long Island Systems, Applications and Technology, pp. 1–6. Farmingdale (2015). https://doi.org/10.1109/LISAT.2015.7160183
38. Sáez, A., Sánchez-Monedero, J., Gutiérrez, P.A., Hervás-Martínez, C.: Machine learning methods for binary and multiclass classification of melanoma thickness from dermoscopic images. IEEE Trans. Med. Imaging. **35**(4), 1036–1045 (2016). https://doi.org/10.1109/TMI.2015.2506270
39. Sánchez-Monedero, J., Sáez, A., Pérez-Ortiz, M., Gutiérrez, P.A., Hervás-Martínez, C.: Classification of melanoma presence and thickness based on computational image analysis. In: Martínez-Álvarez, F., Troncoso, A., Quintián, H., Corchado, E. (eds.) Hybrid Artificial Intelligent Systems: 11th International Conference, HAIS 2016, pp. 427–438. Springer International Publishing, Seville (2016). https://doi.org/10.1007/978-3-319-32034-2_36
40. García, V., Florencia-Juárez, R., Sánchez-Solís, J., Rivera, G., Contreras-Masse, R.: Predicting airline customer satisfaction using k-nn ensemble regression models. Res. Comput. Sci. **148**(6), 205–215 (2019). https://doi.org/10.13053/rcs-148-6-15
41. Pérez-Ortiz, M., Sáez, A., Sánchez-Monedero, J., Gutiérrez, P.A., Hervás- Martínez, C.: Tackling the ordinal and imbalance nature of a melanoma image classification problem. In: International Joint Conference on Neural Networks, pp. 2156–2163. Vancouver (2016). https://doi.org/10.1109/IJCNN.2016.7727466
42. Yang, S., Oh, B., Hahm, S., Chung, K.Y., Lee, B.U.: Ridge and furrow pattern classification for acral lentiginous melanoma using dermoscopic images. Biomed. Signal Process. Control. **32**, 90–96 (2017). https://doi.org/10.1016/j.bspc.2016.09.019
43. Kononenko, I.: Estimating attributes: analysis and extensions of RELIEF. In: Bergadano, F., De Raedt, L. (eds.) European Conference on Machine Learning, vol. 784, pp. 171–182. Springer, Heidelberg (1994)
44. Witten, I., Frank, E., Hall, M., Pal, C.: Data mining. In: Practical Machine Learning Tools and Techniques, 4th edn. Morgan Kaufmann, Burlington (2017). https://doi.org/10.1016/C2015-0-02071-8

45. Han, J., Kamber. M.: Data Mining: concepts and techniques. Morgan Kaufmann, Cambridge (2011)
46. Cristianini, N., Shawe-Taylor, J.: An Introduction to Support Vector Machines and Other Kernel-Based Learning Methods. Cambridge University Press, Cambridge (2000)
47. Chen, T., Guestrin, C.: XGBoost: a scalable tree boosting system. In: ACM SIGKDD International Conference on Knowledge Discovery and Data Mining, pp. 785–794. ACM, New York (2016). https://doi.org/10.1145/2939672.2939785
48. XGBoost: about XGBoost. Royal Society of Chemistry. https://xgboost.ai/about (1999). Accessed 11 Mar 2019
49. Ma, B., Meng, F., Yan, G., Yan, H., Chai, B., Song, F.: Diagnostic classification of cancers using extreme gradient boosting algorithm and multiomics data. Comput. Biol. Med. **121**, 103761 (2020). https://doi.org/10.1016/j.compbiomed.2020.103761
50. Zhang, D., Qian, L., Mao, B., Huang, C., Huang, B., Si, Y.: A data-driven design for fault detection of wind turbines using random forests and XGBoost. IEEE Access **6**, 21020–21031 (2018). https://doi.org/10.1109/ACCESS.2018.2818678
51. Spolaôr, N., Cherman, E.A., Monard, M.C., Lee, H.D.: A comparison of multi-label feature selection methods using the problem transformation approach. Electron. Notes Theor. Comput. Sci. **292**, 135–151 (2013). https://doi.org/10.1016/j.entcs.2013.02.010
52. Pereira, R.B., Plastino, A., Zadrozny, B., Merschmann, L.H.C.: Information gain feature selection for multi-label classification. J. Inf. Data Manag. **6**(1), 48–58 (2015)
53. Oliva, J.T., Lee, H.D., Spolaôr, N., Coy, C.S.R., Chung, W.F.: Prototype system for feature extraction, classification and study of medical images. Expert Syst. Appl. **63**, 267–283 (2016). https://doi.org/10.1016/j.eswa.2016.07.008
54. Chollet, F., Allaire, J.J.: Deep learning in R. Manning publications, Shelter Island (2018)

Modeling Performance of NP-hard Problems by Applying Causal Analysis for the VisTHAA Tool

Claudia Gómez-Santillán, Juan Gerardo Ponce-Najera, Luis Rodolfo García-Nieto, Laura Cruz-Reyes, Nelson Rangel-Valdez, Héctor J. Fraire-Huacuja, and Lucila Morales-Rodríguez

Abstract At present the analysis of the algorithms is a necessary process, especially algorithms that solve difficult problems in the daily life; Since the analysis of algorithms helps to give explanations of the performance and to understand the behavior of the algorithms when the input data change it means, the input instances present inherent characteristics of the environment of the problem to solve, causing a differently behave in the algorithms. In the scientific literature, several researchers have been given the task of working in this area of research, for example, *The Multi-objective Optimization* and *The Analysis and Design of Optimization Algorithms Via Integral Quadratic Constraints* which addresses the analysis of algorithms for optimization problems. This research work proposes the incorporation of a set of modules that allow researchers to redesign algorithms that solve NP-Hard problems belonging to the container family through a causal methodology, allowing a formal explanation of the behavior of algorithms through indexes of problem structure, behavior, trajectory, and algorithmic performance. These modules are incorporated into an analysis tool called VisTHAA.

Keywords Causal models · Algorithmic performance · Portfolio selection problem · Knapsack problem 0/1 · Design of optimization algorithms

1 Introduction

The objective of computational experimentation is to promote that the experiments are relevant, accurate, and reproducible in order to generate knowledge and improve the performance of algorithms [1]. Currently, it is not enough to show that an algorithm is better than others in solving a set of instances, you should take into account both the behavior of the algorithm and performance, some researchers propose the

C. Gómez-Santillán (✉) · J. G. Ponce-Najera · L. R. García-Nieto · L. Cruz-Reyes · N. Rangel-Valdez · H. J. Fraire-Huacuja · L. Morales-Rodríguez
División de Estudios de Posgrado E Investigación, Tecnológico Nacional de México/Instituto Tecnológico de Ciudad Madero, Ciudad Madero 89440, Tamaulipas, México
e-mail: claudia.gomez@itcm.edu.mx

W. Pedrycz et al. (eds.), *Computational Intelligence for Business Analytics*, Studies in Computational Intelligence 953,
https://doi.org/10.1007/978-3-030-73819-8_8

redesign of an algorithm to improve its performance in an instance set, taking as an example the work of Refs. [2, 3].

In this work, it is proposed to characterize algorithms and instances through indexes and tools that promote causal analysis, t o impact on the improvement of algorithm performance. The problems addressed to validate the algorithm redesign methodology are from the container family, and it has been shown that they are NP-hard, and there is a relationship between them [4].

To carry out this was developed a module of the redesign of algorithms that give solution to the mono objective knapsack problem and the portfolio selection problem multi-objective (PSP), through causal techniques, taking as a basis a methodology proposed in the work of Landero [3], this module will be incorporated into the Visualization Tool for Heuristic Algorithms Analysis (VisTHAA).

2 Background

This section presents topics, concepts, and approaches related to the development of this research work, describes the research problems to be analyzed, solving strategies for mono and multi-target optimization problems, techniques allow the realization of the causal redesign of the solution strategies and the incorporation of indicators that allow the characterization of the performance of the solver algorithms to be used. These approaches will be incorporated into the VisTHAA architecture, which will improve its functionality.

2.1 Optimization Problem

Given the importance of optimization problems, various studies have been performed in order to provide solutions that allow cost minimization. However, the required solution process is highly complex, since in most cases, real-world problems belong to a special class of problems called NP-hard [5], which implies that efficient algorithms are not known to solve them exactly in the worst case.

NP-hard problems are of great interest to computer science. One of the characteristics of the problems of this kind is that the exact algorithms used to solve them require an exponential amount of time in the worst case. In other words, these problems are very difficult to solve [5]. When solving NP-hard problems, it is necessary to settle for "good" solutions, that in some cases, may obtain optimal results. Under these conditions, algorithms are used that provide approximate solutions in a reasonable time but do not guarantee to obtain the best solution.

2.1.1 Knapsack Problem 0/1

One of the classic problems of combinatorial optimization because it is classified as an NP-hard problem [6]. It is a combinatorial optimization problem that plays an important role in the theory of computation because it has as a goal the search for the best solution between a finite set of possible solutions to a problem. The problem is to select such a quantity of objects to be carried in the backpack so that the total value is maximized without exceeding the capacity of the knapsack [7].

$$max\ Z = \sum_{i=1}^{n} v_i x_i \tag{1}$$

$$s.t.\ \sum_{i=1}^{n} w_i x_i \leq W \tag{2}$$

$$x_i \in \{0, 1\},\ i \in n = \{1, 2, \ldots, n\} \tag{3}$$

where:

Z is the maximum profit obtained, n is the number of objects available, x_i refers to the object, w_i is the weight of the object i, v_i is the profit of the object i, and W is the maximum capacity of the knapsack.

2.1.2 Portfolio Selection Problem (PSP)

Project selection is a periodic activity that involves meeting an organization's stated objectives without exceeding available resources or violating other constraints [8]; each one is described by estimations of its impacts and resource consumption. In an organization, there are many projects, this is known as a project portfolio, within this set, some projects will be financially supported, and others will not.

One of the most important decision-making problems for both public and private institutions is the correct selection of projects to create a feasible portfolio of projects; therefore, they use a Decision Maker (DM). The DM is a person or a group of people in charge of selecting, in their opinion, and given experiences, the better solutions [9]. In the case of risks, the DM should know the probability of distribution of profits.

The main economic and mathematical models for the PSP, assume that you have a defined set n of projects, each project perfectly characterized with costs and revenues, of which the distribution in the time is known. The mathematical model of the problem is expressed as follows:

$$max\ Z = \sum_{j=1}^{n} c_j x_j \tag{4}$$

where:

The objective function (1) maximizes the total profit associated with Z to a certain portfolio of projects, c_j is the benefit associated with project j and $x_j = 1$ indicates whether project j is part of the portfolio and if $x_j = 0$ necessary opposite. Therefore if $x_j = 1$ then the project j receives the requested funding.

2.2 *Metaheuristic Algorithm*

Heuristic algorithms generally have troubles with stagnation in local optimum because they have not a "searching for a better solution" mechanism. To fix this lack of mechanism there are algorithms called metaheuristics that provide procedures to help heuristic algorithms to avoid getting stuck in local optimum [10].

2.2.1 Population Metaheuristic Algorithm

These types of methods are based on the recombination of candidate solutions of a population, which evolves through genetic mechanisms such as the selection, the cross, and the mutation of the individuals of a population [10].

The concept of solution recombination is one of the Genetic Algorithms (GAs) core contributions. On the other hand, it is also relevant to the explicit difference between the representation of the problem (called genotype), which is usually determined by bits chains known as chromosomes, and the variables of the problem itself (called phenotype). GA represents a simple and intuitive population metaheuristic that is most likely to be the most widely used. A generic GA procedure is Shown in Algorithm 1.

Algorithm 1. Basic GA

```
1. Procedure AG (g)
2. Input t (Population size)
3. Output sol
4. Generate Random Population → p
5. Fitness Function → p
6. For i=1 to g
7. Selection operator
8. Recombination operator
9. Mutation operator
10. End for
11. End procedure
```

2.2.2 Non-Dominated Sorting Genetic Algorithm (NSGAII)

It is a popular genetic algorithm based on non-dominance for multi-objective optimization [11], which has a built-in algorithm for ordering and elitism [12]. The NSGAII algorithm is Shown in Algorithm 2.

Algorithm 2. *fast – non - dominated – sort* (P)

1	for each $p \in P$	
2	$S_p = \emptyset$	
3	$n_p = 0$	
4	for each $q \in P$	p and q are *interval numbers*
5	if$(p \prec q)$	*Then if* p *dominates* q
6	$S_p = S_p \cup \{q\}$	*Add* q *to the set of solutions dominated by* p
7	else if $(q \prec p)$ then	
8	$n_p = n_p + 1$	*Increment the domination counter of* p
9	if $n_p = 0$ then	p *belongs to the first front*
10	$p_{rank} = 1$	
11	$F_1 = F_1 \cup \{p\}$	
12	$i = 1$	*Initialize the front counter*
13	while $F_i \neq 0$	
14	$Q = \emptyset$	*Used to store the members of the next front*
15	for each $p \in F_i$	
16	for each $q \in S_p$	
17	$n_p = n_p - 1$	
18	if $n_p = 0$ then	q *belongs to the next front*
19	$q_{rank} = i + 1$	
20	$Q = Q \cup \{q\}$	
21	$i = i + 1$	
22	$F_i = Q$	

In order to sort the population of size N according to the level of not dominance, each solution must be compared with all the solutions in the population to find out if it is dominated (Lines 4–10), the process is described below. For the set of solutions of the population *P*, on the line 5 comparison is performed by a vector of objectives corresponding to the solution, in turn, to determine if *p* dominates *q* (*p* and *q* are interval numbers) if so, this solution is included in any structure, in order to identify which solutions were dominated by *p*. Otherwise, if *q* dominate *p*, increments the value of n_p, this variable indicates the number of solutions that *p* did not dominates (Lines 7 and 8). In the lines, 9–11 since the evaluation of the solution is known *p* in the previous process, and if there are no solutions that dominate it, the solution *p* will be part of the first front F_0. In order to find the individuals of the following front, the solutions of the first front are temporarily discounted and the process is carried out again (Lines 13 a 22). The procedure is repeated to find the other fronts; the worst case would be that there is only one solution in each front.

2.3 *Causality*

A term that indicates how the world responds to an intervention, such intervention can be represented in causal models which are a generalized representation of knowledge, obtained by finding dependencies that imply relationships of cause and effect.

Causality requires identifying the direct relationship that exists between events or variables. The variables involved in the model are random, and some may have a causal relationship with others.

2.3.1 Causal Modeling

A causal model is a widespread representation of knowledge gained by finding dependencies that involve cause-and-effect relationships. Causation requires identifying the direct relationship between events or variables. The variables involved in the model are naturally random, and some may have a causal relationship with others. In practice, the variables are divided into two sets, the exogenous variables, whose values are determined by factors outside the model, and the endogenous ones, which have values described by a model of structural equations.

Causal relationships are transitive, unreflecting, and antisymmetric. This means that: (1) if A is the cause of B and B is the cause of C, then A is also the cause of C, (2) an event A cannot be caused to itself, and (3) if A is the cause of B then B does not cause of A.

The representation of a causal system is done through a directed acyclic graph (DAG), which shows the causal influences between system variables and helps to estimate the total and partial effects that result from the manipulation of a variable. A causal model can be defined as a causal Bayesian network [13], which consists of a causal model, a set of variables, an acyclic directed graph, and a conditional probability function.

2.3.2 PC Algorithm

Most of the methods used to create causal models are based on conditional independence tests, this defines a set of restrictions that must be satisfied to induce the causal structure [2].

Existing algorithms are based on tests of conditional independence between the variables of the model, with the specific objective of generating graphs that best represent the properties of the model, through a qualitative study.

The PC algorithm aims to find the causal graph for large samples. This graph represents the same conditions of conditional independence of the population under the following assumptions.

- The causal graph in the population is acyclic
- The causal graph in the population is reliable

- The set of causal variables is causally sufficient.

This algorithm is based on tests of independence of variables; therefore, the PC algorithm assumes that it has enough data to initialize and that the statistical tests to be used have no errors.

The PC algorithm is made up of three main stages:

1. Construction of a completely connected graph, in which all the input variables are related.
2. Elimination of indirect relationships, subsistence in tests of conditional dependency.
3. Orient the rest of the relations of the graph, employing conditional operators, without producing cycles.

2.3.3 Casual Modeling Process

This module works with information obtained from the indices of the selection phases of the problem indices and the algorithm (box 1); Furthermore, information is also needed on the set of dominance regions, which is obtained from the Identification phase of the dominance regions (box 2). This information is necessary for the creation of the causal model. Figure 1 shows the phases of the process of creating the causal model used in this work.

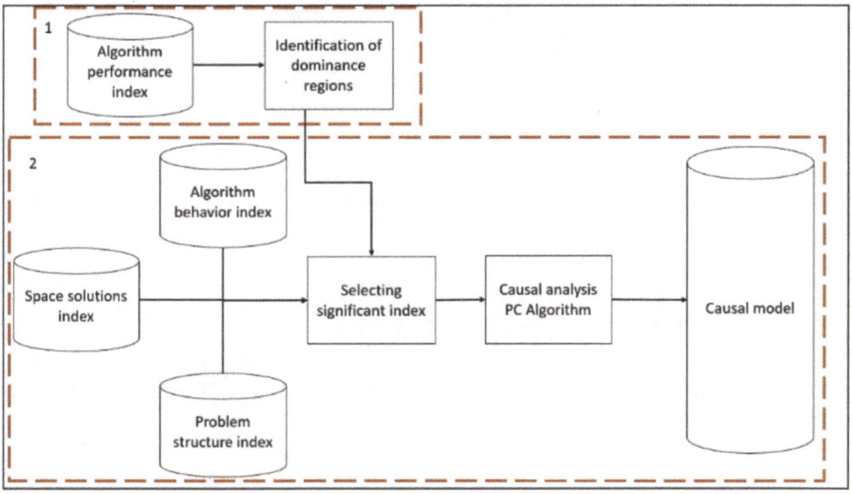

Fig. 1 Causal modeling process

Table 1 Mono objective index performance

Indicator	Description	Formula
MaxBestF	Calculates the maximum aptitude for an instance	$MaxBestF = max(M)$
MinBestF	Calculates the minimum aptitude for an instance	$MinBestF = min(M)$
TeoricalRadio	It indicates the theoretical reason where Z is the best value obtained in the algorithm and Z_{opt} is the best value reported in the literature for that instance	$TeoricalRadio = \frac{Z}{Z_{opt}}$
ErrorRate	Indicates the error percentage of the value obtained from an instance's aptitude	$ErrorRate = \frac{(Z_{opt}-Z)}{Z_{opt}}$

2.4 *Performance Indicators*

In order to measure the performance of the algorithms to be studied, measures are needed to quantify the performance of these algorithms, in this work will be used some basic measures proposed in the literature for mono-objective and multi-objective algorithms.

In this work, it is proposed to characterize algorithms and instances through indexes and tools that promote causal analysis, t o impact on the improvement of algorithm performance. The group of indexes with which we initially work belongs to various groups such as Structure of the problem, space of solutions, behavior, trajectory, and algorithmic performance. However, due to the nature of the problems, only the indexes that provide information to improve the performance of the algorithm will be used, and these are described below.

2.4.1 Mono Objective Performance Indicators

Below, Table 1 will explain some performance indicators applied for mono-objective algorithms proposed in the work of [14].

2.4.2 Multi-Objective Performance Indicators

Table 2 will explain the performance indicators applied for multi-objective algorithms, which are reported in the literature.

2.5 *VisTHAA*

It is a diagnostic tool dedicated to the analysis of heuristic algorithms; this tool allows researchers to extend its functionalities through modules integrated into an

Table 2 Multi-objective index performance

Indicator	Description	Formula
Generational Distance (GD)	It measures the distance of the solutions that are in the set of non-dominated solutions found so far from the optimal Pareto set. Where n is the number of unmastered solutions found and d_i is the Euclidean distance between each of the solutions and the closest member of the Pareto front [15]	$GD = \frac{\sqrt{\sum_{i=1}^{n} d_i^2}}{n}$
Inverted Generational Distance (IGD)	It determines the average of the distances between each solution from the true Pareto front to the nearest Pareto front solution found [16]	$IGD \triangleq \frac{(\sum_{i=1}^{n} d_i^p)^{\frac{1}{p}}}{n}$
Euclidean distance (ED)	It allows knowing the distance that exists between 2 solutions obtained [17]	$ED = \sqrt{\sum_{i=1}^{n} (p_i - q_i)^2}$
Spread	It is a metric of diversity that measures the degree of propagation achieved between the solutions obtained. Where it is the Euclidean distance between N consecutive solutions, it is the average of these distances, and are the Euclidean distances with the extreme solutions (delimiter) of the exact Pareto front in the objective space [14]	$\Delta = \frac{d_f + d_l + \sum_{i=1}^{N-1} \left\| d_i - \overline{d} \right\|}{d_f + d_l + (N-1)\overline{d}}$
Spacing (s)	It is a metric which indicates that so well the solutions are distributed in the discovered front where is the number of members in and is the Euclidean distance (in the domain of the objectives) between the member i in PF_{know} and its close member in PF_{know} [14]	$s = \frac{\left[\frac{1}{n_{PF}} \sum_{i=1}^{n_{PF}} \left(d_i' - \overline{d'}\right)^2\right]^{1/2}}{\overline{d'}}$
Hypervolume (HV)	It is a combined indicator of convergence and diversity that calculates the volume (in the target space) covered by the members of an unmastered set of solutions [18]	$HV = volume\left(\bigcup_{i=1}^{\|Q\|} (v_i)\right)$

(continued)

Table 2 (continued)

Indicator	Description	Formula
The proportion of non-dominated solutions (RNI)	Measures the proportion of non-dominated solutions of a P population. For this, you obtain the Pareto front (*POF*) of this population and divide the cardinality of POF between the cardinality of *P* [19]	$RNI = \frac{\lvert POF \rvert}{\lvert P \rvert}$

architecture for the facilitation of heuristic analyses. The tool has the following modules available [20]:

1 Data entry and preprocessing
2 Instance characterization
3 Viewing instances and algorithm behavior
4 Visualization of the search space in three dimensions and the analysis of the algorithms.

Figure 2 shows the internal architecture of the VisTHAA, and modules that have a dependency on other modules.

Table 3 shows up the modules that currently integrate VisTHAA, along with a brief description of how it works.

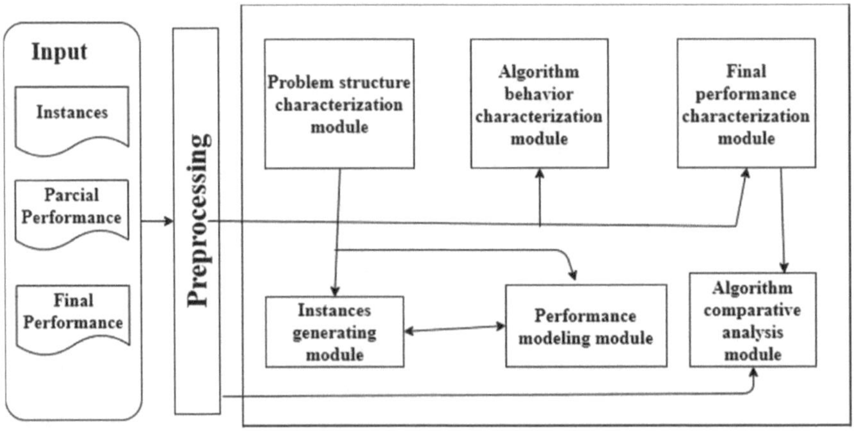

Fig. 2 Architecture of VisTHAA

Table 3 Modules integrated into VisTHAA

Module	Description
Data entry and preprocessing	The input and preprocessing module are intended for researchers to use their formats for reading and processing files, to fulfill this function it is necessary to adjust the input files to an input format required by the tool
Characterization of Instances	After the information upload process is complete, the researcher can apply a characterization process to the set of instances of the problem to be analyzed. This process can be done statistically or visually. Statistical characterization is performed through the application of indexes to the set of instances, which aims to characterize and quantify factors that define the structure of the problem, partial and final performance of the algorithm, allowing the identification of the factors that impact the optimization process. The visual characterization allows showing graphically to the researcher, the results obtained from the indices applied in the characterization process, allowing a better understanding of them, and a comparison between the results obtained and those that they were better
Visualization	The display module aims to be able to graphically visualize the instances loaded in the tool, frequency graphs of the instances, fitness landscape graphs of the algorithm behavior, and statistical graphs
Statistical analysis module	The function of this module allows the performance evaluation of two algorithms; such evaluation is carried out through the Wilcoxon test, which is classified as a nonparametric procedure that determines whether two sets of data show a difference [22]

3 Proposed Solution

Based on the methodology used in the work of [3], this paper shows the improvement in the performance of algorithms that solve problems pertaining to the family of containers through a casual redesign. As can be seen in Fig. 3, the redesign of algorithms that solve mono objective and multi-objective problems.

The necessary elements to execute the methodology shown in Fig. 3 are the set of input instances belonging to the problem to be analyzed and the set of algorithms to be redesigned.

As the first process, the characterization will be performed which intends to quantify the following processes: the structure of the problem, a sample of the solution space, behavior, path, and performance of the algorithm through a set of indexes. The following is a description of the three threads in the characterization process shown in Fig. 4.

Table 4 describes the stages of the casual redesign methodology shown in Fig. 3.

The next section presents the experimentation where the methodology presented in Fig. 3 will be used to solve two optimization problems.

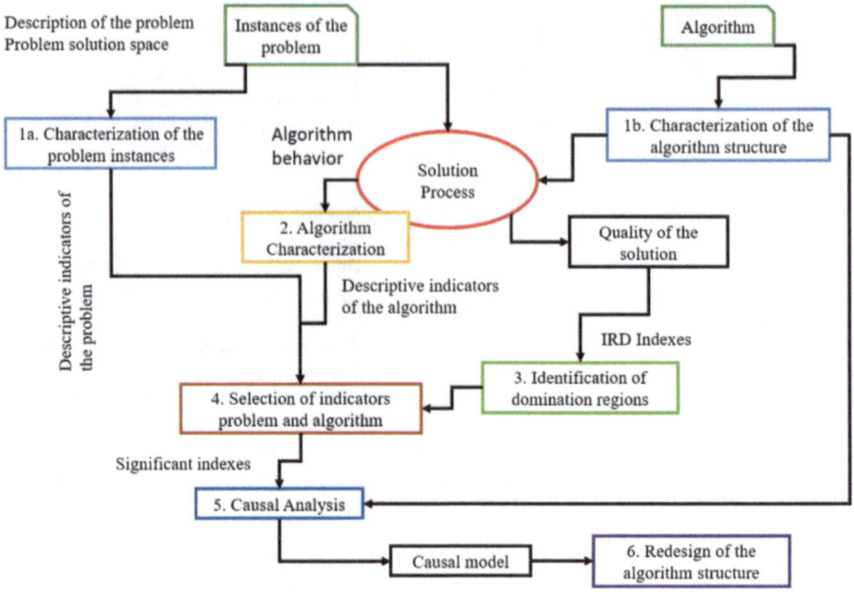

Fig. 3 Causal redesign methodology mono/multi-objective [3]

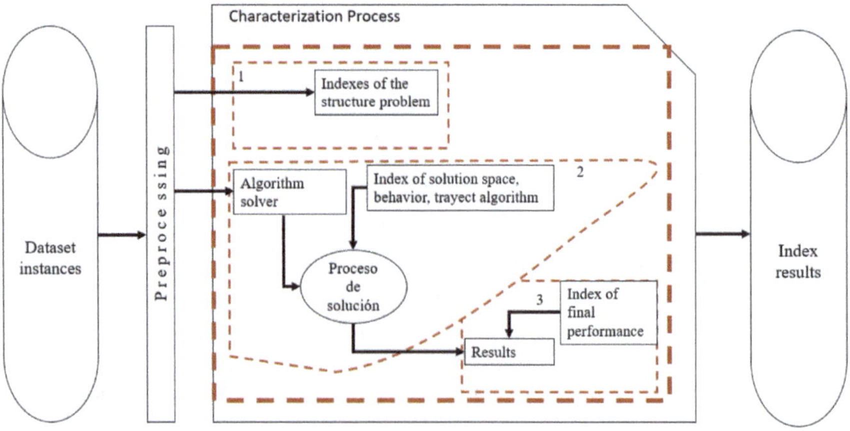

Fig. 4 Characterization process in VisTHAA

4 Experimentation

This section presents the results obtained by the casual redesign methodology explained in the previous section. Applying the methodology to the knapsack 0/1 problem and PSP using evolutionary population metaheuristics algorithms.

Table 4 Casual redesign phases

Phase	Description
Characterization process (part 1)	Applying indexes of the problem structure by taking as input information the set of instances of the problem to be analyzed and allowing to calculate the complexity of the problem structure
Characterization Process (Part 2)	Characterizing the algorithm structure; through indexes that are applied to the problem–solution space sample, taking as input information the set of instances to be analyzed, and the resolver algorithm. Both processes output the results of the indexes applied to the instance structure and the sample of the problem–solution space
Characterization process (part 3)	Characterization of behavior, trajectory, and algorithmic performance through indexes; receiving as input information the set of instances to be processed and a set of variants of the solver algorithm, returning the calculated results from the indexes for the phases of *Identifying the Dominance Regions and Selecting indexes of the problem and the algorithm*
Identification of domination regions	Creates a set of variants belonging to the solver algorithm and identify the variants through limits imposed by the researcher, applied to a set of instances of a problem; returning a set of variants (regions) where the performance obtained is efficient, ending with obtaining conformed dominant regions from the set of variants of the solver algorithm
Selecting indicators problems and algorithm	Identifies relevant indexes of the characterization process, taking as input results obtained from the structure of the problem, solution space, behavior and trajectory of the algorithm, through of a general and/or visual analysis in the form of a graph for its understanding, returning a set of significant indices for the researcher, which will serve as input information for the Causal Analysis phase
Causal analysis	Creates the causal model through the PC algorithm (see Sect. 3.3.2) for algorithmic redesign, taking as input information the significant indices and dominant regions obtained in the previous phases, returning a causal graph with model equations
Redesign of algorithm structure	Applies causal redesign to the set of variants that were not selected in the *Identification of Dominating Regions phase,* based on the causal model suggested in the *Causal Analysis* phase

Table 5 Characteristics of the set of problems analyzed

Problem	Objective	Number of instances	Characteristics	
Knapsack 0/1	Mono	4	Number of objects	15,24,100,1000
			Capacity	750, 6,404,180, 3254, 13,001
PSP	Multi	3	Budget	250,000
			Number of objectives	9
			Number of areas	2
			Number of regions	3
			Number of projects	100

Table 6 Characteristics of set of solving algorithms

Solver algorithm	Parameter	Values
Genetic	Population size	100
	Number of generations	500
	Percentage of crossover	70%
NSGA-II	Percentage of mutation	5%
	Population size	100
	Number of generations	500
	Percentage of crossover	70%
	Percentage of mutation	5%

Table 5 describes the characteristics of the set of instances belonging to the problems analyzed in the causal characterization and redesign process.

Table 6 specifies the configurations of the set of solver algorithms used in the characterization process and the causal redesign process.

4.1 Knapsack 0/1 Problem Experimentation

- Characterization process

Taking as a basis the indices proposed in the theses [2, 3, 14], a subset of indices was selected which were applied in the Knapsack 0/1 problem, Table 7 shows the results obtained from the index used in the characterization process.

- Identification of domination regions

To carry out this phase, an analysis of the genetic algorithm was carried out to identify in which parts of the algorithm different configurations could be applied. To present

Table 7 Example of characterization process indexes for the Knapsack problem 0/1

Index type	Index	Knapsack problem			
		Inst15	Inst24	Inst100	Inst1000
Problem size	*p*	1.87	2.00	1.28	2.33
Relationship between weights and container size	*b*	0.52	0.5	0.66	0.25
	t	0.12	0.08	0.01	0.003
	d	0.02	0.04	0.008	0.002
Central weights trend	*ma*	95.53	533,681.66	48.7	51.28
	mode	0	0	53	32
Weight scattering	*r*	50	919,726	98	99
	dm	14.83	252,395.91	25.24	24.66
	s	16.62	294,478.56	832.73	814.63
	cv	0.17	0.55	0.59	0.55
Shape of weight distribution	*amed*	0.27	0.09	0.17	0.02
	amod	5.74	1.81	-0.14	0.67
	curtosis	23.62	44.76	174.01	1818.98
Sample of the solution space	*vo*	4119.92	301,158,751,713.34	34,777.40	89,460.80
Algorithm behavior	*p_f*	70.06	69.13	55.56	81.26
	p_uf	29.93	30.86	44.43	18.73
	ro	0.53	0.33	0.44	0.35
	delta	2.12	1.49	1.81	1.54
Algorithm trajectory	*nc*	0.95	0.97	0.90	0.98
	nv	1	1	1	1
	pp	0.47	0.48	0.45	0.49
	pn	0.51	0.50	0.53	0.49
Algorithmic performance Mono objective	*Max Best*	1453	13,162,729	4914	14,635
	Average Best	1321.38	11,893,553.1	4564.17	9666.94
	Teorical radio	0.99657064	0.98826091	0.99837464	0.99659517
	Error percentage	0.00342936	0.01173909	0.00162536	0.00340483
	Coefficient of deception	0.23333333	0.21528943	0.07036536	0.12486719

the variety of population configurations on genetic algorithms, there was made Table 8 showing each variant of the algorithm.

Table 8 Genetic algorithm variants

Variant	Initial solution		Crossing method		Mutation method		Local search
	Random	Heuristic	MC1	MC2	MM1	MM2	
V1	X		X		X		X
V2	X		X		X		
V3	X		X			X	X
V4	X		X			X	
V5	X			X	X		X
V6	X			X	X		
V7	X			X		X	X
V8	X			X		X	
V9		X	X		X		X
V10		X	X		X		
V11		X	X			X	X
V12		X	X			X	
V13		X		X	X		X
V14		X		X	X		
V15		X		X		X	X
V16		X		X		X	

- Selecting indicators problems and algorithm

Once each of the possible variants described in Table 8 has been executed, the indices shown in Table 5 are applied, an analysis of the results is made to select the best variants that solve the backpack problem. Algorithm 3 explains the overall process of the significant index used in the characterization process.

Algorithm 3. Significant index process

1. **Start**
2. **Obtaining** the index results of the structure of the problem → indEst
3. **Obtaining** the index results of the sample from the solution space of the problem→ mueEsp
4. **Obtaining** the indexes of the behavior of the algorithm → compAlg
5. **Obtaining** the results of the algorithm path indexes → trayAlg
6. **Application** of general and/or visual analysis to indEst, mueEsp, compAlg, trayAlg
7. **Selection** of significant indices to indEst, mueEsp, compAlg, trayAlg
8. **End**

- Casual analysis

Taking as a basis those variants of the genetic algorithm that have obtained the best results, an analysis is made of all its variables with which a causal graph will be assembled. Figure 1 explains the process of creating the causal model used in this research work, taking as input information the regions of dominance identified in the third phase and the significant indices identified in the fourth phase.

Once the elements for the creation of the causal model are specified, the PC algorithm is executed to the data set which is made up of the indexes of the characterization process, which were mapped with the regions of dominance, at the end of the PC algorithm gets the directed graph with the equations of the causal model.

An example of the graph generated by the PC algorithm applied to the Knapsack 0/1 problem can be seen in Fig. 5 and listed in Table 9 are the set of equations obtained from this causal model.

- Redesign of Algorithm Structure

The results obtained from the process of redesigning the population of the genetic algorithm, focused on improving the profit obtained by variant, are described below. Table 10 shows the percentage of performance improvement of the set of variants of the solver algorithm, comparing the results obtained before and after the redesign.

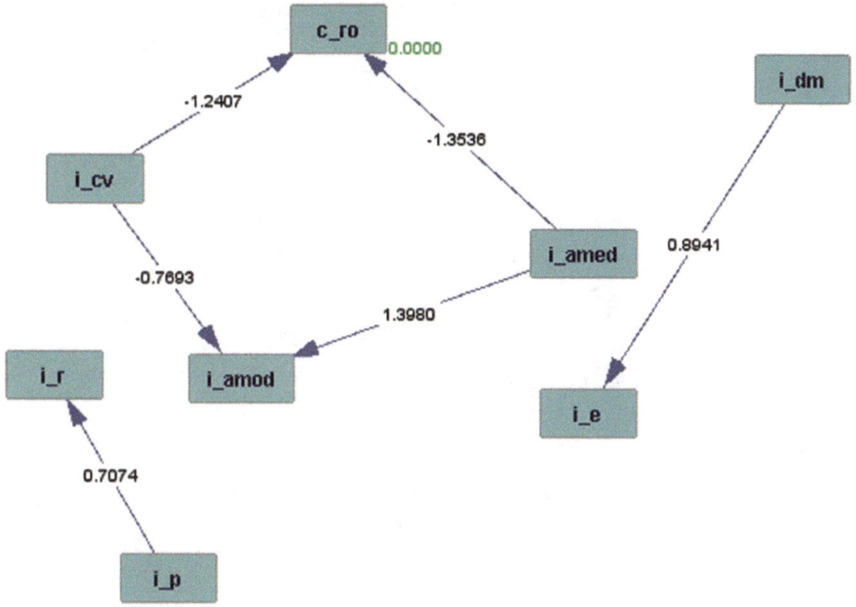

Fig. 5 Causal modeling graph for the metaheuristic algorithm

Table 9 Causal model equations for the metaheuristic algorithm

C_ro = −1.2407 (i_cv) + 1.3536 (i_amed)
I_amod = −0.7693 (i_cv) + 1.3980 (i_amed)
I_e = 0.8941 (i_dm)
I_r = 0.7074 (i_p)

Table 10 Percentage of performance improvement of the genetic algorithm redesign

	V1	V2	V3	V4	V5	V6	V7	V8
Inst15	0.83276	0.97697	1.04312	0.76442	0.62718	0.13879	0.62630	0.34626
Inst24	0.61333	0.24095	0.82003	−2.57089	0.43768	0.60956	2.66525	1.37895
Inst100	−0.63136	−0.22349	−0.26602	0.52887	19.62570	13.17702	−2.49939	0.97769
Inst1000	−0.99751	−2.20840	−0.59503	13.23024	19.77585	32.71535	2.09970	5.23370
	V9	V10	V11	V12	V13	V14	V15	V16
Inst15	0.90909	0.20848	0.62327	1.25436	0.55440	0.41725	1.11343	−2.97990
Inst24	−0.11037	1.06274	−0.00792	0.90091	2.49633	0.83942	1.83511	3.05075
Inst100	−0.85470	−0.63072	4.68544	6.18790	8.13953	15.30000	3.30684	5.11616
Inst1000	−1.11180	−1.10912	14.68765	8.45121	30.95799	−4.56695	5.97941	7.25614

After reviewing Table 10, in the instance of 1000 objects the best improvement percentages were obtained after applying the redesign of the causal algorithm.

4.2 Portfolio Selection Problem Experimentation

- Characterization process

Taking as a basis the indices proposed in the theses [2, 3, 14], a subset of indices was selected which were applied in the PSP, and the results are shown in Table 11 shows the results obtained from the index used in the characterization process.

- Identification of domination regions

To carry out this phase, an analysis of the NSGA-II was carried out to identify in which parts of the algorithm different configurations could be applied. To present the variety of population configurations on genetic algorithms, there was made Table 12 showing each variant of the algorithm.

- Selecting indicators problems and algorithm

Once each of the possible variants described in Table 8 has been executed, the indices shown in Table 12 are applied, an analysis of the results is made to select the best variants that solve the backpack problem. Algorithm 3 explains the overall process of the significant index used in the characterization process.

- Casual analysis

Taking as a basis those variants of the NSGA-II that have obtained the best results, an analysis is made of all its variables with which a causal graph will be assembled. Figure 1 explains the process of creating the causal model used in this research work, taking as input information the regions of dominance identified in the third phase and the significant indices identified in the fourth phase.

Once the elements for the creation of the causal model are specified, the PC algorithm is executed to the data set which is made up of the indexes of the characterization process, which were mapped with the regions of dominance, at the end of the PC algorithm gets the directed graph with the equations of the causal model.

Figure 6 shows an example of the graph obtained from the PC algorithm applied to the PSP and the solver algorithm, and the set of equations obtained from the casual model is shown in Table 13.

- Redesign of Algorithm Structure

The following describes the results obtained from the process of redesigning the NSGA-II algorithm, focused on improving the number of solutions belonging to the

Table 11 Example of characterization process indexes for the PSP

Index type	Index	Portfolio selection problem		
		o9p100_1	o9p100_2	o9p100_3
Relationship between weights and container size	*b*	0.3305	0.3326	0.3330
	t	0.0302	0.0300	0.0300
	d	0.0302	0.0300	0.0300
Central weights trend	*ma*	7563.8	7515.3	7505.8
	moda	8010	5410	8435
Weight scattering	*r*	4905	4845	4980
	dm	1302.19	1362.7	1263.85
	s	1495.55	1533.16	1467.82
	cv	0.1977	0.2040	0.1955
The shape of weight distribution	*curtosis*	171.02	161.66	184.94
Sample of the solution space	*vo*	8,023,320	5,840,849	4,096,121
Algorithm behavior	*p_f*	100	100	100
	p_uf	0	0	0
	ro	0	0	0
	delta	1	1	1
Algorithm trajectory	*nc*	0.98	0.98	0.98
	nv	0.06	0.04	0.04
	pp	0.5	0.54	0.48
	pn	0.48	0.44	0.50
Algorithmic performance Multi-objective	*Non dominated solution ratio*	0.42	0.64	0.32

(continued)

Table 11 (continued)

Index type	Index	Portfolio selection problem		
		o9p100_1	o9p100_2	o9p100_3
	Hypervolume	513,231,298,226.517	513,278,774,235.003	513,599,570,205.834
	Spread	39.7311811933996	68.3242860079803	27.6662580350649
	Spacing	0.000104680462973524	0.000254480650303477	0.0000772010656403887

Table 12 NSGA-II algorithm variants

Variant	Initial solution		Crossing method		Mutation method	
	Random	Heuristic	MC1	MC2	MM1	MM2
V1	X		X		X	
V2	X		X			X
V3	X			X	X	
V4	X			X		X
V5		X	X		X	
V6		X	X			X
V7		X		X	X	
V8		X		X		X

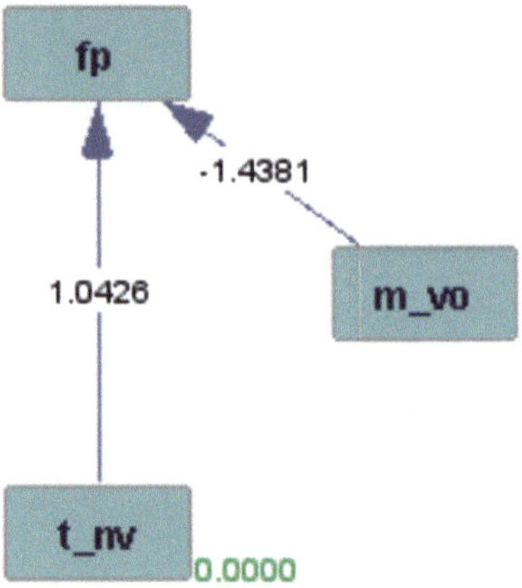

Fig. 6 Causal modeling graph for the NSGA-II algorithm

Table 13 Causal model equations for the NSGA-II algorithm

fp = −1.4381 (m_vo)
fp = 1.0426 (t_nv)

front of Pareto and improving the number of solutions that make up the optimal Pareto front. Table 14 shows the results obtained from the calculation of the number of solutions per variant that belong to the Pareto optimal front without applying the causal redesign.

Table 15 shows the results obtained from the calculation of the number of solutions that belong to the Pareto optimal front, applied to the set of variants of the solver algorithm after the execution of the causal redesign.

As can be seen in Tables 14 and 15, all versions of the NSGA-II algorithm obtained improvements in the number of solutions that belong to the Pareto optimal frontier in the o9p100_1 instance after applying the causal redesign. Also, it is observed that

Table 14 Results obtained from the number of solutions belonging to the pareto optimal front without applying the causal redesign

Instances	Versions							
	V1	V2	V3	V4	V5	V6	V7	V8
o9p100_1	12	7	2	7	28	27	15	13
o9p100_2	47	30	8	9	9	29	28	21
o9p100_3	37	24	30	13	13	35	40	15

Table 15 Results obtained from the number of solutions belonging to the pareto optimal front after the application of the causal redesign

Instances	Versions							
	V1	V2	V3	V4	V5	V6	V7	V8
o9p100_1	41	17	29	22	34	47	24	20
o9p100_2	43	23	25	14	49	29	24	26
o9p100_3	29	27	20	10	41	38	19	25

for the o9p100_2 instance, it only found improvement in **50%** of the versions after the causal redesign. And finally, for the o9p100_3 instance, it is observed that it only had an improvement in **25%** of the versions.

5 Conclusions

In this work, algorithms and instances were characterized through indexes and tools that support causal analysis, to impact the improvement of algorithm performance.

The group of indexes with which we initially work belongs to various groups such as Structure of the problem, space of solutions, behavior, trajectory, and algorithmic performance. However, due to the nature of the problems, only a subset of the original set was applied.

The casual modeling applied to the analysis of algorithms allows a statistical explanation, of how the structure of problems and algorithms design affect its performance.

The main contributions of this work are the following:

1. The incorporation of a methodology of a causal redesign of mono/multi-objective algorithms to VisTHAA.
2. Incorporation of more indicators of the description of the problem, the behavior of the algorithm, and the performance for algorithms mono/multi-objective.
3. Causal modeling of the performance for genetic and NSGAII algorithms.
4. A strategy for the generation of formal explanations of the relationships between the elements influencing the algorithm performance.

Acknowledgements The authors thank the support from CONACYT projects: (a) A1-S-11012-“Análisis de Modelos de NO-Inferioridad para incorporar preferencias en Problemas de Optimización Multicriterio con Dependencias Temporales Imperfectamente Conocidas”; (b) project 3058 from the program Cátedras CONACyT; and, (c) and from project 312397 from Programa de Apoyo para Actividades Científicas, Tecnológicas y de Innovación (PAACTI 2020-1). Also thank the support from TecNM project no. 5797.19-P and Laboratorio Nacional de Tecnologias de Información (LaNTI) del TecNM/Campus ITCM.

References

1. McGeoch, C.: Experimental analysis of algorithms. In: Pardalos, P.M., Romeijn, H.E. (eds.) Handbook of Global Optimization, vol. 2, pp. 489–513 (2000)
2. Pérez, V.: Modelo Causal de Desempeño de Algoritmos Metaheurísticos en Problemas de Distribución de Objetos. Ph.D. thesis, Instituto Tecnológico de Ciudad Madero (2007)
3. Landero, V.: Desarrollo de un Método Formal que Muestre la Interrelación en-tre las Características de un Conjunto de Casos y las de un Algoritmo que los Resuelve Eficientemente, para el Problema de Distribución de Objetos en Contenedores. Ph.D. tesis, Centro Nacional de Investigación y Desarrollo Tecnológico (2008)
4. Gómez, C., Fernández, E., Cruz-Reyes, L., Bastiani, S., Rivera, G., Ruız, V.: Memetic algorithm for solving the problem of social portfolio using out-ranking model. In: Castillo, O., Melin, P., Kacprzyk, J. (eds.) Recent Advances on Hybrid Intelligent Systems. Studies in Computational Intelligence, vol 451. Springer, Heidelberg (2013). https://doi.org/10.1007/978-3-642-33021-6_27
5. Garey, M., Johnson, D.: Computers and Intractability a Guide to the Theory of NP-Completeness. Freeman, New York (1979)
6. Cruz, L.: Caracterización de Algoritmos Heurísticos Aplicados al Diseño de Bases de Datos Distribuidas. Tesis, Centro Nacional de Investigación y Desarrollo Tecnológico (2004)
7. Martello, S., Toth, P.: Knapsack Problems: Algorithms and Computer Implementations. Wiley, New York (1990)
8. Rivera, G., Gómez, C., Cruz, L., García, R., Balderas, F.A., Fernández, E.R., López, F.: Solution to the social portfolio problem by evolutionary algorithms. Int. J. Comb. Optim. Probl. Inform. **3**(2), 21–30 (2012)
9. Rivera, G., Gómez, C.G., Fernández, E.R., Cruz, L., Castillo, O., Bastiani, S.: Handling of synergy into an algorithm for project portfolio selection. In: Castillo, O., Melin, P., Kacprzyk, J. (eds.) Recent Advances on Hybrid Intelligent Systems, vol. 451, pp. 417–430. Springer, Heidelberg (2013). https://doi.org/10.1007/978-3-642-33021-6_33
10. Ochoa, A., Rivera, G., Gómez-Santillán, C., Sánchez, B.: Handbook of research on metaheuristics for order picking optimization in warehouses to smart cities. IGI Global, (2019). https://doi.org/10.4018/978-1-5225-8131-4
11. Srinivas, N., Deb, K.: Multiobjective Optimization using nondominated sorting in genetic algorithms. Evol. Comput. **2**(3), 221–248 (1994). https://doi.org/10.1162/evco.1994.2.3.221
12. Deb, K., Pratap, A., Agarwal, S., Meyarivan, T.A. M.: A fast and elitist multiobjec-tive genetic algorithm: NSGA-II. IEEE Trans. Evol. Comput. **6**(2), 182–197 (2002). https://doi.org/10.1109/4235.996017
13. Spirtes, P., Glymour, C., Scheines, R.: Causation, prediction, and search, 2nd edn. MIT Press, Cambridge (2000). https://doi.org/10.1002/sim.1415
14. Quiroz, M.: Caracterización de Factores de Desempeño de Algoritmos de Solución De BPP. Tesis, Instituto Tecnológico de Cd. Madero (2009)
15. Van Veldhuizen, D.A., Lamont, G.: Multiobjective evolutionary algorithm research: a history and analysis. Technical report TR-98–03, Department of Electrical and Computer Engineering,

Graduate School of Engineering, Air Force Institute of Technology, Wright-Patterson AFB (1998)

16. Jiang, S., Ong, Y., Zhang, J., Feng, L.: Consistencies and contradictions of performance metrics in multiobjective optimization. IEEE Trans. Cybern. **44**(12), 2391–2404 (2014). https://doi.org/10.1109/TCYB.2014.2307319
17. Wang, L., Zhang, Y., Feng, J.: On the euclidean distance of images. IEEE Trans. Pattern Anal. Mach. Intell. **27**(8), 1334–1339 (2005). https://doi.org/10.1109/TPAMI.2005.165
18. Zitzler, E., Deb, K., Thiele, L.: Comparison of multiobjective evolutionary algorithms: empirical results. Evol. Comput. **8**(2), 173–195 (2000). https://doi.org/10.1162/106365600568202
19. Tan, K., Lee, T., Khor, E.: Evolutionary algorithms for multi-objective optimization: performance assessments and comparisons. Artif. Intell. Rev. **17**(4), 251–290 (2002). https://doi.org/10.1023/A:1015516501242
20. Castillo, N.: Evaluación de Estrategias de Mejora del Desempeño de Metaheurísticos Aplicados a BPP Vía Diagnóstico Visual. Tesis, Instituto Tecnológico de Cd. Madero (2001)
21. Rivera, G., Rodas-Osollo, J., Bañuelos, P., Quiroz, M., Lopez, M.: A genetic algorithm for surgery scheduling optimization in a mexican public hospital. Recent advances in artificial intelligence research and development. In: Aguiló, I., Alquézar, R., Angulo, C., Ortiz, A. (eds.) Frontiers in Artificial Intelligence and Applications, vol. 300, pp. 269–274. IOS Press, Amsterdam (2017). https://doi.org/10.3233/978-1-61499-806-8-269
22. Mendenhall, W., Sincich, T.: Probabilidad y Estadística para Ingeniería y Ciencias, 4th edn. Prentice-Hall Hispanoamérica, D.F. (1997)

Fuzzy Logic to Measure the Legal and Socioeconomic Effect of the Debtors Declared in the Canton of Pastaza

Diego Vladimir Garcés Mayorga, Danilo Rafael Andrade Santamaría, and Luis Rodrigo Miranda Chávez

Abstract As part of bad economic planning carried out by people, they sometimes declare themselves debtors. Debts are a legal effect that causes a socio-economic impact. The present work proposes a method to measure the legal and socioeconomic effect in debtors declared in default. The method bases its operation on fuzzy logic through user experience. A case study was implemented in the canton of Pastaza with the aim of measuring the socioeconomic legal effect from which a working tool for decision-making is obtained.

Keywords Fuzzy logic · User experience · Legal and socioeconomic effect

1 Introduction

As a consequence of a low level in terms of financial culture, many people are declared as debtors. Debts entail facing a set of legal actions that can end with the auction of your assets or in the declaration of insolvency that generates legal and economic repercussions.

Insolvency is not only a judicial consequence but also social and obviously economic as a result of the lack of economic income. In Ecuador, an individual is declared insolvent when he/she incurs in the breach of an obligation or contract and is submitted to the jurisdictional decision, in which he is sentenced to pay the amount owed.

Insolvency has constituted one of the most imprecise legal concepts in its meaning, which includes relatively heterogeneous meanings: equity insufficiency, inability to pay, lack of liquidity, over-indebtedness, assets less than liabilities, among others [1, 2].

Insolvency represents a recurring feature linked to an economic situation in which a debtor finds himself unable to satisfy his creditor, consequently causing an inability

D. V. Garcés Mayorga (✉) · D. R. Andrade Santamaría · L. R. Miranda Chávez
Universidad Regional Autónoma de Los Andes (UNIANDES), Puyo 160150, Pastaza, Ecuador
e-mail: up.diegogarces@uniandes.edu.ec

W. Pedrycz et al. (eds.), *Computational Intelligence for Business Analytics*,
Studies in Computational Intelligence 953,
https://doi.org/10.1007/978-3-030-73819-8_9

to pay. The moratorium causes an injury to the credit right with the consequent origin of liability of the debtor.

In accounting or financial terms, it does not seem very difficult to determine who is solvent and who is not, since arithmetically, a conclusive answer is obtained not in purely legal terms since for what is decisive is the capacity and the realization value of your assets [3, 4].

The declaration of insolvency is the consequence of the implementation of a judicial process that determines in a sentence the payment of what is owed. The procedures established in this regard are contemplated in the General Organic Code of processes; these procedures can be ordinary and executive.

The procedure is considered as the action to proceed before the judicial authority by means of the respective demand to obtain effective judicial protection, being a system of actions or set of acts that include the orderly development of judicial proceedings [5, 6].

The procedures used are the methods established by law to give life to the process; it is clear then that it is not the same process as a procedure because the process has several elements that make it up, according to our General Organic Code of Processes.

Insolvency has been approached in the scientific literature from different perspectives, affecting different sectors [7, 8]. The Constitution of Ecuador, in article 168 states: "The administration of justice, in the fulfillment of its duties and in the exercise of its powers, will apply the following principles: The substantiation of the processes in all matters, instances, stages, and procedures will be carried out through the oral system, in accordance with the principles of concentration, contradiction, and device" [9, 10].

Based on the aforementioned problems, this research defines a solution in which it proposes a method to measure the legal and socioeconomic effect of debtors that bases their operation using fuzzy logic. The research is structured in materials and methods where the main theoretical references of the research are presented and the inference process is described. The results and discussions present an example of the implementation of fuzzy logic based on user experience to measure the legal and socioeconomic effect on debtors.

2 Materials and Methods

Fuzzy logic is a mode of reasoning that applies multiple truth or confidence values to restrictive categories during problem solving [11]. The set is a collection of objects that can be classified thanks to the characteristics they have in common. It is defined in two ways: by an extension ({a, e, i, o, u}) or by understanding.

A boolean set A is an application of a referential set S in the set $\{0, 1\}$, $A : S \to \{0, 1\}$, and it is defined with a characteristic function:

$$\mu_A(x) = \begin{cases} 1 \; if \; x \in A, \\ 0 \; otherwise. \end{cases} \tag{1}$$

Fuzzy sets give a quantitative value to each element, which represents the degree of belonging to the set [12, 13].

A fuzzy set A is an application of a referential set S in the interval [0, 1].

$A : S \to [0, 1]$, and it is defined by means of a membership function: $0 \leq \mu_A(x) \leq 1$.

The closer the value is to 0, the less we can assure the membership of an element to a set [12–16]. On the contrary, the closer the value is to 1, the more we can assure the element's membership to the set [17].

Can be represented as a set of ordered pairs of a generic element x, $x \in A$ and their degree of belonging $\mu_A(x)$:

$$A = \{(x, \mu_A(x)), \mu_A(x) \in [0, 1]\} \quad (2)$$

Work with fuzzy logic can be represented using linguistic variables to improve the interpretability of the data. Linguistic variables are those of natural language characterized by the fuzzy sets defined in the universe of discourse in which they are defined [18–20].

To define a set of linguistic terms, the granularity of the uncertainty of the set of linguistic labels with which you are going to work must be previously established [21]. The granularity of uncertainty is the cardinal representation of the set of linguistic labels used to represent the information [22]. Figure 1 shows a set of language labels.

The degree of membership of an element $M(x)$ in a fuzzy set will be determined by membership functions. The typical membership functions most addressed in the scientific literature are Ahsan and Kayacan [23, 24]:

Triangular Function, Trapezoidal Function, Gaussian Function.

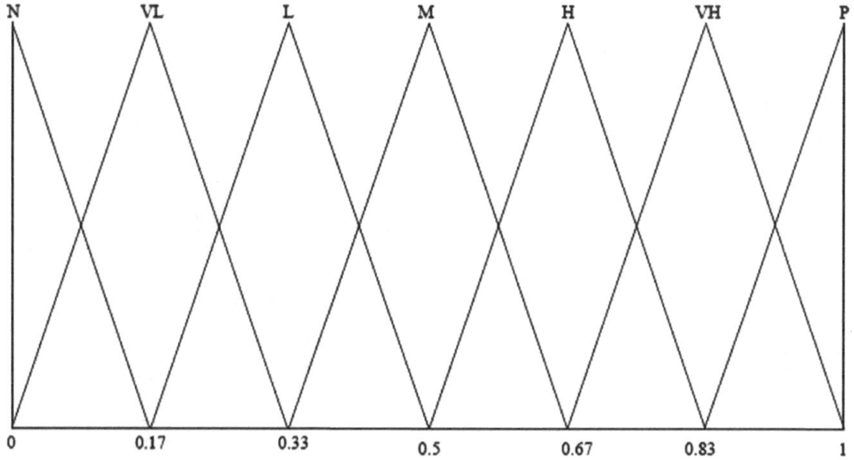

Fig. 1 Linguistic labels set

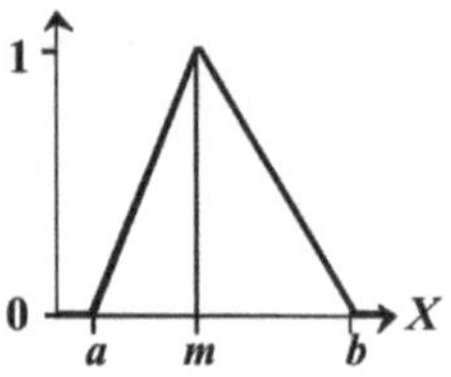

Fig. 2 Triangular function

Triangular Function: Defined by its lower limits a and upper b, and the modal value m, such that $a < m < b$ [25, 26] (Fig. 2).

$$A(x) = \begin{cases} 0 & if\ x \le a, \\ \frac{(x-a)}{(m-a)} & if\ a < x \le m, \\ \frac{(b-x)}{(b-m)} & if\ m < x < b, \\ 0 & otherwise. \end{cases} \tag{3}$$

Trapezoidal Function: Defined by its lower limits a and upper d, and limits b and c, corresponding to the lower and upper respectively of the plateau [26, 27] (Fig. 3).

$$A(x) = \begin{cases} 0 & if\ x \le a\ or\ x \ge d, \\ \frac{(x-a)}{(b-a)} & if\ a < x < b, \\ 1 & if\ b \le x \le c, \\ \frac{(d-x)}{(d-c)} & otherwise. \end{cases} \tag{4}$$

Gaussian function. It is defined by its mean value m and the value $k > 0$. It is the typical Gaussian bell (greater k, narrower the bell) [28, 29] (Fig. 4).

$$A(x) = e^{-k(x-m)^2} \tag{5}$$

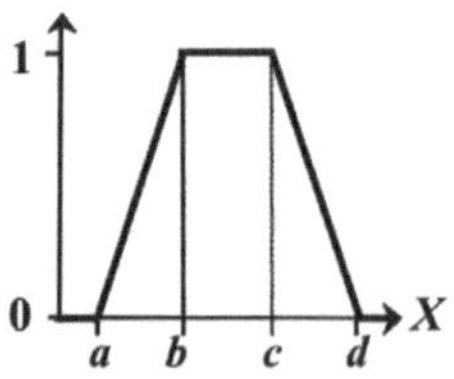

Fig. 3 Trapezoidal function

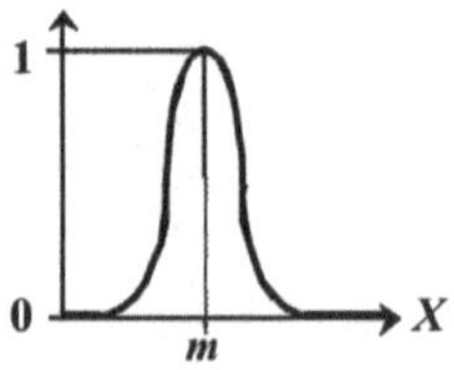

Fig. 4 Gaussian function

Logical operations of intersection (conjunction), union (disjunction), and complement (negation) can be performed on fuzzy sets. To carry out these operations, the T-Norms and the S-Norms can be used. The T-Standards specify the conditions that operations must meet to intercept assemblies, and the S-Standards do so for unions.

Intersections occur at conjunctions and contributions; similarly, unions occur at disjunctions and the global. These operations are performed in expert systems to calculate the certainty factors of the production rules. According to the T-Norms and S-Norms, these operations meet the following conditions:

It is a T-norm operation if it meets the following properties:

$$\textit{Commutative} \quad T(x, y) = T(y, x) \tag{6}$$

$$\textit{Associative} \quad T(x, T(y, z) = T(T(x, y), Z) \tag{7}$$

$$\textit{Increasing monotone} \quad T(x, y) > T(x, y)\, si\, x \geq x' \cap y \geq y' \tag{8}$$

$$\textit{Neutral element} \quad T(x, 1) = x \tag{9}$$

It is an S-Norms operation if it meets the following properties:

$$\textit{Commutative} \quad S(x, y) = T(y, x) \tag{10}$$

$$\textit{Associative} \quad S(x, S(y, z) = S(S(x, y), Z) \tag{11}$$

$$\textit{Increasing monotone} \quad S(x, y) > T(x, y)\, si\, x \geq x' \cap y \geq y' \tag{12}$$

$$\textit{Neutral element} \quad S(x, 1) = x \tag{13}$$

In a system expressed by fuzzy logic, there are linguistic variables, their labels, the label membership functions, the production rules, and the certainty factors associated with these rules. As input data to the system, we have the numerical values that the linguistic variables take.

The input values are converted to fuzzy tag membership values that are equivalent to the certainty factors. This process is called Fuzzyfication since it converts numeric values to fuzzy.

From the values obtained in the Fuzzyfication process, the certainty propagation process occurs using the defined production rules. This is the Fuzzy Inference process, in which the functions of the T-Norms and S-Norms are used [30, 31]. As a result, certainty values are obtained that refer to the membership of the output sets. From the values of belonging to the output linguistic variables, the numerical values of these must be obtained, and this process is called Defuzzification. DeFuzzyfication

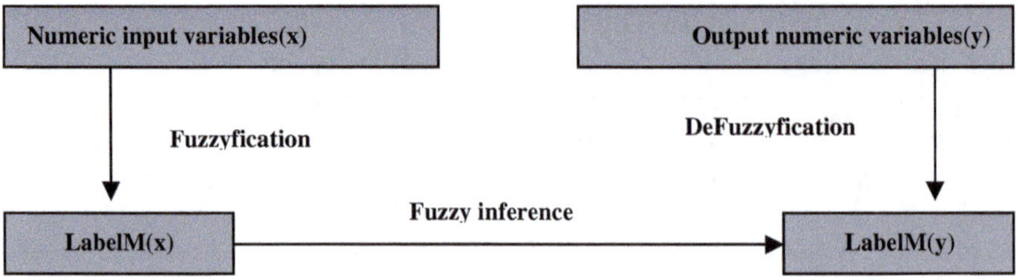

Fig. 5 Scheme of a system expressed by fuzzy logic

of the variables can be done by the Centroid Method, which is the most used for this process [32, 33]. Figure 5 shows a schematic of a system expressed by fuzzy logic.

2.1 Fuzzy Logic to Measure the Legal and Socioeconomic Effect of Debtors

The proposed method for measuring the legal and socioeconomic effect of debtors bases its operation on fuzzy logic. It uses the inference process based on the Centroid or Center of Gravity (GOC) in the DeFuzzyfication.

GOC-based inference guarantees that you do not have to adjust any coefficient; it is only necessary to know the membership functions of each of the defined labels. To infer with GOC, we start from the values of belonging to each of the labels associated with the variable that we want to Defuzzify. For each fuzzified output variable, the maximum value of the membership function of each label is truncated, starting from the value obtained during the inference.

The effect of truncation is shown in Fig. 6. The process is carried out in the same way for each label. Each tag is truncated based on the inferred certainty value. Graphics of inferred labels with lower value are guaranteed to be more truncated.

Then the result of the truncation of all these functions is combined, and the center of gravity is obtained [34]. For that, Eq. 14 is used:

$$GOC = \frac{\int M(x)xdx}{\int M(x)dx} \tag{14}$$

Fig. 6 Truncation of the fuzzy set according to the certainty factor of the output variable

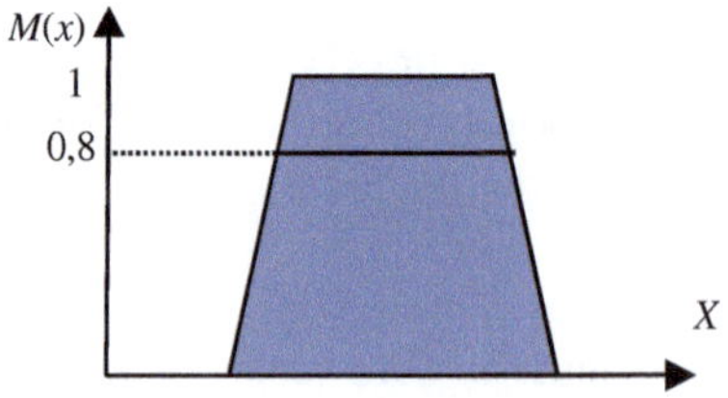

Table 1 Impact of the output variable labels

Label	Impact
Low	Debts can be paid in several months
Half	Debts can be paid in one month
High	Debts can be paid in one week
Excellent	Debts can be paid immediately

where $M(x)$ represents the degree of belonging of the element X that will take values in the discourse universe, using a defined step. The smaller this step, the more accurate the GOC result will be.

To measure the legal and socioeconomic effect of debtors using fuzzy logic, the indicators defined in Table 1 will be taken as linguistic variables. These indicators are job stability, history of previous debts, and seriousness in payments of previous debts. The legal and socioeconomic evaluation of debtors is the output variable. It was defined that each of these input or output variables will have associated the labels of Low, Medium, High, and Excellent. To assess the impact that linguistic labels have on the output variable, see Table 1.

For the Baja label, the associated membership function will be the triangular function, such that $\langle 0, 4, 5 \rangle$. The first value represents where the function begins, the second where 1 is made, the third where it begins to decrease, and the fourth where 0 is made.

For the Media tag, using the PI function, we have $\langle 4, 5, 6, 7 \rangle$. For the high label, with a Gaussian distribution function, it will be $\langle 6, 7, 8, 9 \rangle$. For the Excellent tag, using the trapezoid function, it can be represented through $\langle 8, 9, 10, 10 \rangle$. Figure 7 shows the membership functions of the linguistic labels of the input variables.

Using the assessment of experts in the field, the production rules were defined. These rules guarantee that the evaluation of the legal and socio-economic effect of debtors is always largely determined by the lowest evaluation obtained in the input indicators. The production rules and the certainty factors (FC) associated with each one were defined as follows:

- **R1**: If job stability is low (LE) or past debt history is low (PDA), or seriousness of past debt payments is low (ADP), then the legal and socioeconomic effect of debtors is low (DSJE). (FC = 1)
- **R2**: If job stability is medium (LE) or past debt history is medium (PDA), or the seriousness of past debt payments is medium (ADP), then the legal and socioeconomic effect of debtors is medium (DSJE). (FC = 0, 8)
- **R3**: If job stability is high (LE) or past debt history is high (PDA), or the seriousness of past debt payments is high (ADP), then the legal and socioeconomic effect of debtors is high (DSJE). (FC = 0, 6)
- **R3**: If the job stability is excellent (LE) or the previous debt history is excellent (PDA), or the seriousness of past debt payments is excellent (ADP), then the legal and socio-economic effect of debtors is excellent (DSJE). (FC = 0, 4).

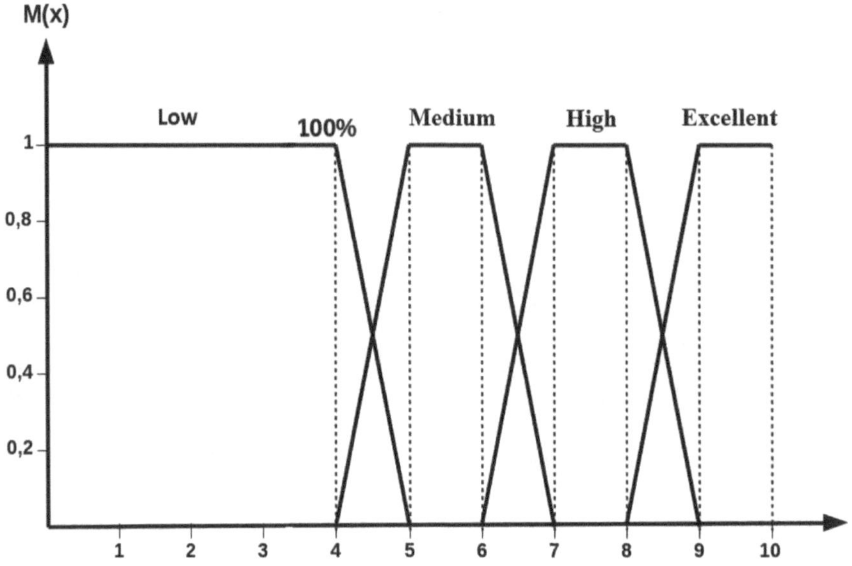

Fig. 7 Membership functions of the linguistic labels of the input variables

Once these data are available, the Fuzzyfication of the input variables can be carried out. The universe discourse is the same for all the input variables that have been defined, so all the input variables have the same linguistic labels and membership functions.

After calculating the certainty factors for each of the labels of the input variables, we will proceed to the Fuzzy Inference phase. This will calculate the factors associated with the labels of the output variables. From the four defined production rules, the necessary DISY and CTR will be calculated, following the Minimum–Maximum pair of T-Standards and S-Standards.

In the third phase, the Desfuzzyfication will be carried out using the Centroid Method. The step will be 1, since x will go from X1 to X10, to gain in accuracy to the extent of the legal and socioeconomic effect on debtors. The labels of the output variable legal and socioeconomic effect on debtors will be the same ones used for the input variables, as well as their membership functions [35, 36]. Figure 8 shows the general scheme of fuzzy logic to measure the legal and socioeconomic in debtors.

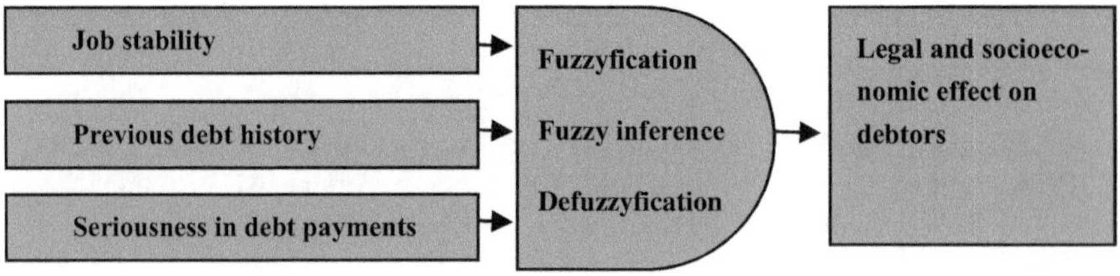

Fig. 8 General scheme of the fuzzy logic of the proposed method

3 Results and Discussions

To evaluate the results of the present investigation, experimentation will be carried out. The main objective of the experiment will be to demonstrate the applicability of fuzzy logic based on user experience to measure the legal and socioeconomic effect on debtors.

To do this, the process will be tested based on the location of the Pastaza canton. Once this process is completed, the results of the experimentation will be discussed.

There are the input values [2, 3, 5] for the indicators of job stability, history of previous debts, and seriousness in debt payments. In the Fuzzyfication process, the certainty factors of each of the input variables are calculated for each of its labels. By applying Fuzzyficación to the input variables, having the numerical values associated with each one, the results of Table 2 are obtained. The calculation of the degrees of belonging is performed according to the typical membership functions.

The Fuzzy Inference process is performed through the defined rules, using the Minimum–Maximum pair of T-Norms and S-Norms. Once this process has been carried out, the values shown in Table 3 are obtained for the output variable legal and socioeconomic effect.

The Fuzzy Inference shows that the degree of belonging of the output variable is 1 for the low label and 0.8 for the medium label; therefore for the input values of the legal and socioeconomic effect, it will below. To Desfuzzyficar, the output variable, the Centroid Method is applied, with which the value shown in Table 4 is obtained.

Figure 9 shows the degree of membership of the output variable (legal and socioeconomic effect) with a value of 3.94 for the input values [2–4] corresponding. Here it is observed that for the value of this variable, the degree of belonging to the low linguistic label is 100%, which means that actions are required to mitigate the legal and socioeconomic effect, as stated in Table 1.

Table 2 Degrees of membership of input values to fuzzy sets

Linguistic variables	Label low	Label medium	Label high	Label excellent
1-Job stability		1		
2-Previous debt history	1			
3-Seriousness in debt payments	1			

Table 3 Degrees of membership of input values to fuzzy sets

Linguistic variables	Label low	Label medium	Label high	Label excellent
Legal and socioeconomic effect	1	0.8	0	0

Table 4 Results of the application of the centroid method

X_i	$M(x_i)$	$x_iM(x_i)$
1	1	1
2	1	2
3	1	3
4	0.8	3
5	0	3
6	0	3
7	0	0
8	0	0
9	0	0
10	0	0
Sum	3.8	15
GOC	3.94	

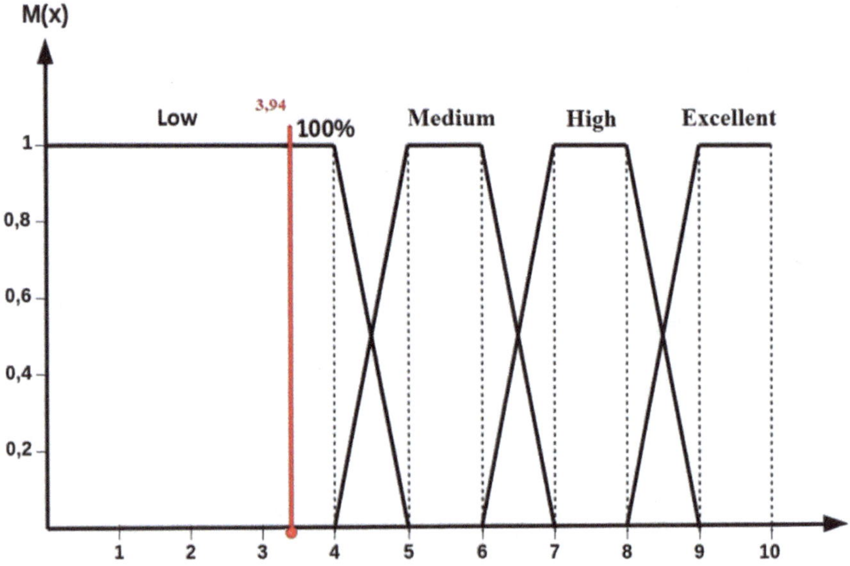

Fig. 9 Degree of belonging of the variable legal and socioeconomic effect

4 Conclusions

The fuzzy logic theory applied to perform the analysis and evaluation of the Socioeconomic and Legal Effect generates and delivers accurate data compared to other qualitative methods. This gives the main managers in charge of administrative and economic management the possibility of a better interpretation, free of other subjectivities.

Once the research results have been analyzed, a method of evaluating the legal and socioeconomic effect is obtained, contributing a tool for the analysis of the phenomenon using fuzzy logic capable of quantifying the variable under study.

Fuzzy logic guarantees an alternative for the development of systems and tools that support business and administrative decision-making for the analysis of different phenomena.

References

1. Usman, A.K.: Theory and Practice of International Economic Law. Malthouse Press, Lagos (2017)
2. Bermeo, E.: Determinants of Financial Inclusion: Results of Multilevel Analyses. University of Bristol, Bristol (2019)
3. Yuniarto, P.R.: Indonesian migration industry in Taiwan: some socio-economic implications and improvement challenges. Jurnal Kajian Wilayah **6**(1), 17–33 (2016)
4. Goldmann, M.: The Law and Political Economy of Mozambique's Odious Debt (2019). https://doi.org/10.2139/ssrn.3513651
5. Moncayo, A.L., Granizo, G., Grijalva, M.J., et al.: Strong effect of Ecuador's conditional cash transfer program on childhood mortality from poverty-related diseases: a nationwide analysis. BMC Public Health **19**(1), 1132 (2019). https://doi.org/10.1186/s12889-019-7457-y
6. Vasco, C., Sirén, A.: Correlates of wildlife hunting in indigenous communities in the Pastaza province, Ecuadorian Amazonia. Anim. Conserv. **19**(5), 422–429 (2016). https://doi.org/10.1111/acv.12259
7. Fonseca, S.J.: Régimen de insolvencia empresarial: propuesta de unificación de los privilegios concursales para los países miembros de la comunidad andina de naciones. Estado del arte. Civilizar. Ciencias Sociales y Humanas **7**(13), 173–191 (2007). https://doi.org/10.22518/16578953.772
8. Ortíz, E., Noboa, P.: Propuestas societarias y concursales para mitigar el impacto económico del covid-19 en Ecuador. X-pedientes Económicos **4**(8), 38–48 (2020). https://doi.org/10.2139/ssrn.3568267
9. Lupien, P.: The incorporation of indigenous concepts of plurinationality into the new constitutions of Ecuador and Bolivia. Democratization **18**(3), 774–796 (2011). https://doi.org/10.1080/13510347.2011.563116
10. Fine-Dare, K.S.: The claims of gender: indigeneity, Sumak Kawsay, and horizontal women's power in Urban Ecuador under the 2008 political constitution. Soc. Dev. Issues **36**(3), 18–33 (2014)
11. Solís, P.Y.J., et al.: Compensatory fuzzy logic model for impact. Neutrosophic Sets Syst. **26**, 40 (2019)
12. Lumba, L.A., Khayam U., Lumba, L.S.: Application of fuzzy logic for partial discharge pattern recognition. In: 2019 International Conference on Electrical Engineering and Informatics (ICEEI), Bandung (2019). https://doi.org/10.1109/ICEEI47359.2019.8988844
13. Ricardo, J.E., Llumiguano-Poma, M.E., Arguello-Pazmiño, A.M., Albán-Navarro, A.D.: Neutrosophic model to determine the degree of comprehension of higher education students in Ecuador. Neutrosophic Sets Syst. **26**(1) (2019). https://doi.org/10.5281/zenodo.3244297
14. Chang, M., Kim, K., Jeon, D.: Research on terrain identification of the smart pros-thetic ankle by fuzzy logic. IEEE Trans. Neural Syst. Rehabil. Eng. **27**(9), 1801–1809 (2019). https://doi.org/10.1109/TNSRE.2019.2933874
15. Soesanti, I., Syahputra, R.: A fuzzy logic controller approach for controlling heat exchanger temperature. J. Electr. Technol. UMY **3**(4), 206–213 (2020). https://doi.org/10.18196/jet.3462

16. Zhang, S., Huang, X., Min, J., Chu, Z., Zhuang, X., Zhang, H.: Improved fuzzy logic method to distinguish between meteorological and non-meteorological echoes using C-band polarimetric radar data. Atmos. Meas. Tech. **13**(2), 537–537 (2020). https://doi.org/10.5194/amt-13-537-2020
17. Hernandez, N.B., Ruilova-Cueva, M.B., Mazacón, B.N., et al.: Prospective analysis of public management scenarios modeled by the fuzzy Delphi method. Neutrosophic Sets Syst. **26**(1), 17 (2019)
18. Ye, J.: Multiple attribute group decision making based on interval neutrosophic uncertain linguistic variables. Int. J. Mach. Learn. Cybern. **8**(3), 837–848 (2017). https://doi.org/10.1007/s13042-015-0382-1
19. Liu, P., Teng, F.: An extended TODIM method for multiple attribute group decision-making based on 2-dimension uncertain linguistic variable. Complexity **21**(5), 20–30 (2016). https://doi.org/10.1002/cplx.21625
20. Fan, J., et al.: Research on multi-objective decision-making under cloud platform based on quality function deployment and uncertain linguistic variables. Adv. Eng. Inform. **42**, 100932 (2019). https://doi.org/10.1016/j.aei.2019.100932
21. Li, C., Yuan, J.: A new multi-attribute decision-making method with three-parameter interval grey linguistic variable. Int. J. Fuzzy Syst. **19**(2), 292–300 (2017). https://doi.org/10.1007/s40815-016-0241-6
22. Ponce-Ruiz, D.V., Albarracín-Matute, J.C., et al.: Softcomputing in neutrosophic linguistic modeling for the treatment of uncertainty in information retrieval. Neutrosophic Sets Syst. **26** (2019). https://doi.org/10.5281/zenodo.3244320
23. Ahsan, R., et al.: Prediction of autism severity level in Bangladesh using fuzzy logic: FIS and ANFIS. In: Choroś K., Kopel M., Kukla E., Siemiński A. (eds.) International Conference on Multimedia and Network Information System 2018. Advances in Intelligent Systems and Computing, vol. 833. Springer, Cham (2018). https://doi.org/10.1007/978-3-319-98678-4_22
24. Kayacan, E., et al.: Elliptic membership functions and the modeling uncertainty in type-2 fuzzy logic systems as applied to time series prediction. In: 2017 IEEE International Conference on Fuzzy Systems (FUZZ-IEEE). IEEE, Naples (2017). https://doi.org/10.1109/FUZZ-IEEE.2017.8015457
25. Li, Y., Tong, S.: Adaptive fuzzy control with prescribed performance for block-triangular-structured nonlinear systems. IEEE Trans. Fuzzy Syst. **26**(3), 1153–1163 (2017). https://doi.org/10.1109/TFUZZ.2017.2710950
26. Kreinovich, V., Kosheleva, O. Shahbazova, S.N.: Why triangular and trapezoid membership functions: a simple explanation. In: Shahbazova S., Sugeno M., Kacprzyk J. (eds.) Recent Developments in Fuzzy Logic and Fuzzy Sets, vol. 391. Springer, Cham (2020). https://doi.org/10.1007/978-3-030-38893-5_2
27. Mustafa, S., Asghar, S., Hanif, M.: Fuzzy logistic regression based on least square approach and trapezoidal membership function. Iran. J. Fuzzy Syst. **15**(6), 97–106 (2018). https://doi.org/10.22111/IJFS.2018.4369
28. Azimi, S., Miar-Naimi, H.: Designing programmable current-mode Gaussian and bell-shaped membership function. Analog Integr. Circ. Sig. Process **102**(2), 323–330 (2020). https://doi.org/10.1007/s10470-019-01567-y
29. Tolga, A.C., Parlak, I.B., Castillo, O.: Finite-interval-valued type-2 Gaussian fuzzy numbers applied to fuzzy TODIM in a healthcare problem. Eng. Appl. Artif. Intell. **87**, 103352 (2020). https://doi.org/10.1016/j.engappai.2019.103352
30. Motylska-Kuźma, A., Mercik. J.: Fuzzyfication of repeatable trust game. In: Nguyen, N., Jearanaitanakij, K., Selamat, A., Trawiński, B., Chittayasothorn, S. (eds.) Asian Conference on Intelligent Information and Database Systems, vol. 12033. Springer, Cham (2020). https://doi.org/10.1007/978-3-030-41964-6_12
31. Kankaras, M., Cristea, I.: Fuzzy reduced hypergroups. Mathematics **8**(2), 263 (2020). https://doi.org/10.3390/math8020263
32. Kolekar, K., et al.: Fuzzy logic modelling to predict residential solid waste generation: a case study of Baranagar. In: Ghosh S. (ed.) Waste Management and Resource Efficiency, pp. 1155–1166. Springer, Singapore (2019). https://doi.org/10.1007/978-981-10-7290-1_95

33. Shrivastav, U., Singh, S.K., Khamparia, A.: A nobel approach to detect edge in digital image using fuzzy logic. In: Luhach, A., Kosa, J., Poonia, R., Gao, X.Z., Singh, D. (eds.) First International Conference on Sustainable Technologies for Computational Intelligence, vol. 1045. Springer, Singapore (2020). https://doi.org/10.1007/978-981-15-0029-9_6
34. García-Jacas, C.R., et al.: Smoothed spherical truncation based on fuzzy membership functions: application to the molecular encoding. J. Comput. Chem. **41**(3), 203–217 (2019). https://doi.org/10.1002/jcc.26089
35. Mar, O., Ching, I., González, J.: Operador por selección para la agregación de información en Mapa Cognitivo Difuso. Revista Cubana de Ciencias Informáticas, **14**(1), 20–39 (2020)
36. Ortega, R.G., et al.: Pestel analysis based on neutrosophic cognitive maps and neutro-sophic numbers for the sinos river basin management. Neutrosophic Sets Syst. **26**(1), 16 (2019). https://doi.org/10.5281/zenodo.3244633

Predictive Analytics

A Look at Artificial Intelligence on the Perspective of Application in the Modern Education

Ana Carolina Borges Monteiro, Reinaldo Padilha França, Rangel Arthur, and Yuzo Iano

Abstract Artificial intelligence (AI) has the objective to produce devices that reproduce human capacity in the spheres of reasoning, awareness, guidance, decision making, and collaboration in problem-solving. Its operation is based on a stimulus from the outside world centered on data analysis; it is usually related to Machine Learning. AI in Education also expands the teachers' capabilities, allowing them to focus on his most important task, i.e., accompanying students individually and supporting the teaching–learning process more effectively. In this sense, the greater the volume of interactions in the digital environment, the greater the system's ability to update its information based on student interactions. Through the performance analysis of students and classes, it is possible to articulate new approaches and educational actions through the data collected concerning learning. Since the student, when engaged, learns very well with technology, and so, the system can, for example, know in which areas a student does best, how they reason to solve problems and what they seek to complement his learning. However, AI makes it possible for the teacher to make more assertive decisions in supporting student learning, it is still worth considering that technology does not replace the teacher, but helps them to perfect and optimize his classes. Therefore, this chapter aims to provide an updated overview of AI in Education, showing its context on the horizon in teaching, addressing and approaching its branch of application potential, with a concise bibliographic background, synthesizing the potential of technology.

Keywords Artificial intelligence · Machine learning · Learning · Education · Technologies

A. C. Borges Monteiro · R. Padilha França (✉)
School of Electrical and Computer Engineering (FEEC), University of Campinas—UNICAMP, Av. Albert Einstein, Barão Geraldo, Campinas, SP 400, Brazil
e-mail: padilha@decom.fee.unicamp.br

R. Arthur · Y. Iano
Faculty of Technology (FT), University of Campinas—UNICAMP, Paschoal Marmo Street, Jardim Nova Italia, Limeira 1888, Brazil

W. Pedrycz et al. (eds.), *Computational Intelligence for Business Analytics*, Studies in Computational Intelligence 953,
https://doi.org/10.1007/978-3-030-73819-8_10

1 Introduction

Artificial intelligence (AI) is related to the Computer Science area, in which the objective is to produce devices that reproduce human capacity in the spheres of reasoning, awareness, guidance, decision making, and collaboration in problem-solving. Its operation is based on a stimulus from the outside world centered on data analysis [1–4].

The Fourth Industrial Revolution or Industry 4.0, is marked by AI as a new revolution, which has been transforming the way society relates, lives daily, and produces. This revolution is beneficial in terms of collective responsibility in designing a future in which technology and innovation are at the service of people; the same is designed for Education 4.0, a term that is linked to the technological revolution that includes computational language, AI, Internet of things (IoT) and includes "learning by doing" related to learning through experimentation, projects, experiences, and hands-on experience of the students themselves, supported by intelligent technology [5–8].

It is in this sense that AI can help with the creation of personalized teaching and learning environments, allowing interactive platforms and intelligent tutors to guide students on their development journey [9, 10].

There is no ready-to-apply model that breaks old paradigms imposed over the years in a decontextualized education, based on knowledge transmission and environments that are not conducive to the learning process. Assessing that the immersion of technology in Education makes it possible to carry out teaching ruled in creativity and inventiveness, linked to the model based on the maker culture, i.e., "do it yourself", as one of the ways, using various AI resources and relying on an environment based on experimentation with the student at the center of the learning process. Thus, Education 4.0 considers models based on AI technology that overcome challenges by transforming the learning experience through technology, but it is essential that they are accompanied by pedagogical practices that enable meaningful experiences [5–10].

However, although it is still a distant reality for many developing countries, it is possible to observe nowadays many movements to create more personalized, flexible, inclusive, and interactive teaching that goes through the application of AI [5–10].

It is important that, from an early age, children believe in their ability to solve challenges to learn mathematics. Allied to your needs to explore a good repertoire of problems that allow advancing in the concept formation process. It is important that this happens in a personalized way and that each child has their pace respected. This trend is expected to grow even more in the coming years since AI makes it possible to amplify human intelligence at its best [11, 12].

AI-based systems are produced by programming and provide content that is previously curated by teachers, so that students have access to these contents, interacting with the platform. The use of adaptive platforms has been gaining strength in Education, as they can propose individual learning paths. The platform collects data on each user experience, analyzes, and offers a different path based on their initial knowledge [13, 14].

AI in Education also expands the teacher's skills, allowing them to focus on his most important task: accompanying students individually and supporting the teaching–learning process more effectively. Since the student, when engaged, learns very well with technology, the problem is to overcome the engagement stage. The best solution occurs when it is combined with the teacher with the system AI; after all, the biggest driver o f engagement is the teacher [15, 16].

The adaptive platform integrates the classroom and offers a range of benefits, such as reports that serve as input for the teacher to decide student learning. In this sense, the greater the volume of interactions in the environment, the greater the system's ability to update its information based on student interactions. The system can, for example, know in which areas a student does best, how they reason to solve problems and what they seek to complement his learning [15–18].

Using game-based learning strategies makes this challenge much more engaging and also helps teachers detect which are the points of greatest difficulty for their students. The most complex behavioral patterns emerge when techniques as Big Data, Machine Learning, and AI are used instead of simple statistical analysis [15–20].

Thus, according to students participating in the activities proposed in the classroom, a large number of data is generated. This information is later used as analysis tools to detect each student's learning difficulties. In this context, the main object is not the amount of data generated, but the tool and type of analysis that this information will be submitted [5–20]. In this context, education can be seen as a movement to take AI to the classroom, through games, computer programs, security applications, robotics, devices for handwriting recognition, and voice recognition [15–21].

Therefore, this chapter aims to provide an updated overview of AI in education, showing its context on the horizon in Teaching, addressing and approaching its branch of application potential, with a concise bibliographic background, synthesizing the potential of technology.

2 Methodology

This research was developed endorsed on the analysis of scientific papers and scientific journal sources referring to AI towards Modern Education, concerning evolution and fundamental concepts of technology aiming to gather pertinent information regarding thematic. Thus, also enabling to boost more academic research through the background provided through this study.

This chapter is motivated to provide a major scientific contribution related to an overview of AI and discussion of the areas in education in which it can be applied. It also highlights the benefits and challenges of using AI in this field.

Addressing their key points and their importance, which are a complex and heterogeneous concept but which involves the role of AI in modern perspectives in education.

3 Artificial Intelligence Concept

The definition and explanation of the basic concepts of this theme are of fundamental importance since they will elucidate the clear and necessary integration of this technology, together with the area of education, which is one of the oldest sciences in the world.

Currently, there is a strong tendency to integrate emerging technologies such as AI, Deep Learning, Machine Learning, Big Data, cybersecurity, and the internet of things to traditional branches of study and human activity. Examples of this type are abundant, such as autonomous cars, robotic surgeries, telemedicine, virtual assistants, among others. In this context, we understand that the next part to be hit hard by these modernizations is education.

The generation of people born between the 1980s and 1990s literally grew up with the internet, computers, and smartphones; however, these technologies were often inaccessible to these children, especially those from emerging countries.

However, children born between 2000 and 2010 have already been born amid low-cost technologies, and consequently, since childhood, they have had access to smartphones, interactive, colorful, and ergonomic.

This generation and the ones that follow show us that traditional learning methodologies such as pencil, paper, pen, whiteboard, and exhaustively theoretical classes are no longer compatible with a generation born amid interactivity. Thus, AI can make the learning process more attractive, interactive, and personalized to each of the students because each person presents his own pace and learning time.

A definition of AI is to make computers think like humans or to be as intelligent as a person. Thus, the ultimate goal of research on this topic is to be able to develop a machine that can simulate some human skills and replace them in some activities. The term AI refers to computer systems capable of performing tasks that usually require human knowledge. AI, however, is closer; it is usually related to Machine Learning or even Deep Learning (Fig. 1) [22, 23].

AI is part of Computer Science studies, considering programs that use the same language as conventional systems, in which, in some cases, the intelligent system works with simple logic, in other cases, such as studies in neural networks. The machine tries to reproduce the functioning of human neurons, in which information is transmitted from one cell to another and is combined with other data to arrive at a solution [22–25].

AI on Education, it emerges what it means for a system to be intelligent, relating how students and teachers can benefit from this technology. Since the teacher can select online the questions to develop in class, asking students to watch the video-lessons to prepare for the lesson and, after it, complete the exercises also via the internet. Through AI, it is expected to obtain a set of statistical tools displaying graphs indicating the level of understanding of the class, what percentage completed the exercises correctly, and what were the main flaws, that is, that creates more knowledge the more students use it [19–25].

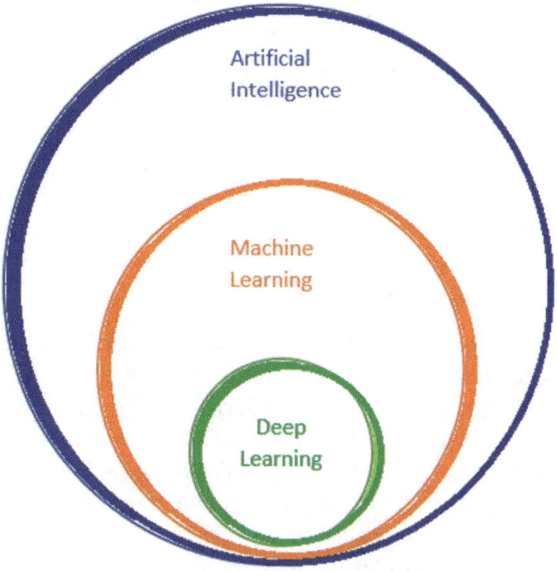

Fig. 1 Artificial learning technologies

Artificial Neural Networks are a computer concept that aims to work in data processing in a similar way to the human brain since this ability is known that biological neurons are five to six orders of magnitude slower than electronic neurons, even though the brain is considered to be a highly complex processor that performs parallel processing. Even if the human brain performs its necessary processing at an extremely high speed, there is no computer currently capable of doing what the human brain does. Synapses, starting from the principle of contact between neurons, is a transmission of nerve impulses from one cell to another in the human brain that occurs, it is given by chemical reactions [26–28].

In artificial neural networks, the principle is to carry out information processing based on the organization of neurons in the brain. Evaluating that they are created from algorithms designed for a certain purpose, still pondering the impossibility of creating such an algorithm without having knowledge of mathematical models that simulate the learning process of the human brain. The human brain is capable of learning and making decisions based on learning. Based on this, artificial neural networks can be interpreted as a processing scheme capable of storing knowledge based on learning (experience) and making that knowledge available for the application in question [26–30].

Basically, a neural network resembles the brain at two points, (1) knowledge is obtained through learning stages and (2) synaptic weights that are used to store knowledge. A synapse is a name given to the connection between neurons, where values are assigned, which are called synaptic weights (assigned value). In artificial neural networks, a series of artificial neurons are formed and connected to each other, forming a network of processing elements [26–30].

There are 3 types of learning in artificial neural networks:

- Supervised Learning: type of AI where the neural network receives a set of standardized inputs and their respective output patterns. In this process, adjustments in the synaptic weights occur until the error between the output patterns generated by the network has the desired value.
- Unsupervised learning consists of the neural network working with the data in order to determine some properties of the data sets. Based on this, learning is built.
- Hybrid learning is the type where there is a mixture of supervised and unsupervised types, whereas one layer can work with one type while another layer works with the other type [31–37].

Another type of emerging industry is electronic games, nowadays, being considered even sports, also called e-game. As stated earlier, the generation born from the 2000s onwards naturally has a greater affinity for this type of modality when compared to previous generations. Considering this characteristic, the union of electronic games, education, and AI can be proposed. Learning through electronic games makes the process easier and more natural, making it lose the aspect of something mandatory and massive. So to keep the student engaged, if the game/activity is very difficult and the player/student is not winning, the AI-based system itself will reduce the level of difficulty so that he does not lose interest and continue practicing in order to improve their skills and understand the content.

Following the same principle, if the game/activity is very easy, and the player/student always starts to win, the level of difficulty increases to make them feel challenged [27–38].

Another example of AI's performance in people's daily lives is cameras that automatically focus on people's faces or take a picture when they encounter a smile. Another example is the spell checkers of computer word processors, where an intelligent system is needed to detect that there is a syntax problem in the sentence and offer a possible correction. This can also be seen in recognition of images from patterns, being able to recognize letters, figures, people, and an infinity of elements of neural networks that can be of fundamental importance for each niche of the industry, among them the education of children and young [32–40].

Still thinking about the large volume (Big Data) of data that is generated on an educational platform accessed by thousands of students, it is possible to create a Data Model that serves to structure the data captured and stored by the system. Through computational processing, relevant information from students can be extracted, such as grades, class attendance, and student and behavior. However, the technique still does not show satisfactory results for understanding emotions, replicating the intellect of a good teacher. In this context, it is important to note that Big data is related to the large volume of data that needs to be processed and stored, which nourishes AI. A context can be illustrated according to Fig. 2 [41–46].

Based on this information, AI platforms can adapt to the needs of students, but also help the teacher to understand the behavior of students, offering them potential recommendations on how to alleviate or reduce the difficulties encountered, which can prevent school dropout, for example [38, 41–46].

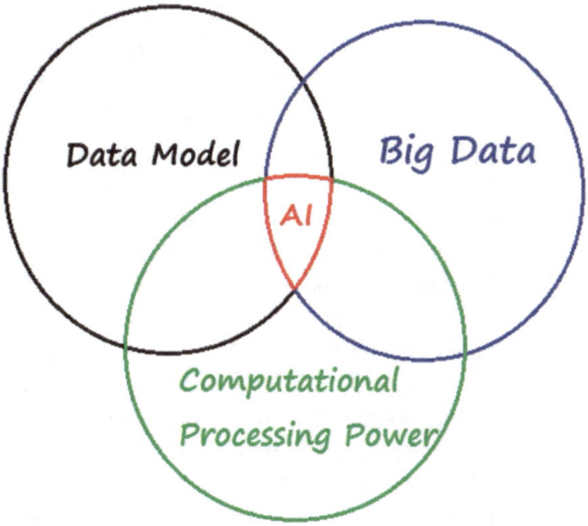

Fig. 2 AI operation

4 Artificial Intelligence in Education

Historically the school does not usually walk at the same speed as technological advances, considering that they are usually behind when it comes to the use of information and communication technology in school life. In this context, some solutions and technologies aimed at AI are already available and are known in the educational environment, such as chatbot. This tool assists in distance communication in writing, functioning as a personal assistant capable of answering questions automatically. Predictive Analysis acts as a solution capable of predicting future scenarios such as the evasion of a student from the school through a historical database. Another tool that can be used is the simultaneous translation of voice into any language, overcoming the barrier of access to other content produced in different languages. Another possibility is the application of intelligent tutoring to assist each student individually in their learning process. It is important to note that each of these scenarios has AI as the basis of the process [41, 47–54].

China has advanced a lot in the area of focused education, as it has focused its efforts on the use of AI. The United States is also taking steps to put AI in the classroom. Recent studies have pointed out 34 h in the Duolingo app are equivalent to a complete university semester of language teaching. Companies like Carnegie Learning and Fuel Education apply AI to elementary and high school. One of EdTech's most popular platforms, McGraw Hill's ALEKS [52].

Another example of the use of AI in the Teaching Platform can be seen by the multinational VIA Technologies, Inc., which recently launched the VIA Pixetto vision sensor, an intuitive and engaging platform for teaching AI and Machine Learning to students of 12 years or older. VIA Pixetto comes with an integrated set of tools that makes it easy to understand the basic principles and technologies underlying AI and ML and apply them to your own AI vision, manufacturer, and robotics projects. The tools include pre-built models of object recognition, shapes, colors, faces, and manuscripts that students can use to configure the vision sensor; coding blocks for beginners, integrated with the popular Scratch platform, to teach students basic programming; accelerated machine learning platform for students to create templates for their projects and support for advanced coding using Python and TensorFlow Lite [53]. These tools, in turn, refer to elementary and high school and even level content universities. These tools mentioned exemplify and align with the focus of the approach of the present study.

The learning of mathematics is one of the tasks considered most difficult by children and adolescents who attend elementary school since the discipline has a language composed of symbols that requires logical and deductive reasoning. It is necessary to insert in the students' daily lives more playful ways of assimilating and abstracting content and stimulating thinking through games that can transform what once seemed a burden, with the performance of AI in fun, stimulating learning [41, 49–54]. It is important that the child believes in his ability to solve challenges concerning learning mathematics. Given this, AI has a field of exploration, and a repertoire of problems that allow it to advance in the concept formation process in a personalized way and that each child has their pace respected [54].

The techniques of Big Data and Machine Learning, that is, AI, through more complex behavioral patterns, it is possible to measure engagement, perseverance, performance, and time spent on a task. AI algorithms allow outlining a profile of each student and their personalized experiences [55, 56].

Through the performance analysis of students and classes, at any desired moment, it is possible to articulate new approaches and educational actions, through the data collected concerning learning, in the treatment of the heterogeneity of groups, and the delivery of performance and interaction information of students. Still evaluating the possibility of observing where students show improvement in both performance and engagement in certain subjects, composing indicators of efficiency of the platform based on AI as non-traditional learning [46, 55, 56].

Finally, the use of platforms and techniques of joint intelligence contribute to the development of learning. Whereas what AI is not about automating processes per se, but what is valid and important for education is the combination of human intelligence and machine intelligence, showing more effective results.

4.1 AI Applications in Education

Around the world, different modalities of AI technology in education have already been applied in search of advances in the learning process. Considering that to create an intelligent tutor system, it is necessary to build an algorithm to teach the computer to deal with information from three different sources based on the content that will be taught (1) domain model; the way that content will be taught, (2) pedagogical model (3) and the knowledge that the student already has (student model). Fostering

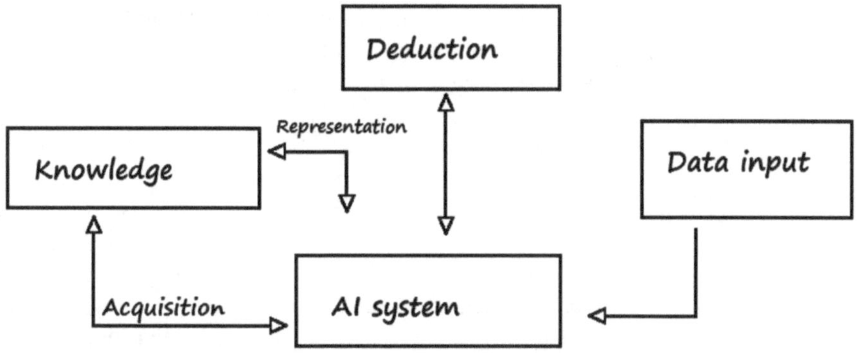

Fig. 3 A conceptual illustration of AI systems

that this initial information is the models that will feed the computer system based on AI (Fig. 3) [55–57].

Considering that from innovative software to tablets, much has been tried in terms of technology in the classroom, however, not always with significant impacts on learning [43–46, 55–57]. Another way to assist in the learning process is the use of a teaching platform adapted to each student, containing a "playlist" of videos, texts, and exams prepared according to their preferences and pedagogical deficiencies adapted according to such deficiencies. Another alternative is the use of programs that create a database over the years, based on various educational assessments, if used in education networks, to help teachers accurately identify what students' needs are [43–46, 57].

Still considering the use of techniques aimed at online education aimed at expanding and improving the offer of courses, which consist of distance learning modality, used to define learning by electronic means. Another alternative is to virtual tutoring adapted for each student, which seeks to improve the learning of specific subjects. This task is performed from the analysis of thousands of hours of previous classes available [10, 55–58].

Societies and cultures around the world have remained unable to provide a teacher for each student, but it will be possible to simulate this reality with the tools of AI. The creation of personalized teaching and learning environments for each student, using platforms that employ technologies such as intelligent tutoring systems, are already able to do this [25, 55–61].

It is important to emphasize that technology does not dispense with the teacher, but that individual ceases to be the owner of knowledge and becomes a mediator, making his pedagogical role in teaching the student to be a good researcher. Because one of the potentials of technology is to allow the student not to depend so much on the physical availability of the teacher. The intelligent system, or AI solution, will not yet be equivalent to a private teacher, but simulates that teacher at a lower cost.

Thus, in this perspective, it is also necessary to take into account the limits of current technology, which, at least for the time being, is not efficient in assessing nuances, such as students' emotional intelligence or their writing skills [46, 55–61].

5 Discussion

Currently, i t is possible to observe many movements to create more personalized, flexible, inclusive, and interactive teaching through the application of AI, which will allow that is maximized human intelligence. Assessing that this trend should grow even more in the coming years as easily as the people do with the use of technology for the most routine activities of daily life, it must be embraced by the school and the teachers.

In Education, it shouldn't be different, since teachers and students can benefit a lot from learning platforms, which capture learning difficulties and provide data for the teacher to determine the best intervention. Implying that for effective use and obtaining a real pedagogical impact, despite being a very different experience for students, they must familiarize themselves with the technologies.

The movements to create more personalized, flexible, inclusive and interactive teaching involve the application of AI, helping children and young people to find new ways to build learning. Going through the facilities of AI and enabling students to advance in their discoveries by building their own knowledge with small achievements, exercising protagonism.

Evaluate that the innovations brought by the advancement of technology affect all sectors, from the most basic forms of contact in human relations to the education systems and health services. In this context, it is important to note that technology can help create personalized teaching and learning environments, allowing interactive platforms and smart tutors to guide students on their development journey.

When interacting with a platform of this type, students have access to the contents available and can interact on the platform itself, since the greater the volume of interactions in the digital environment, the greater the system's ability to update information based on the interactions of students. Considering that the system has properties of knowing in which areas the student has more competence, assessing how he reasoned to solve problems, and even seeking to complement his learning. If the student demonstrates mastery of a particular topic, the platform suggests more complex subjects. In the same sense, that if the student shows difficulties in activities, the AI refers them to previous subjects that are prerequisites for understanding that subject; that is, the AI "understands" the student's profile better and generates personalized study plans.

The infinity of data generated by these AI solutions, which can be integrated with Big Data, refer to the large volume of data generated, so from these data and reports, it is possible to accurately indicate the level of knowledge of each student, the teacher can provide personalized service, even in large classes. The information contained in these platforms can adapt to the real needs of students, as teachers can make an analysis of student behavior on the platform to offer content with the potential to advance learning and deal with problem-solving.

Furthermore, the student's motivation is an important point because, from the student's point of view, a school is interesting, with different classes, with maker space, information technology, and derivatives, which makes this student fall in love and be motivated by teaching and by the place. From another perspective, the student, when engaged, learns very well with technology, emphasizing that the problem is to overcome the stage of engagement. Since the biggest driver of engagement is the teacher, note that better models occur when the teacher is combined with the system. Thus, for better use and understanding of the benefits of AI in Education, it is necessary to invest in the training and digital qualification of managers, the education network, and teachers.

In Education, there is still a movement to take AI to the classroom, through games, computer programs, security applications, robotics, devices for handwriting recognition and voice recognition, and derivatives. Since rule games usually represent a "fair fight", because when playing, opponents are usually sought within the same level of competitiveness (even if it is an AI), as in the case of electronic games in which the opponent is the computer. Emphasizing that the objective is the same and the conditions are the same, that is, the victory is for those who play best in that match. This perspective in education encourages and includes that the student must be focused, have discipline, exercise a lot mentally (reasoning), and be skilled, which is important for good performance. Considering that these characteristics are variables that can be trained and stimulated (by these AI solutions), which are very important in today's world, and thus guaranteeing that 'gamification' is a powerful learning strategy.

In this sense, it is possible to adjust the amount of work and motivation of each student profile, through the possibility of personalizing learning trails with games and guidelines developed according to the curricular parameters and disciplinary matrices, allowing the teacher to create personalized scripts with the games for the whole room, for groups and even students individually. What makes the use of adaptive AI platforms gain space in Education, as they are able to propose these individual learning paths. Since the platform itself collects data for each user experience (student), it analyzes and offers a different path based on its initial knowledge.

Adaptive AI platform integrates the classroom offering a range of benefits, such as reports with usefulness for teachers to make a decision regarding student learning. Highlighting the role of the teacher reflecting its essentiality in the process for allowing interactions, carrying out the planning that best suits the needs of students, and even mediating the cognitive process through personalized teaching using AI and virtual classes to complement the work done by the teacher in a classroom. Considering that AI allows education to be beyond the curve since technology can be defined as a computer system that simulates the human capacity to solve problems.

The advancement of AI offers possibilities of care for teachers' mental and emotional health that will have an indirect impact on Education, reflecting that generally, teachers around the world report levels of anxiety, stress, loneliness, and withdrawal from their duties due to issues related to mental health. Pondering the impact of this new digital age on the profession and the possibilities that technology presents for care related to well-being, still measuring the focus on problem-solving and as resources that deal with emotional issues. Which whether due to lack of time, financial resources, lack of professionals to meet all the demand in public health services, or the stigma of seeking help, many teachers and educators do not receive the psychological support they need.

Assessing that these services of information and psychological reception through AI via chatbots still allow the recording of audios for the user to talk about feelings, and from that, they are able to map keywords that may indicate risk profiles. Still considering the scenario of less care concerning health, the technologies capable of recognizing what a person feels from their facial expression, through AI applications that differentiate the main mental health conditions via interaction with the individual, and even personalized integrated services with guidelines and techniques for reducing stress and anxiety.

The need for investment in public policies in order to overcome the infrastructure obstacle, given the need for schools to have quality internet and good equipment for all students, still considering those located in a region far from the city center, the which are the first major barriers to working with technology. In this type of reality, there are many Brazilian schools that, depending on the location, only have dial-up internet, which impacts a lot in more technological scenarios. However, having good infrastructure and connectivity is important, but these factors alone are not enough; that is, high technological resources do not guarantee to learn.

In the face of world scenarios where the poor quality of the mobile internet is still glaring, provided that little by little, the tendency is for these bottlenecks to be overcome, in the face of the tendency that technology helps to democratize education, even if its use is more subtle than is still imagined. Assessing that sometimes, technologies are a little invisible to end-users and that they are being incorporated into the education routine. Weighing on adaptive teaching, which has characteristics that are shaped for each student, even systems capable of recommending reliable materials for students in certain areas, examples that follow a common identity of incorporating the analysis of the data of its users to identify deficiencies from the students.

Thus, it is necessary to see that AI technology is not an end in itself; that is, it is a driver of learning, also considered as a driver of innovation, creativity, and inventiveness through experimentation, and its use must always be accompanied by reflection and ethics. For this, in addition t o proposals and clear objectives in the use of digital tools based on AI, the actions should involve students in actions that emphasize the use of active methodologies, which allow the student to get out of passivity and assume the center of the learning process, allowing them to be protagonists, authors, and builders of their own learning.

6 Trends

AI applied to Education is a multi and interdisciplinary research area, the use of these technologies in systems whose objective is teaching and learning is related to those that use technologies aimed at Affective Intelligent Tutoring Systems (STIs), the Learning Management Systems (LMSs), Intelligent Educational Robotics and even Massive Open Online Course (MOOCs), concerning Learning Analytics (LA). Still pondering that each of these applications makes use of AI technologies in different ways, causing a great impact on the teaching–learning processes [45, 62–70].

Intelligent tutorial systems (ITS) are computer programs based on AI and independent of the constant intervention of the teacher. Seen its application as adaptive teaching technologies, capable of shaping teaching according to the students' level of knowledge [71–73].

Personalized teaching is a trend, but also an evolving knowledge, evaluating that adaptive teaching programs are able to identify students' interests and facilities, proposing personalized learning paths for each one. Still pondering the analysis, it is done in real-time from the experience of using virtual learning platforms [74, 75].

The use of learning analytics, which is the interpretation of a wide range of data produced by students. Making it possible to assess academic progress, predicting performance, and even detecting possible problems in learning, detecting the points of greatest difficulty in understanding the content, or even the tendency to drop out [76, 77].

In education, the AI systems combined with machine learning behind the chatbots, consist of virtual attendance services that offer tutoring, assistance, and feedback in different niches of society. Which favor personalized teaching and facilitate the learning process, the educational chatbot is an excellent instrument for inclusion and awareness, which are used to deepen knowledge on a virtual platform that centralizes the activities of students and teachers [78, 79].

The increasingly recurrent use of resources such as natural language processing (PLN), in which a computer is able to understand and interpret human language. Allowing a teacher of another nationality and not fluent in a given language spoken in a given country, to teach a specific language class to native students without knowledge of the specific language. Considering that in education, the PLN will contribute to the exchange between students of different nationalities and to the real-time transmission of classes in different languages, which will be translated to students [80–84].

7 Conclusions

AI technology will allow society to be able to move millions of people in autonomous vehicles in large metropolises, causing an avalanche of changes in other areas. Causing changes in the way people interact with the world, reformulating standards and relationships in society. This will occur because AI technology, i.e., robots, algorithms, machines, sensors, systems, and other derivatives, will cause jobs that were previously exclusively human to change, making solid and traditional careers no longer make sense. From this technological strand AI, society is also experiencing a revolution in education, in which the youngest constantly charging teachers and schools for novelty and greater interaction. What from this stimulus, it is necessary to rethink the existing tools, also creating online and offline possibilities, developing platforms that provide a panoramic view, so that schools can take advantage of the existing technology, adapting the reality of the units.

The possibility of seeing performance standards by a class of students, or even teaching unit, investigating why a class or school in a given education network has higher performance in a specific discipline such as Physics, and others have deficits in Mathematics or other specific topics, such as derivatives and integrals. In this sense, making it possible to discover and develop successful practices to replicate them with a greater chance of effectiveness, still evaluating the point of view of students who use an AI virtual platform to perform the activities indicated by the teachers, since it is possible to monitor their own performance in each discipline and classify the contents by their degree of difficulty. Still evaluating the moment the student completes the activities, the AI system identifies, via intelligent algorithms, how much this student understood and abstracted knowledge of each discipline and subject, indicating which classes should attend again to answer his doubts regarding a specific subject.

Still measuring and considering the other point of view, i.e., teachers, they have the possibility to measure the learning of each student and each class, passing complimentary classes, and even making the automatic correction of the exercises. Allowing students to use the AI tool, attending classes, and answering questionnaires; and teachers receive the data and compare it to specific models of teaching levels, understanding what students have learned and what their difficulties have been. The most important thing is that the use of this collected data can be applied to any area, still considering the use of the data walk with subsidies for macro and micro decision making by education network managers, as in general, in most countries, there are Education departments. Assessing the need to integrate technology into the student's curriculum, exploring its potential, and promoting conversation with the areas of knowledge.

This type of context opens scenarios for another possibility, such as the creation of the student's digital identification (ID), since, in general, universities already have the Student Registry (RA), i.e., a unique digital disciplinary record. Giving teachers and teaching managers access to the student's digital trajectory, enabling customization and personalization for this student, enabling their performance in all tests and activities performed to be counted from the software with AI generating reports of their weaknesses and strengths throughout school life. This allows the student to be heard and is a vitally active part of the educational processes since they leave passivity and learns to become an effective actor in your learning cycle. Assessing the properties and characteristics that it has through AI solution to identify, through tests, in which topics of the subject have more difficulty, offering activities and support material to overcome the deficiencies. This results, from the point of view of a high school that receives a student, does not have to face an illustrious stranger, having to know them during his stay in that place. On the contrary, these AI solutions will allow these schools to set up individual planning before classes start to help the students along the learning path.

The AI technology experience at the school is an example of how this can be applied in education as its global trend is still full of challenges and opportunities. Assessing that to achieve this success, it is necessary to explore active methodologies when working with projects, research, problem-solving, production of digital narratives, and development of activities, transforming digital tools into language that students understand and abstract. Finally, there is no single model to follow, and something ready should not be expected, the process of Education based on AI tools is being created. However, technology should not be considered as perfumery, it should be at the center of pedagogical management, and the essential thing is to emphasize the message that technology is at the center of the process of modernizing teaching for the twenty-first century, and it's potential for transforming education. Education should not be underestimated.

References

1. Mohamed, E.: The relation of artificial intelligence with internet of things: a survey. J. Cybersecurity Inf. Manage. **1**(1), 24–30 (2020). https://doi.org/10.5281/zenodo.3686810
2. Goksel, N., Bozkurt, A.: Artificial intelligence in education: crrent insights and future perspectives. In: Handbook of Research on Learning in the Age of Transhumanism, pp. 224–236. IGI Global (2019). https://doi.org/10.4018/978-1-5225-8431-5.ch014
3. Malik, G., Devendra, K.T., Sonakshi, V.: An analysis of the role of artificial intelligence in education and teaching. In: Recent Findings in Intelligent Computing Techniques, pp. 407–417. Springer, Singapore (2019)
4. Estevez, J., Gorka, G., Manuel, G.: Gentle introduction to artificial intelligence for high-school students using scratch. IEEE Access **7**, 179027–179036 (2019)
5. Frank, A.G., Dalenogare, L.S., Ayala, N.F.: Industry 4.0 technologies: implementation patterns in manufacturing companies. Int. J. Prod. Econ. **210**, 15–26 (2019). https://doi.org/10.1016/j.ijpe.2019.01.004
6. Ardito, L., et al.: Towards industry 4.0. Bus. Process Manage. J. (2019)
7. Almeida, F., Simoes, J.: The role of serious games, gamification and Industry 4.0 tools in the education 4.0 paradigm. Contemp. Educ. Technol. **10**(2), 120–136 (2019). https://doi.org/10.30935/cet.554469
8. Salmon, G.: May the fourth be with you: creating education 4.0. J. Learn. Dev.-JL4D **6**(2) (2019)
9. Schmid, R., Petko, D.: Does the use of educational technology in personalized learning environments correlate with self-reported digital skills and beliefs of secondary-school students? Comput. Educ. **136**, 75–86 (2019). https://doi.org/10.1016/j.compedu.2019.03.006
10. Peng, H., Ma, S., Spector, J.M.: Personalized adaptive learning: an emerging pedagogical approach enabled by a smart learning environment. Smart Learn. Environ. **6**(1), 9 (2019). https://doi.org/10.1186/s40561-019-0089-y

11. Mukhanov, S.A., Arkhangelsky, A.I., Mukhanova, A.A.: Differentiated and indi-vidualised teaching mathematics to students of technical universities. In: 1st International Scientific Conference "Modern Management Trends and the Digital Economy: from Regional Development to Global Economic Growth" (MTDE 2019). Atlantis Press (2019)
12. Cai, W., et al.: MathBot: a personalized conversational agent for learning math (2019)
13. Luckin, R., Cukurova, M.: Designing educational technologies in the age of AI: a learning sciences-driven approach. Br. J. Educ. Technol. **50**(6), 2824–2838 (2019). https://doi.org/10.1111/bjet.12861
14. How, M.-L.: Future-ready strategic oversight of multiple artificial superin-telligence-enabled adaptive learning systems via human-centric explainable AI-empowered predictive optimizations of educational outcomes. Big Data Cogn. Comput. **3**(3), 46 (2019). https://doi.org/10.3390/bdcc3030046
15. Holmes, W., Bialik, M., Fadel, C.: Artificial Intelligence in Education. Center for Curriculum Redesign, Boston (2019)
16. Goksel, N., Bozkurt, A.: Artificial intelligence in education: current insights and future perspectives. In: Handbook of Research on Learning in the Age of Transhumanism, pp. 224–236. IGI Global (2019)
17. Karsenti, T.: Artificial intelligence in education: the urgent need to prepare teachers for tomorrow's schools. Formation et profession **27**(1), 112–116 (2019). https://doi.org/10.18162/fp.2019.a166
18. Khosravi, H., Kitto, K., Williams, J.J.: Ripple: a crowdsourced adaptive platform for recommendation of learning activities (2019)
19. Henderson, N.L., et al.: 4D affect detection: improving frustration detection in game-based learning with posture-based temporal data fusion. In: International Conference on Artificial Intelligence in Education. Springer, Cham (2019)
20. Monteiro, A.C.B., et al.: Methodology of high accuracy, sensitivity and specificity in the counts of erythrocytes and leukocytes in blood smear images. In: Brazilian Technology Symposium. Springer, Cham (2018)
21. Monteiro, A.C.B., et al.: A comparative study between methodologies based on the Hough transform and watershed transform on the blood cell count. In: Brazilian Technology Symposium. Springer, Cham (2018)
22. Chen, Z.-H., Lu, H.-D., Chou, C.-Y.: Using game-based negotiation mechanism to enhance students' goal-setting and regulation. Comput. Educ. **129**, 71–81 (2019). https://doi.org/10.1016/j.compedu.2018.10.011
23. Bailey, L.W.: New technology for the classroom: mobile devices, artificial intelligence, tutoring systems, and robotics. In: Educational Technology and the New World of Persistent Learning, pp. 1–11. IGI Global (2019)
24. Arrieta, A.B., et al.: Explainable artificial intelligence (XAI): concepts, taxonomies, opportunities and challenges toward responsible AI. Inf. Fusion **58**, 82–115 (2020). https://doi.org/10.1016/j.inffus.2019.12.012
25. Lu, Y.: Artificial intelligence: a survey on evolution, models, applications and future trends. J. Manage. Analytics **6**(1), 1–29 (2019). https://doi.org/10.1080/23270012.2019.1570365
26. Horie, Y., et al.: Diagnostic outcomes of esophageal cancer by artificial intelligence using convolutional neural networks. Gastrointest. Endosc. **89**(1), 25–32 (2019). https://doi.org/10.1016/j.gie.2018.07.037
27. Walczak, S.: Artificial neural networks. In: Advanced Methodologies and Technologies in Artificial Intelligence, Computer Simulation, and Human-Computer Interaction, pp. 40–53. IGI Global (2019)
28. Shahid, N., Rappon, T., Berta, W.: Applications of artificial neural networks in health care organizational decision-making: a scoping review. PLoS ONE **14**(2), e0212356 (2019). https://doi.org/10.1371/journal.pone.0212356
29. Mohandes, S.R., Zhang, X., Mahdiyar, A.: A comprehensive review on the application of artificial neural networks in building energy analysis. Neurocomputing **340**, 55–75 (2019). https://doi.org/10.1016/j.neucom.2019.02.040}

30. Gonzalez-Fernandez, I., et al.: A critical review on the use of artificial neural networks in olive oil production, characterization and authentication. Crit. Rev. Food Sci. Nutr. **59**(12), 1913–1926 (2019). https://doi.org/10.1080/10408398.2018.1433628
31. Amer, M., Maul, T.: A review of modularization techniques in artificial neural networks. Artif. Intell. Rev. **52**(1), 527–561 (2019). https://doi.org/10.1007/s10462-019-09706-7
32. Al-Mubayyed, O.M., Abu-Nasser B.S., Abu-Naser S.S.: Predicting overall car performance using artificial neural network (2019)
33. Bevilacqua, V., et al.: A performance comparison between shallow and deeper neural networks supervised classification of tomosynthesis breast lesions images. Cogn. Syst. Res. **53**, 3–19 (2019). https://doi.org/10.1016/j.cogsys.2018.04.011
34. Alshehhi, R., et al.: Supervised neural networks for helioseismic ring-diagram in-versions. Astron. Astrophys. **622**, A124 (2019). https://doi.org/10.1051/0004-6361/201834237
35. Zhang, Y., et al.: Bayesian graph convolutional neural networks for semi-supervised classification. In: Proceedings of the AAAI Conference on Artificial Intelligence, vol. 33 (2019)
36. Jiang, B., et al.: Semi-supervised learning with graph learning-convolutional networks. In: Proceedings of the IEEE Conference on Computer Vision and Pattern Recognition (2019). https://doi.org/10.1109/CVPR.2019.01157
37. Pehlevan, C., Chklovskii, D.B.: Neuroscience-inspired online unsupervised learning algorithms: artificial neural networks. IEEE Signal Process. Mag. **36**(6), 88–96 (2019). https://doi.org/10.1109/MSP.2019.2933846
38. Hou, T., Huang, H.: Statistical physics of unsupervised learning with prior knowledge in neural networks. Phys. Rev. Lett. **124**(24), 248302 (2020). https://doi.org/10.1103/PhysRevLett.124.248302
39. Mao, X., Yang, H., Huang, S., Liu, Y., Li, R.: Extractive summarization using supervised and unsupervised learning. Expert Syst. Appl. **133**, 173–181 (2019). https://doi.org/10.1016/j.eswa.2019.05.011
40. Crigger, E., Khoury, C.: Making policy on augmented intelligence in health care. AMA J. Ethics **21**(2), 188–191 (2019). https://doi.org/10.1001/amajethics.2019.188
41. Jian, L., et al.: Combining unmanned aerial vehicles with artificial intelligence technology for traffic-congestion recognition: electronic eyes in the skies to spot clogged roads. IEEE Consum. Electron. Mag. **8**(3), 81–86 (2019). https://doi.org/10.1109/MCE.2019.2892286
42. Monteiro, A.C.B., et al.: Development of a laboratory medical algorithm for simultaneous detection and counting of erythrocytes and leukocytes in digital images of a blood smear. In: Deep Learning Techniques for Biomedical and Health Informatics, pp. 165–186. Academic Press (2020)
43. How, M.-L., Hung, W.L.D.: Educational stakeholders' independent evaluation of an artificial intelligence-enabled adaptive learning system using Bayesian network predictive simulations. Educ. Sci. **9**(2), 110 (2019). https://doi.org/10.3390/educsci9020110
44. How, M.-L., Hung, W.L.D.: Harnessing entropy via predictive analytics to optimize outcomes in the pedagogical system: an artificial intelligence-based Bayesian networks approach. Educ. Sci. **9**(2), 158 (2019). https://doi.org/10.3390/educsci9020158
45. Samuel, Y., George, J., Samuel, J.: Beyond stem, how can women engage big data, analytics, robotics and artificial intelligence? An exploratory analysis of confidence and educational factors in the emerging technology waves influencing the role of, and impact upon, women (2020)
46. Jules, T.D., Salajan, F.D. (eds).: The Educational Intelligent Economy: Big Data, Artificial Intelligence, Machine Learning and the Internet of Things in Education. Emerald Group Publishing (2019)
47. Helbing, D., et al.: Will democracy survive big data and artificial intelligence? In: Towards Digital Enlightenment, pp. 73–98. Springer, Cham (2019)
48. Duan, Y., Edwards, J.S., Dwivedi, Y.K.: Artificial intelligence for decision making in the era of big data–evolution, challenges and research agenda. Int. J. Inf. Manage. **48**, 63–71 (2019). https://doi.org/10.1016/j.ijinfomgt.2019.01.021

49. Jalal, A., Mahmood, M.: Students' behavior mining in e-learning environment using cognitive processes with information technologies. Educ. Inf. Technol. **24**(5), 2797–2821 (2019). https://doi.org/10.1007/s10639-019-09892-5
50. Gams, M., et al.: Artificial intelligence and ambient intelligence. J. Ambient Intell Smart Environ. **11**(1), 71–86 (2019). https://doi.org/10.3233/AIS-180508
51. Smutny, P., Schreiberova, P.: Chatbots for learning: a review of educational chatbots for the facebook messenger. Comput. Educ. 103862 (2020). https://doi.org/10.1016/j.compedu.2020.103862
52. Lexalytics. AI in education: where is it now and what is the future. www.lexalytics.com/lexablog/ai-in-education-present-future-ethics (2020). Accessed 10 Sep 2020
53. Terra. Nova plataforma acelera ensino de inteligência artificial nas escolas. www.terra.com.br/noticias/dino/nova-plataforma-acelera-ensino-de-inteligencia-artificial-nas-escolas,661d6edbba3b21b49ecd0182bd5d3904709p4101.html (2020). Accessed 10 Sep 2020
54. Rădulescu, A.: Algorithmic textual practices: improving fluency and word order in neural machine-translation output. Linguist. Philos. Invest. **18**, 126–132 (2019). https://doi.org/10.22381/LPI1820198
55. Almasri, A., et al.: Intelligent tutoring systems survey for the period 2000–2018 (2019)
56. Yang, J., Zhang, B.: Artificial intelligence in intelligent tutoring robots: a systematic review and design guidelines. Appl. Sci. **9**(10), 2078 (2019). https://doi.org/10.3390/app9102078
57. Saltz, J., et al.: Integrating ethics within machine learning courses. ACM Trans. Comput. Educ. (TOCE) **19**(4), 1–26 (2019)
58. Sekeroglu, B., Dimililer, K., Tuncal, K.: Student performance prediction and classification using machine learning algorithms. In: Proceedings of the 2019 8th International Conference on Educational and Information Technology (2019)
59. Loftus, M., Madden, M.G.: A pedagogy of data and artificial intelligence for student subjectification. Teach. High. Educ. **25**(4), 456–475 (2020). https://doi.org/10.1080/13562517.2020.1748593
60. Afzal, S., et al.: The personality of AI systems in education: experiences with the watson tutor, a one-on-one virtual tutoring system. Child. Educ. **95**(1), 44–52 (2019). https://doi.org/10.1080/00094056.2019.1565809
61. Cukurova, M., Kent, C., Luckin, R.: Artificial intelligence and multimodal data in the service of human decision-making: a case study in debate tutoring. Br. J. Educ. Technol. **50**(6), 3032–3046 (2019). https://doi.org/10.1111/bjet.12829
62. Cabada, R.Z., et al.: Hyperparameter optimization in CNN for learning-centered emotion recognition for intelligent tutoring systems. Soft Comput. **24**(10), 7593–7602 (2020). https://doi.org/10.1007/s00500-019-04387-4
63. de Oliveira, W.C., Gottardo, E., Pimentel, A.R.: Changes of Affective States in Intelligent Tutoring System to Improve Feedbacks through low-cost and open Electroencephalogram and Facial Expression. International conference on intelligent tutoring systems. Springer, Cham (2020)
64. Winstone, N., et al.: "Check the grade, log out": students' engagement with feedback in learning management systems. Assess. Eval. High. Educ. 1–13 (2020). https://doi.org/10.1080/02602938.2020.1787331
65. Bervell, B., Umar, I.N.: Blended learning or face-to-face? Does tutor anxiety prevent the adoption of learning management systems for distance education in Ghana? Open learning. J. Open, Distance, e-Learning **35**(2), 159–177 (2020). https://doi.org/10.1080/02680513.2018.1548964
66. Moro, M., Alimisis, D., Iocchi, L.: Educational robotics in the context of the maker movement. Springer International Publishing (2020)
67. Jansen, R.S., et al.: Supporting learners' self-regulated learning in massive open online courses. Comput. Educ. **146**, 103771 (2020). https://doi.org/10.1016/j.compedu.2019.103771
68. de Jong, P.G.M., et al.: Twelve tips for integrating massive open online course content into classroom teaching. Med. Teach. **42**(4), 393–397 (2020). https://doi.org/10.1080/0142159X.2019.1571569

69. Knight, S., Gibson, A., Shibani, A.: Implementing learning analytics for learning impact: taking tools to task. Internet High. Educ. **45**, 100729 (2020). https://doi.org/10.1016/j.iheduc.2020.100729
70. Gutiérrez, F., et al.: LADA: a learning analytics dashboard for academic advising. Comput. Hum. Behav. **107**, 105826 (2020). https://doi.org/10.1016/j.chb.2018.12.004
71. Chabay, R.W., Larkin, J.H. (eds.): Computer-assisted Instruction and Intelligent Tutoring Systems: Shared Goals and Complementary Approaches. Routledge (2020)
72. Reiser, B.J., et al.: Knowledge Representation and Explanation in GI L, An Intelligent Tutor for Programming. Computer-assisted instruction and intelligent tutoring systems: shared goals and complementary approaches, vol. 111 (2020)
73. González-Hernández, F., et al.: Recognition of learning-centered emotions using a convolutional neural network. J. Intell. Fuzzy Syst. **34**(5), 3325–3336 (2018)
74. Almarzooq, Z., Lopes, M., Kochar, A.: Virtual learning during the COVID-19 pandemic: a disruptive technology in graduate medical education (2020)
75. Ehrlich, H., McKenney, M., Elkbuli, A.: We asked the experts: virtual learning in surgical education during the COVID-19 pandemic—shaping the future of surgical education and training. World J. Surg. **1** (2020). https://doi.org/10.1007/s00268-020-05574-3
76. Shibani, A., Knight, S., Shum, S.B.: Educator perspectives on learning analytics in classroom practice. Internet High. Educ. **46**, 100730 (2020). https://doi.org/10.1016/j.iheduc.2020.100730
77. Herodotou, C., et al.: The scalable implementation of predictive learning analytics at a distance learning university: insights from a longitudinal case study. Internet High. Educ. **45**, 100725 (2020). https://doi.org/10.1016/j.iheduc.2020.100725
78. Yang, H.-G., Lee, T.-W.: Development of AI Chatbot Education Based on Maker-Education. Proceedings of the Korean society of computer information conference. Korean society of computer information (2020)
79. Sands, S., et al.: Managing the human–chatbot divide: how service scripts influence service experience. J. Serv. Manage. (2020)
80. Lucy, L., et al.: Content analysis of textbooks via natural language processing: findings on gender, race, and ethnicity in Texas US history textbooks. AERA Open **6**(3), 2332858420940312 (2020). https://doi.org/10.1177/2332858420940312
81. Pikhart, M.: Natural Language Processing and Deep Learning For Blended Learning as an Aspect of Computational Linguistics. International conference on remote engineering and virtual instrumentation. Springer, Cham (2020)
82. Haldorai, A., Murugan, S., Ramu, A.: Evolution, challenges, and application of intelligent ICT education: an overview. Comput. Appl. Eng. Educ. (2020)
83. Kolleck, N., Yemini, M.: Environment-related education topics within global citizenship education scholarship focused on teachers: a natural language processing analysis. J. Environ. Educ. 1–15 (2020). https://doi.org/10.1080/00958964.2020.1724853
84. Denisov, P., Vu, N.T.: Pretrained semantic speech embeddings for end-to-end spoken language understanding via cross-modal teacher-student learning. arXiv preprint (2020)

Knowledge Discovery by Compensatory Fuzzy Rough Predicates

Rafael Alejandro Espin-Andrade, Erick González, Rafael Bello, and Witold Pedrycz

Abstract Compensatory Fuzzy Logic (CFL) are fuzzy logic systems, which satisfy axiomatic properties of bivalent logic and Decision Theory simultaneously. There is a coherence between CFL and other theories like classical logic, t-norm, and t-conorm based fuzzy logic, mathematical statistics, and decision theory. Those properties are the basis of transdisciplinary interpretability in relation to natural language. Hence, CFL has the advantage to model easily the problems by using natural and professional language. The main objective of this paper is to propose a method, inspired by rough sets theory (RST), to approximate decision classes by means of two clusters, defined by logic predicates formed on condition attributes. The importance of the method is mainly that it is a compliment, and not a substitute, of other methods for forecasting, in the sense that its results are more useful in the way of linguistic values than in numerical values.

Keywords Compensatory fuzzy logic · Decision theory · Bivalent logic · Rough sets theory

1 Introduction

Fuzzy Logic [1] and Rough Sets (see [2, 3]) are two theories used for modeling problems in Soft computing. They are able to deal with different kinds of uncertainty existing in natural language and knowledge.

R. A. Espin-Andrade (✉)
Universidad Autónoma de Coahuila, Saltillo 25280, Coahuila, México

E. González
Technological University of Havana, 19390 La Habana, Cuba

R. Bello
Universidad Central Marta Abreu de Las Villas, 54830 Villa Clara, Santa Clara, Cuba

W. Pedrycz
University of Alberta, Edmonton, Alberta T6G 2R3, Canada

W. Pedrycz et al. (eds.), *Computational Intelligence for Business Analytics*,
Studies in Computational Intelligence 953,
https://doi.org/10.1007/978-3-030-73819-8_11

Membership functions of fuzzy sets allow representing gradual membership of an element to a certain set. Linguistic variables [4] are an effective manner to calculate by means of words, not only by numbers [5]. Therefore, these two concepts are very important to represent the knowledge and its intrinsic fuzziness.

On the other hand, rough sets are based on the delimitation of the knowledge by unambiguous borders, which are represented by a data table and an indiscernibility relation. Those relations can be, for example, equivalence relations or similarity relations. Here, the approximation is obtained by determining two sets, called lower and upper approximations. Usually, even for a small number of elements in the table, it is possible to apply the method of knowledge discovery.

One classical way t o joint both theories is to consider fuzzy rough sets and rough fuzzy sets [6], where there are two options; one of them is the fuzzification of rough sets and the other one is the conversion of one fuzzy set into a rough set.

Very commonly, classic Zadeh max–min operators are used in fuzzy rough sets and rough fuzzy sets, even though they are not continuous operators, in the sense that one change in a single truth value of a simple predicate changes the truth value of the compound predicate.

Approaches to RST, fuzzy logic, and the interpretability of logical propositions in natural language have been made last years. Some of them can be found in [7–12].

This hybrid approach exhibits several advantages. It is a practical way to resolve decision problems, as the data are summarized in a table of attributes; even if the number of data is small, the method is able to be applied; this method can be used for every type of data, even if they are mixed data. Classical RST works only with continuous data, but some extension of it includes the work with mixed data.

A general method of Knowledge Discovery has been created by using Compensatory Fuzzy Logic [13, 14]. The method that we introduce in this paper is part of this approach to generalize the Knowledge Discovery, here based on rough sets. One advantage of this generalization is that results can be used directly in decision making.

This paper aims to develop a method to find fuzzy predicates inspired in RST, that is to say, beginning from a decision table we will obtain two predicates, and not sets, called lower and upper approximation predicates, such that the truth value of the decision attribute is obtained from an approximation of these two predicates evaluated in the condition attributes. Hence, this method allows to predict decision attributes or making decisions based on the discovered predicates.

A linguistic interpretation from the data is built in the form of two clusters a sufficient condition and a necessary condition with truth values associated. Here Zadeh max–min system is substituted by CFL systems [15] with the objective to get the interpretable results we are looking for.

It is essential to remark that our main purpose is to introduce a method for knowledge discovery, but where this knowledge can be expressed with words and not only numerically. That is to say; its results are more useful if they are understood in the way of linguistic values than in numerical values. Therefore, it is a complementary method for forecasting and not a substitute respects to the others.

The advantage in the management is that it allows overcoming the traditional incomprehension between the language used by strategic decision makers or users, which usually are not specialized in this kind of methods, and the tactic decision makers, which usually provide to the primers with numerical values. The natural language is comprehensible for each of them.

2 Basic Concepts of Compensatory Fuzzy Logic and Rough Sets Theory

Compensatory Fuzzy Logic (CFL) [15] is a multivalued logic axiomatic approach different from the one based on t-norms and t-conorms. They satisfy characteristics of the descriptive approach of Decision making and the normative approaches of the decision making.

This is based on four logic operators (c, d, n, o). Here c is the conjunction operator, d is the disjunction operator, n is the negation operator, and o is a fuzzy-strict ordering.

The following axioms are postulated.

I. **Compensation Axiom** $\min(x_1, x_2, \cdots, x_n) \leq c(x_1, x_2, \cdots, x_n) \leq \max(x_1, x_2, \cdots, x_n)$.
II. **Symmetry or Commutativity Axiom** $c(x_1, x_2, \ldots, x_i, \ldots, x_j, \ldots, x_n) = c(x_1, x_2, \ldots, x_j, \ldots, x_i, \ldots, x_n)$.
III. **Strict Growth Axiom** If $x_1 = y_1, x_2 = y_2, \ldots, x_{i-1} = y_{i-1}, x_{i+1} = y_{i+1}, \ldots, x_n = y_n$ are different to zero and $x_i > y_i$ then $c(x_1, x_2, \ldots, x_n) > c(y_1, y_2, \ldots, y_n)$.
IV. **Veto Axiom** If $x_i = 0$ for any i then $c(\mathbf{x}) = 0$.
V. **Fuzzy Reciprocity Axiom** $o(\mathbf{x}, \mathbf{y}) = n\,[o(\mathbf{y}, \mathbf{x})]$.
VI. **Fuzzy Transitivity Axiom** If $o(\mathbf{x}, \mathbf{y}) \geq 0.5$ and $o(\mathbf{y}, \mathbf{z}) \geq 0.5$, then $o(\mathbf{x}, \mathbf{z}) \geq \max(o(\mathbf{x}, \mathbf{y}), o(\mathbf{y}, \mathbf{z}))$.
VII. **De Morgan's Laws:**$n(c(x_1, x_2, \ldots, x_n) = d(n(x_1), n(x_2), \ldots, n(x_n))$ $n(d(x_1, x_2, \ldots, x_n) = c(n(x_1), n(x_2), \ldots, n(x_n))$.

Implications can be defined in different ways:

- S-implication: $S(x, y) = d(n(x), y)$, where d and n are the disjunction and negation operators, respectively.
- R-implication: $R(x, y) = \sup\{z \in [0, 1]: c(x, z) \leq y\}$, where c is the conjunction operator.
- QL-implication, $IQL(x, y) = d(n(x), c(x, y))$.
- A-implication: The operator satisfies a group of axioms, which implicitly associate it with the conjunction, disjunction, and negation operators. For example, the Law of Importation $(x \wedge y \to z) \leftrightarrow (x \to (y \to z))$ is one of its axioms, where the symbol $\leftrightarrow$ is the logic equivalence.

The quasi-arithmetic means, which include for example the geometric mean, are operators of the form represented as shown below:

$$M_f(x_1, x_2, \ldots, x_n) = f^{-1}\left(\frac{1}{n}\sum_{i=1}^{n} f(x_i)\right) \tag{1}$$

where f is a strictly monotone continuous function which is extended to non-defined points by using the corresponding limit. These operators satisfy Axioms I-III.

In addition, it is easy to prove that axiom IV is also satisfied.

Then, taking

$$d(x_1, x_2, \ldots, x_n) = 1 - f^{-1}\left(\frac{1}{n}\sum_{i=1}^{n} f(1 - x_i)\right) \tag{2}$$

$$n(x) = 1 - x \tag{3}$$

and

$$o(\boldsymbol{x}, \boldsymbol{y}) = 0.5\left[C(\boldsymbol{x}) - C(\boldsymbol{y})\right] + 0.5 \tag{4}$$

We have a family of CFLs referred to as Quasi Arithmetic Mean based Compensatory Logic (QAMBCL).

$$\forall_{i\in\{1,2,\cdots,n\}} i\, p(x_i) = \bigwedge_{i\in\{1,2,\cdots,n\}} p(x_i) = f^{-1}\left(\frac{1}{n}\sum_{i=1}^{n} f(p(x_i))\right) \tag{5}$$

is the universal proposition,

$$\exists_{i\in\{1,2,\cdots,n\}} i\, p(x_i) = \bigvee_{i\in\{1,2,\cdots,n\}} p(x_i) = 1 - f^{-1}\left(\frac{1}{n}\sum_{i=1}^{n} f(1 - p(x_i))\right) \tag{6}$$

is the existential proposition.

An even more particular system of QAMBCL is the Geometric Mean based Compensatory Logic (GMBCL), where conjunction and disjunction operators are expressed by (7) and (8), respectively.

$$c(x_1, x_2, \ldots, x_n) = \sqrt[n]{\prod_{i=1}^{n} x_i} = \exp\left(\frac{1}{n}\sum_{i=1}^{n} \ln(x_i)\right) \tag{7}$$

Disjunction is the dual of the conjunction:

$$d(x_1, x_2, \ldots, x_n) = 1 - \sqrt[n]{\prod_{i=1}^{n}(1 - x_i)} = 1 - \exp\left(\frac{1}{n}\sum_{i=1}^{n} \ln(1 - x_i)\right) \tag{8}$$

CFL is a multivalued generalization of the classical bivalent logic; therefore, it maintains the properties of veto, symmetry, idempotency, and the De Morgan Laws. The main differences between logic systems based on t-norm and t-conorm are the idempotency and compensation axioms, and also the continuity of the compensatory operators.

Because of these axioms, CFL is a logic theory coherently based on two sources, the bivalent logic and decision theory, with the advantage that we can model many problems with the use of natural language.

A rough set [2] is defined as a pair formed by one universe set U, which is the set of some kind of objects and an indiscernible relation $R \subseteq U \times U$, which can be defined as an equivalence relation.

If X is a subset of U, then, a characterization of X with respect to R is the following below:

- The lower approximation of a set X with respect to R is the set of all objects which can be classified necessarily contained in X with respect to R.
- The upper approximation of a set X with respect to R is the set of all objects which can be classified possibly contained in X with respect to R.
- The boundary region of a set X with respect to R is the set of all objects, which can be classified neither as X nor as not-X with respect to R.
- A set X is crisp if the boundary region of X with respect to R is empty.
- A set X is rough if the boundary region of X with respect to R is non-empty.
- A lower approximation of the set X with respect to R can be defined as Eq. 9:

$$R_*(x) = \bigcup_{x \in U} \{R(x):\mathrm{R}(x) \subseteq X\} \tag{9}$$

An upper approximation of the set X with respect to R is defined in the following form

$$R^*(x) = \bigcup_{x \in U} \{R(x):\mathrm{R}(x) \cap X \neq \varphi\} \tag{10}$$

The boundary region of the set X is expressed in the following form

$$\mathrm{RN}_R(x) = R^*(x) \backslash R_*(x) \tag{11}$$

The necessary information for the application of RST to decision making is a table containing the data, which is called decision table (see Table 1).

Every row represents an object O_i (m rows), and every column represents an attribute A_j of the object (n columns). The n-1 first attributes are called condition attributes, and the last one is called the decision attribute.

Every datum d_{ij} placed in row i and column j of the table represents an actual value of the object i for the j attribute.

Besides, there exist definitions which link fuzzy set and rough set notions; they are the rough fuzzy sets and the fuzzy rough sets, [16].

Table 1 A generic decision table

Object\Attribute	A_1	A_2	...	A_n
O_1	d_{11}	d_{12}	...	d_{1n}
O_2	d_{21}	d_{22}	...	d_{2n}
⋮	⋮	⋮	⋮	⋮
O_m	d_{n1}	d_{n2}	...	d_{mn}

Given a fuzzy set F, the result of the approximation is the pair of fuzzy sets on the quotient set U/R:

$\mu_{qaprR}(F)([x]_R) = \inf\{\mu_F(y)|\ y \in [x]_R\}$ and $\mu_{\overline{qprR}}(F)([x]_R) = \sup\{\overline{\mu_F}(y)|\ y \in [x]_R\}$.

Or equivalently: $\mu_{qaprR}(F)(x) = \inf\{\max\{\mu_F(y), \mu_R(x, y)\}|\ y \in U\}$ and $\mu_{\overline{qprR}}(F)(x) = \sup\{\min\{\overline{\mu_F}(y), 1 - \mu_R(x, y)\}|\ y \in U\}$.

These are the definitions for rough fuzzy sets.

On the other hand, a fuzzy rough set is characterized by a crisp set and two fuzzy sets:

- The reference set: $A \subseteq U$,
- The lower approximation: $\mu_{\underline{apr}}(A)(x) = \inf\{1 - \mu_{\Re}(x, y)|\ yA\}$.
- The upper approximation: $\mu_{\overline{apr\Re}}(A)(x) = \sup\{\mu_{\Re}(x, y)|\ y \in A\}$.

Here $\Re$ is a fuzzy similarity relation.

Let us note that the definitions above use the Zadeh max–min logic system.

3 Compensatory Fuzzy Rough Predicates

Our approach to rough sets and fuzzy sets is different from the classic definitions for fuzzy rough set and rough fuzzy sets. We don't fuzzify rough sets neither convert fuzzy sets into rough sets, but our goal is to revisit the general idea of rough sets like an inspiration for creating two types of predicates. These predicates approximate the knowledge of certain tables of data.

This method is a hybridization of rough sets and fuzzy predicates with a semantic meaning. Because of the use of CFL and the advantages explained above, it is possible to predict and infer with the method.

In this section, we define the concepts used in this paper. Besides, the application of these concepts to Knowledge Discovery by the method created by us is exposed.

Definition 1 *A CFL-Cluster is a disjunctive predicate of the CFL, where every element in the disjunction is called a class.*

A special type of CFL-Cluster is a predicate in the conjunctive normal form, and in this case, the classes are all the conjunction elements.

Definition 2 *A fuzzy rough predicate is a pair ($C_1(x)$,$C_2(x)$) of two CFL-Clusters, where the set of classes into $C_1(x)$ belongs the set of classes into $C_2(x)$.*

For example, if there are three condition attributes in the decision table, let us call them I, GIP, P (see example of Sect. 5), according to Definition 2 $C_1(\mathbf{x})$ can be the proposition $C_1(\mathbf{x}) = I(x_1) \vee GIP(x_2) \vee P(x_3) \vee (I(x_1) \wedge GIP(x_2))$ which is a normal form and classes I, GIP, P, and I∧GIP, with certain membership functions for every attribute.

Then, $C_2(\mathbf{x})$ repeats the same classes of $C_1(\mathbf{x})$ with the same membership functions, and its disjunction with other or the same classes with other memberships functions, for example, $C_2(\mathbf{x}) = I(x_1) \vee GIP(x_2) \vee P(x_3) \vee (I(x_1) \wedge GIP(x_2)) \vee (I(x_1) \wedge P(x_2))$, let us note that we added the class $I(x_1) \wedge P(x_2)$. Here x_i are the values of the object for the attribute i. This pair ($C_1(\mathbf{x})$, $C_2(\mathbf{x})$) is a fuzzy rough predicate.

Definition 3 *A fuzzy rough predicate is said to be a λ-approximation of the fuzzy predicate A(y) if the universal propositions according to formula (12) and(13)have at least the truth value λ, calculated with a CFL.*

$$\mathrm{LA}(\boldsymbol{x}, \boldsymbol{y}) = \forall(\boldsymbol{x}, \boldsymbol{y}) \forall i \; C_{1\mathrm{i}}(\boldsymbol{x}) \rightarrow A_i(\boldsymbol{y}) \tag{12}$$

$$\mathrm{UA}(\boldsymbol{x}, \boldsymbol{y}) = \forall(\boldsymbol{x}, \boldsymbol{y}) \forall i \; A_i(\boldsymbol{y}) \rightarrow C_{1\mathrm{i}}(\boldsymbol{x}) \tag{13}$$

x is the notation for the vector of the condition attributes in Table 1, **y** is the name of the decision attribute, and i is the index of the object in Table 1.

Continuing the example for Definition 2, if D is the decision attribute, then, according to formulas (12) and (13), two predicates for LA and UA are the universal quantifier applied for every value of, $I(x_1) \vee GIP(x_2) \vee P(x_3) \vee (I(x_1) \wedge GIP(x_2)) \rightarrow D(y)$ and $D(\mathbf{y}) \rightarrow I(x_1) \vee GIP(x_2) \vee P(x_3) \vee (I(x_1) \wedge GIP(x_2)) \vee (I(x_1) \wedge P(x_2))$, respectively.

We impose that LA and UA take at least the value λ.**y** is the notation for the values of the attribute D in the decision table. Hence, in case that the truth values of LA and UA be at least equal to λ then, the proposed pair ($C_1(\mathbf{x})$, $C_2(\mathbf{x})$) of this example will be the fuzzy rough predicate λ-approximation of the fuzzy predicate D(**y**).

The λ-approximation is established in the following definition:

Definition 4 *A value $y_0$of y is a λ_1-λ_2-approximation of the fuzzy rough predicate ($C_1(x)$,$C_2(x)$)from the decision table for some values ofx_0, if$y_0$maximizes the expressions ($LA(x_0,y) \leftrightarrow \lambda_1) \wedge (UA(x_0,y) \leftrightarrow \lambda_2$); where LA and UA are the formulas (12) and(13)respectively.*

Below, we offer a description of a method of Knowledge Discovery using fuzzy rough predicates.

First at all, we define the Multistate Membership Function (MMF). Let us recall that the sigmoidal membership function equipped with parameters β and γ sigm(x, [b, g]) is defined as follows:

$$sigm(x, [\beta, \gamma]) = \frac{1}{1 + e^{-\alpha(x-\gamma)}} \tag{14}$$

where $\alpha = \frac{\ln(0.99) - \ln(0.01)}{\gamma - \beta}$, let us note that sigm $(\gamma,[\beta,\gamma]) = 0.5$, which means 'as true as false' or 'indifference' and sigm $(\beta,[\beta,\gamma]) = 0.01$, which means 'almost false'.

Definition 5 *The Multistate Membership Function (MMF) is defined as*

$$MMF(x, [\beta, \gamma, m]) = \frac{sigm(x, [\beta, \gamma])^m \cdot sigm(x, [\gamma, \beta])^{(1-m)}}{M} \tag{15}$$

where, $m \in [0,1]$ and $M = \begin{cases} m^m \cdot (1-m)^{(1-m)}, & \text{if } m \in (0,1) \\ 1, & \text{otherwise} \end{cases}$.

It is easy to prove that (15) satisfies the conditions $0 \le \mathrm{MMF}(x, [\beta, \gamma]) \le 1$.

If $m = 1$, (15) is a sigmoidal membership function, if $m = 0$, it is the negation of the sigmoidal membership function. (15) generalizes the notion of the state of the attribute. When it is a sigmoidal membership function, it can be interpreted linguistically as "the value of the attribute is high", see Fig. 1. When it is the negation of the sigmoidal membership function, it can be interpreted linguistically as "the value of the attribute is low", see Fig. 2. There are other intermediate cases where the state is near or far from those two extreme cases depending on the values of the parameter m. Especially if the parameter m assumes $\frac{1}{2}$, then the interpretation can be "the value of the attribute is medium", see Fig. 3.

Therefore, there is a continuum of linguistic interpretations of the possible states of the attributes, and they can be defined by the values of the parameters β, γ and m.

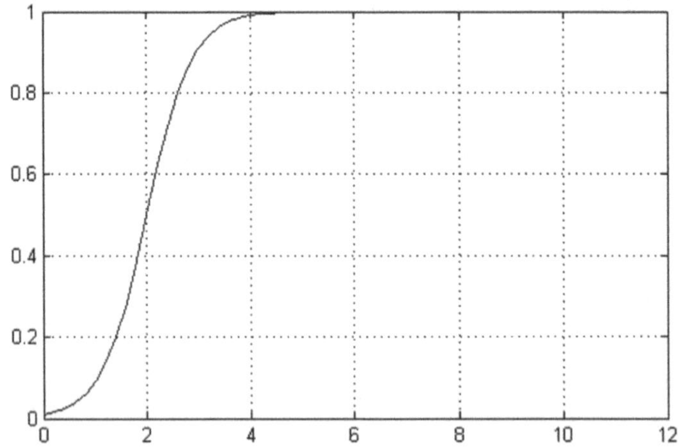

Fig. 1 Multistate membership function with parameters $\beta = 0$, $\gamma = 2$ and $m = 1$, a sigmoidal membership function. It can be linguistically understood as: "the value of the attribute is high"

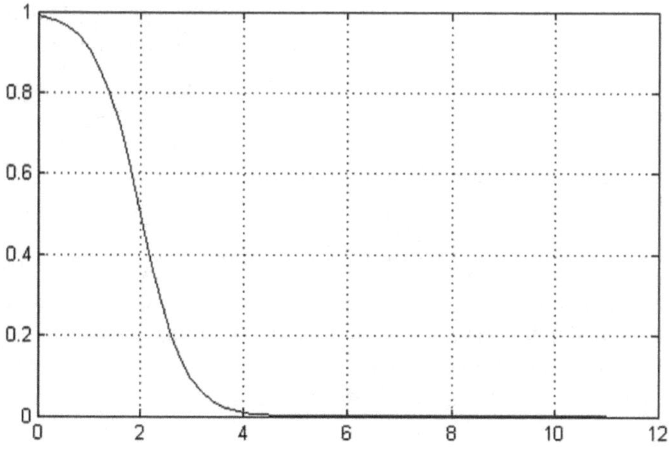

Fig. 2 Multistate membership function with parameters $\beta = 0$, $\gamma = 2$, and $m = 0$, a negation of the sigmoidal membership function. It can be linguistically understood as: "the value of the attribute is low"

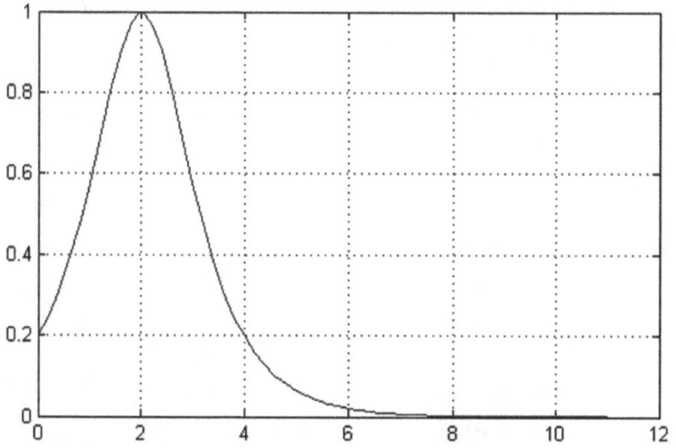

Fig. 3 Multistate membership function with parameters $\beta = 0$, $\gamma = 2$ and $m = 1/2$, an intermediate membership function, nor a sigmoidal and nor a negation of the sigmoidal. It can be linguistically understood as: "the value of the attribute is medium"

4 Knowledge Discovery

Now we describe a new method for Knowledge Discovery based on fuzzy rough predicates.

The characteristics of our method are the following:

1. It's basically a curve fitting method.

2. The method fits the three parameters of each MMF (see Eq. 15), where every MMF represents one linguistic value for one of the linguistic variables. Therefore, the results of our method are membership functions with a meaning in the natural language.
   That is possible because of the flexibility of the MMF.
   The parameters are estimated from the measure data.
3. The objective function is a logic predicate for preserving the semantic meaning. It is based on the predicates in Eq. (12–13).
4. The user selects the level of exactitude and fuzziness that he/she prefers, taking into account that more is exactly the result less it is fuzzy and vice versa.
   The value of λ in the λ-approximation controls the fuzziness.
   If $\lambda = 1$, the result is unique and crisp and if $\lambda = 0$, the result is completely fuzzy. These two extremes values of λ are not recommendable.
5. It's possible to control the number of variables the user wants to use to prevent the impracticability of the method, computationally speaking.
6. Every algorithm of optimization can be used to fit the parameters. In our example, we use the genetic algorithm method of optimization. Genetic algorithms are popular metaheuristic approaches because of their efficiency in addressing complex real-world problems [17–19].

The algorithmic details of the method are the following:

1. **Input elements**:

- Decision table (see Table 1), where the condition attributes (n attributes) are well differentiated from the decision attribute.
- One decision attribute which is fuzzified previously with a MMF with certain parameters β, γ and m.
- N is the maximum number of elements allowed in the classes, $N \leq n$.

For example, I, GIP, P are the condition attributes, and D is the decision attribute used to exemplify the concepts in Sect. 3. In this way, we used three classes; the maximum of number of elements per classes was 2, $I(x_1) \wedge P(x_2)$. Hence, $N = 2$.

This restriction prevents to create a too complex algorithm in the calculus.

2. **Elements of the algorithm**:

The CFL-Cluster is the disjunction of classes, which each of them has the following structure:

- A set of n classes, where each of them is formed by the MMFs of the N condition elements and parameters β, γ and m, to be estimated.
- Another set of n(n-1)/2 classes, each of them formed by all the possible 2-elements conjunctions with MMF into the n attributes.

Also, the parameters β, γ and m associated with every attribute in the classes have to be estimated.

- One third set of $n(n-1)(n-2)/6$ classes formed by all the possible 3-elements conjunctions with MMF into the n attributes.

The parameters β, γ and m have to be estimated.

- So on, until the number of elements of conjunctions is formed by every N-elements conjunctions with MMF into the n attributes.

One simple example is the set of two generic condition attributes, A and B, where each of them has one MMF as a membership function, and the parameters β, γ and m are going to be estimated; thus, one class is a MMF (A, [β_{A1}, γ_{A1}, m_{A1}]), the second one is MMF(B, [β_{B1}, γ_{B1}, m_{B1}]), where 1 means that this is the first set. And also, a third class: MMF (A, [β_{A2}, γ_{A2}, m_{A2}]) ∧ MMF (B, [β_{B2}, γ_{B2}, m_{B2}]).

All the twelve parameters (β_{A1}, γ_{A1}, m_{A1}, β_{B1}, γ_{B1}, m_{B1}, β_{A2}, γ_{A2}, m_{A2}, β_{B2}, γ_{B2} and m_{B2}) have to be estimated.

Finally, the CFL-Cluster is formed by all these classes.

In the example above, we used the classes of only one element I, GIP and P, and the class with two elements I∧P. Because N = 2, we have to add two more classes $I(x_1) \wedge GIP(x_2)$ and $GIP(x_2) \wedge P(x_3)$ for the disjunction.

We should establish the parameter and the membership function for D(y).

3. **Objective functions**:
- Firstly, for estimating the low approximation, see (12),

$$\forall i \; C_{1i}(\boldsymbol{x}) \wedge LA(\boldsymbol{x}, \boldsymbol{y}) \tag{16}$$

where i is the index for the set of objects in the decision table, **x** is the vector of the values of the condition attributes, and y is the value of the decision attribute in the table. Let call this *OF1* (*First Objective Function*).

- Secondly, for estimating the upper approximation Eq. (13) is used. Let call this *OF2* (*Second Objective Function*).

Let us note that the approximation is completed on the space of parameters of the membership functions of every condition attribute which appeared in LA and UA.

4. **Problems**
- Problem1: Maximize the truth-value of the OF1 into the space of parameters.
- Problem2: Maximize the truth-value of the OF2 into the space of parameters.

5. **Optimization**

Every algorithm which can be used for optimization could be considered here.

6. **Selection of the final classes**
- Every class, such that when it is removed from the predicate (16), then the truth value of 12 increases, is eliminated from the Eq. (12).
- This is $C_1(\mathbf{x})$ standing in Definition 2.
- Also, every class, such that, when it is removed from the predicate 13, then the truth value of the equation increases is eliminated from the Eq. (13).
- Finally, the classes of $C_1(\mathbf{x})$ are added to the classes obtained in the point above, and then we obtain $C_2(\mathbf{x})$.

This allows to improve the approximation and to reduce the number of attributes necessary for doing that.

Hence, we obtain one fuzzy rough predicate. Let us note that this is a way to reduce the number of attributes relevant for describing the problem; that is to say, the method act as a reducer of the number of attributes necessary for solving the problem.

7. **Prediction**

Let us use a $\lambda_1 - \lambda_2$-approximation of the fuzzy rough predicate obtained ($C_1(\mathbf{x})$, $C_2(\mathbf{x})$) (see Definition 4), and the method doesn't establish the optimization tool that should be used and then we will predict the value of the decision attribute.

5 An Illustrative Example

We shall detail step by step, every aspect of the method applied to this example as an illustration about how to use it.

Let us study the example of the Mexican economy [20], which we introduced above.

- Input elements:

There are three condition attributes: Inflation (I), Gross Internal Product (GIP), and Parity peso/dollar (P). The decision attribute is the Demand (D).

Decision Table 2 visualizes the results of these attributes for seven consecutive years.

Let us consider the initial hypothesis from an expert criterion for fuzzification that 'a high demand in the Mexican economy' is obtained by a sigmoidal membership function with parameters $\beta = 25{,}000$ and $\gamma = 32{,}000$, or equally the MMF with parameters $\beta = 25{,}000$, $\gamma = 32{,}000$ and $m = 1$.

Table 2 Mexican economy data for seven consecutive years

Year/Attribute	I	GIP	P	D
1	11.01	2	2.3	32,350
2	7.06	4.5	4.8	31,305
3	52	−6.2	5.6	28,083
4	27.7	5.2	6.5	33,408
5	15.7	7	6.95	44,987
6	18.8	4.1	10.05	54,344
7	17.2	2.9	11.4	56,830

- Elements of the algorithm:

The cluster that should be used according to the method for approximation is I∨GIP∨P∨(I∧GIP)∨(I∧P)∨(GIP∧P)∨(I∧P∧GIP), it is formed as the disjunction of all possible conjunctions of the three condition attributes.

Let us note that the first group of classes mentioned in the algorithm is formed by the attributes, I, GIP and P, the second group is formed by I∧GIP, I∧P, and GIP∧P; and the third is formed only by I∧P∧GIP.

Here n = 3 and N = 3, according to the notation used in the algorithm. Every attribute in the predicate is fuzzified with Multistate Membership Functions with their parameters.

We will use the GMBCL, see formulas (7) and (8); and the natural implication, $i(x, y) = d(n(x), y)$.

- Objective function 1 (OF1):

Let us apply formula 16 or OF1 to the decision Table 2, where the parameters of the MMF have to be estimated.

- Problem1:

Genetic Algorithm coded in MATLAB was used for the optimization, and the results were: The truth-value for the OF1 is: 0.6735, the parameters estimated are shown in Table 3.

The truth-value of LA resulting from (12) is 0.5300.

Table 3 Parameters estimated for the attributes by every class for the lower approximation

Class	Parameters		
	β	γ	m
I	47.45	5.51	0.05
GIP	0.0086	10.97	0.228
P	16.99	3.73	0.151
I∧GIP	59.71	5.44	0.060
	0.036	10.95	0.194
I∧P	39.24	5.092	0.0817
	25.52	4.881	0.0302
GIP∧P	0.026	10.86	0.1827
	19.354	1.944	0.0846
I∧GIP∧P	58.0029	5.628	0.0469
	0.0421	10.73	0.1443
	19.18	3.229	0.0765

Table 4 Parameters estimated for the attributes by every class for the upper approximation

Class	Parameters		
	β	γ	m
I	59.53	5.628	0.0239
GIP	0.3920	7.789	0.0961
P	29.811	3.354	0.0157
I∧GIP	53.0678	8.317	0.0240
	0.1783	9.851	0.2073
I∧P	58.1393	7.053	0.0221
	26.3424	6.230	0.0119
GIP∧P	0.0740	4.521	0.0067
	29.5052	1.041	0.0231
I∧GIP∧P	55.6834	5.713	0.0265
	0.0095	9.863	0.1932
	29.1546	3.103	0.0375

If we maintain just the classes GIP, I∧GIP, and GIP∧P, then, the truth-value of LA increases to 0.5523. Hence, to reduce the number of classes, we should use the CFL-Cluster GIP∨(I∧GIP)∨(GIP∧P) for prediction.

- Objective function 2 (OF2):

On the other hand, we must estimate the UA of formula (13).

- Problem2:

The aim of problem 2 is to estimate the true value of UA. The truth-value estimated is 0.9723, and the results of the parameters estimation is resumed in Table 4.

- Optimization:

In both cases, for estimating LA and UA, we used genetic algorithm coded in MATLAB.

- Selection of final classes:

Even though the truth-value obtained for the UA is big, it can be improved by eliminating some classes. The truth-value is 0.9805 when the classes of the UA are only I, GIP, and P. Therefore, the CFL-Cluster I∨GIP∨P is used for the prediction.

The fuzzy rough predicate is formed by $C_1(\mathbf{x})$ and $C_2(\mathbf{x})$, where, $C_1(\mathbf{x})$ = GIP∨(I∧GIP)∨(GIP∧P), with their corresponding parameters in Table 2, and $C_2(x)$ = $C_1(\mathbf{x})$∨I∨GIP∨P and their corresponding parameters in Table 4.

The pair ($C_1(\mathbf{x})$, $C_2(\mathbf{x})$) is a fuzzy rough predicate, according to Definition 3.

Therefore, there are two logical conditions for the proposition 'high demand'; one is the sufficient condition $C_1(\mathbf{x})$, and the other one is the necessary condition $C_2(\mathbf{x})$. Both of them can be expressed in the natural language.

For expressing $C_1(\mathbf{x})$ and $C_2(\mathbf{x})$ in natural language, the first step is to assign linguistic values by experts to every attribute forming $C_1(\mathbf{x})$ and $C_2(\mathbf{x})$, according to the parameters calculated and summarized in Tables 3and 4.

The second step is to express every linguistic value of the attributes according to the operators AND and OR appeared in $C_1(\mathbf{x})$ and $C_2(\mathbf{x})$.

We shall illustrate this part for $C_1(\mathbf{x})$:

The MMFs for GIP as a class, GIP as member of the class $I \wedge GIP$ and GIP as member of the class $GIP \wedge P$, see Table 3, are plotted in Figs. 4, 5 and 6, respectively; and the graphs o f Iand P are plotted in Figs. 7 and 8, respectively.

Evidently, the GIP in the three figures above can be expressed in words with the statement "the GIP is very high".

Also, the I showed in Fig. 7 can be expressed as: "the inflation is very high".

And the P showed in Fig. 8 can be expressed as: "the parity is medium".

Nevertheless, this step should be defined by experts.

Then, we can say the knowledge in natural language in the following way: IF GIP is very high OR GIP is very high AND inflation is very high OR GIP is very high AND parity Peso/dollar is medium THEN Demand is high. Let us note that this statement makes sense.

This interpretation in natural language is an advantage to express the sufficient and necessary conditions for decision making in a more comprehensible manner to experts and users.

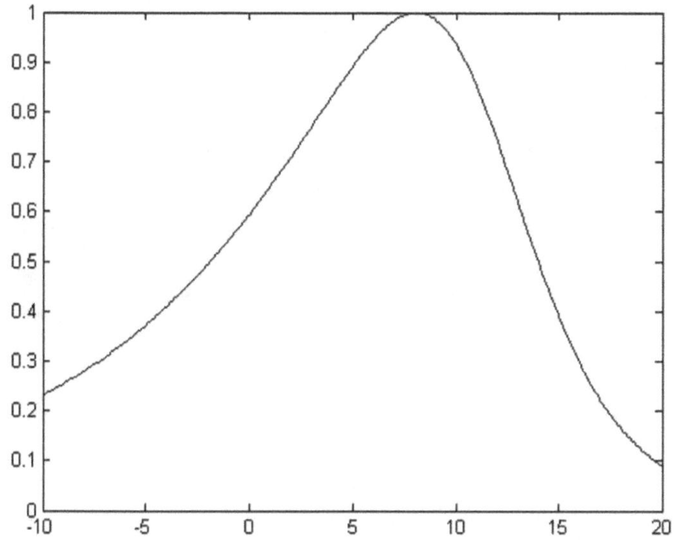

Fig. 4 Graph of the MMF corresponding to GIP as a single class in C1(x)

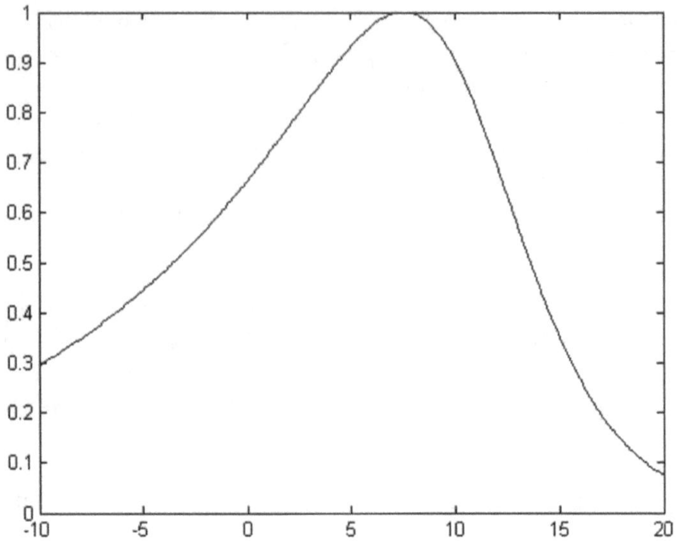

Fig. 5 Graph of the MMF corresponding to GIP as member of the class I∧GIP in C1(x)

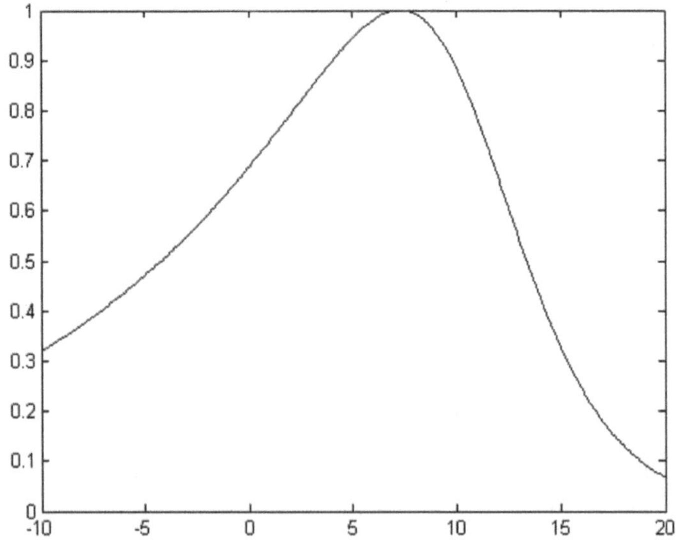

Fig. 6 Graph of the MMF corresponding to GIP as member of the class GIP∧P in C1(x)

Therefore, decision makers can take measures to improve the results of decision attributes. In this example, the decision maker can act over condition attributes to increase the Demand.

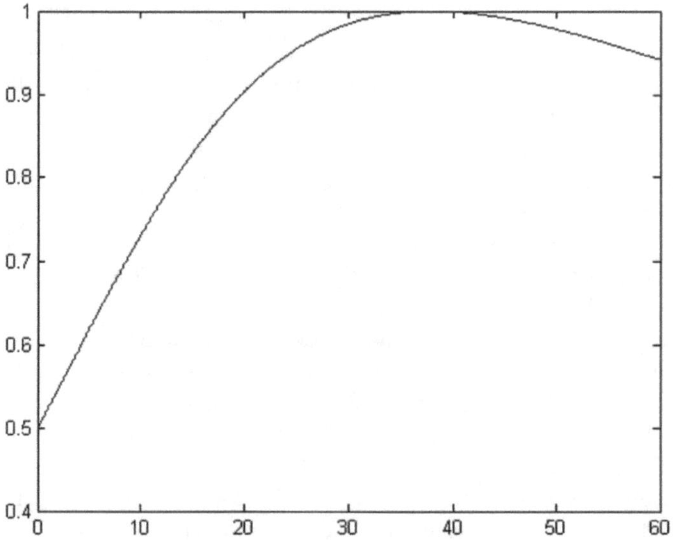

Fig. 7 Graph of the MMF corresponding to I as member of the class I∧GIP in C1(x)

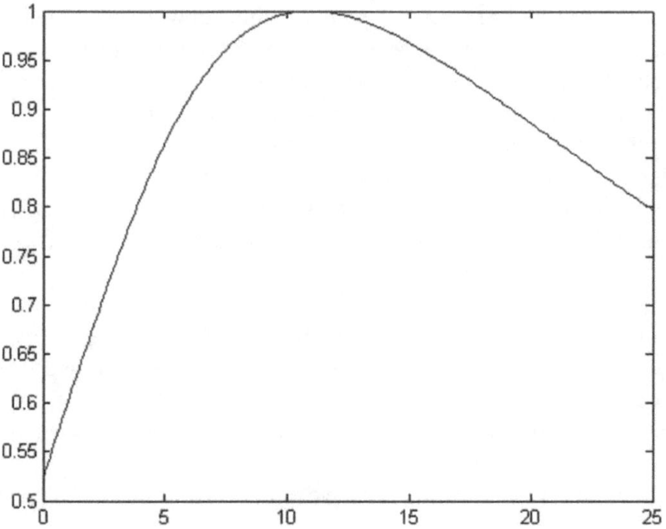

Fig. 8 Graph of the MMF corresponding to P as member of the class GIP∧P in C1(x)

This is the main result that we propose by this method, i.e., for every decision attribute associated with a linguistic value, it is possible to obtain necessary and sufficient conditions expressed in the natural language.

Besides, the method allows forecasting as we do below:

- Prediction

For predicting, we use Definition 4, and we have to find the value y_0 of the demand. If for a certain year in the Mexican Economy the values of Inflation, GIP and Parity are: 3.97, 1.2 and 7.99 respectively, then for predicting the Demand, we must use the expression $(LA(x_0, \mathbf{y}) \leftrightarrow \lambda_1) \wedge (UA(x_0, \mathbf{y}) \leftrightarrow \lambda_2)$, for $\lambda_1 = 0.5523$ and $\lambda_2 = 0.5523 \vee 0.9805 = 0.9066$.

Applying the Genetic Algorithm of MATLAB, we obtain that 57,031.446 is a λ_1-λ_2-approximation of the fuzzy rough predicate $(C_1(\mathbf{x}), C_2(\mathbf{x}))$ for the predicate ´high demand´. The truth-value for this estimation is 0.6368.

Because we used a CFL, the predicates can be interpreted by using the natural language according to the expert sense of the concepts. For example, the truth-values of lambda 0.5523 and 0.9066, in spite of they were calculated from complex predicates, they can be understood as 'more true than false' and 'almost true', respectively. See [15] for more detail about the meaning of interpretability.

Comparing the solution of the problem in Weka, the well-known Environment for Knowledge Analysis, by using the linear regression, the solution is the linear equation $D = -271.9445 * I + 3391.6905 * P + 22{,}930.0107$, where the attribute GIP was eliminated. The Relative absolute error is 28.3931%.

The predicted value for the Demand by using regression is approximately equal to 48,950. Its relative error respects the value estimated with our method is equal to 16.51%. Therefore, both estimations are enough closed each other.

This method is a calculation with numbers and not with words; hence it doesn't establish qualitative relationships between the attributes, only that there is an inverse relationship between Demand and inflation, also, that there is a direct relationship between the tax of exchange Peso/Dollar and the Demand.

The method of regression needs of enough large quantity of cases. Let us note it is not strange that this problem has just a few numbers of cases. This is another disadvantage over our method.

One more closed approach to our method is the decision tree of Weka. Let us call 'low demand' if the Demand datum evaluated in the membership function is in the interval [0, 0.4), 'middle demand' if the interval is [0.4, 0.6] and finally, 'high demand' if the interval is (0.6, 1]. Hence, Table 2 becomes in the following:

For calculating truth values of Demand in Table 5, we used (15) with parameters $\beta = 25{,}000$, $\gamma = 32{,}000$ and $m = 1$.

For resolving the problem with Weka, we ran the algorithm J48, and we obtained the following decision tree's rules:

If $P < \ = 5.6$ then Demand is low (3.0/1.0). Here three cases over four were classified correctly.

Else if $P > 5.6$ then high (4.0). Here four cases over four were classified correctly.

For the prediction by using a decision tree, where $P = 7.99$, we obtain that the Demand is high. By evaluating 57,031.446 for the membership function of ´high demand´ obtained with our method, the truth value is 1; hence, the results for both methods are similar.

Table 5 Mexican economy data for seven consecutive years, where the data to be predicted is nominal

Year/Attribute	I	GIP	P	D (truth value)
1	11.01	2	2.3	0.5572 or 'middle'
2	7.06	4.5	4.8	0.3879 or 'low'
3	52	−6.2	5.6	0.0710 or 'low'
4	27.7	5.2	6.5	0.7159 or 'high'
5	15.7	7	6.95	0.9998 or 'high'
6	18.8	4.1	10.05	1.0000 or 'high'
7	17.2	2.9	11.4	1.0000 or 'high'

Let us note that with this solution, the attributes I and GIP were not included. The values of the True Positive (values well classified per class) were for 'middle demand' a 0%, and for 'high demand' and 'low demand' 100%.

It is important to point out that with this method, a bigger number of cases is necessary for obtaining better results, too. It is a more qualitative method of classification than regression; here, we have a division in a finite number of cases to analyze; it is n't a relationship between continuous variables.

But the division in two crisp intervals is not fuzzy, and hence, we can't say what these disjoint intervals mean in natural language, especially the value $P = 5.6$, which represents a very disrupt and unnatural way to pass from one extreme of Demand to another. See that the case of 'middle demand' is not taken into account.

6 Concluding Remarks

In this study, we introduced and discussed the concept of fuzzy rough predicates. The proposed definition of the two fuzzy predicates was by the Lower and Upper approximations present in the theory of rough sets. These two approximations form the fuzzy rough predicate, which supports prediction.

The definition was used to establish the method of Knowledge Discovery. The importance of the approach is at least two-fold. First, it helps us predict the decision attribute from a decision table data. Secondly, it facilitates to reduce the variables used in the prediction process.

It is worth stressing that the approach is sound as being built on the basis of fuzzy logic and the axiomatic framework of normative decision making. Hence the use of CFL becomes advantageous in modeling problems expressed with the use of natural language.

The novelty of the method into the family of Knowledge Discovery methods is its interpretability in the natural language, that is to say, every logical proposition obtained and the results of its calculus can be expressed in a phrase of the natural language. This method is part of one more general approach to Knowledge Discovery.

The importance of the method is that it can help to decision makers to make decisions in the comprehensible manner of communication in natural language. Also, it is a compliment, and not a substitute, of other methods for forecasting, in the sense that its results are more useful in the way of linguistic values than in numerical values.

References

1. Zadeh, L.A.: Fuzzy sets. Inform. Control **8**, 338–353 (1965). https://doi.org/10.2307/2272014
2. Pawlak, Z.: Rough sets. Int. J. Comput. Inform. Sci. **11**(5), 341–356 (1982). https://doi.org/10.1007/BF01001956
3. Pawlak, Z.: Rough Sets: Theoretical Aspects of Reasoning About Data. Kluwer, Springer Science and Business Media (1991)
4. Zadeh, L.: The concept of a Linguistic variable and its application to approximate reasoning-I. Inf. Sci. **8**(3), 199–249 (1975). https://doi.org/10.1016/0020-0255(75)90036-5
5. Zadeh, L.: From computing with numbers to computing with words-from manipulation of measure to manipulation of perceptions. Int. J. Appl. Math. Comput. Sci. **3**, 307–324 (2002). https://doi.org/10.1007/978-3-7908-1792-8_5
6. Dubois, D., Prade, H.: Rough fuzzy sets and fuzzy rough sets. Int. J. Gen Syst. **17**(2–3), 191–209 (1990). https://doi.org/10.1080/03081079008935107
7. Agarwal, M., Palpanas, T.: Linguistic rough set. Int. J. Mach. Learn. Cybern. **7**, 953–966 (2014). https://doi.org/10.1007/s13042-014-0297-2
8. Ahmadi, F., Maghooli, K.: Missing data analysis: a survey on the effect of different K-means clustering algorithms. Am. J. Signal Process. **4**(3), 65–70 (2014)
9. Chen, M., Ludwig, S.: Fuzzy decision tree using soft discretization and a genetic algorithm based feature selection method. 2013 World Congress on Nature and Biologically Inspired Computing, pp. 238–244. IEEE, Fargo (2013). https://doi.org/10.1109/NaBIC.2013.6617869
10. Liang, D., Liu, D., Pedrycz, W., Hu, P.: Triangular fuzzy decision-theoretic rough sets. Int. J. Approximate Reasoning **54**(8), 1087–1106 (2013). https://doi.org/10.1016/j.ijar.2013.03.014
11. Moewes, C., Kruse, R.: Evolutionary fuzzy rules for ordinal binary classification with monotonicity constraints. In: Soft Computing: State of the Art Theory, vol. 291. Springer, Heidelberg (2013). https://doi.org/10.1007/978-3-642-34922-5_8
12. Slowinski, R., Greco, S., Matarazzo, B.: Rough sets in decision making. In: Meyers R. (eds.). In Computational Complexity. Springer, New York (2012). https://doi.org/10.1007/978-1-4614-1800-9_168
13. Espín, R., González, E., Fernández, E., Martínez, M.: Compensatory fuzzy logic inference. In: Soft Computing for Business Intelligence, pp. 25–43. Springer, Heidelberg (2014)
14. Martínez, M., Espín, R., López, V., Rosete, A.: Discovering knowledge by fuzzy predicates in compensatory fuzzy logic using metaheuristic algorithms. In: Soft Computing for Business Intelligence, pp. 161–174. Springer, Heidelberg (2014). https://doi.org/10.1007/978-3-642-53737-0_11
15. Espín, R., Fernández, E., González, E.: Compensatory fuzzy logic: a frame for reasoning and modeling preference knowledge in intelligent systems. In: Espín, R., Pérez, R., Cobo, A., Marx, J., Valdés, A. (eds.). Soft Computing for Business Intelligence, pp. 3–23. Springer, Heidelberg (2014). https://doi.org/10.1007/978-3-642-53737-0_1
16. Yao, Y.: Combination of rough and fuzzy sets based on α-level sets. In: Rough Sets and Data Mining: Analysis for Imprecise Data, pp. 301–321. Springer, Boston (1997). https://doi.org/10.1007/978-1-4613-1461-5_15
17. Rivera, G., Cisneros, L., Sánchez-Solís, P., Rangel-Valdez, N., Rodas-Osollo, J.: Genetic algorithm for scheduling optimization considering heterogeneous containers: a real-world case study. Axioms **9**(1), 27 (2020). https://doi.org/10.3390/axioms9010027

18. Alvarado, O., Castro, B., González, L., Rivera, G., Rodas-Osollo, J., Sánchez-Solís, J.: Metaheuristic-based optimization of treated water distribution in a Mexican City. Aplicaciones Recientes en la Investigación de Operaciones. Pp. 19–30. Universidad Autónoma de Coahuila, Coahuila (2020)
19. Rivera, G., Rodas-Osollo, J., Bañuelos, P., Quiroz, M., Lopez, M.: A genetic algorithm for surgery scheduling optimization in a Mexican public hospital. Recent advances in artificial intelligence research and development. In: Aguiló, I., Alquézar, R., Angulo, C., Ortiz, A. (eds.) Frontiers in Artificial Intelligence and Applications, vol. 300, pp. 269–274. IOS Press, Amsterdam (2017). https://doi.org/10.3233/978-1-61499-806-8-269
20. Rosete, A., Ceruto, T., Espín, R. Marx, J.: A general method for knowledge discovery using compensatory fuzzy logic and metaheuristics. In: Espín, R., Marx, J., Racet, A. (eds.). Towards a Transdisciplinary Technology for Business Intelligence, pp. 240–268. Shaker (2011)

Nonparametric Tests for Comparing Forecasting Models

Dmitriy Klyushin

Abstract The creation of computer systems for analyzing and forecasting business processes involves choosing the effective tools for assessing the quality of the results obtained and comparing predictive models. Direct comparison of the predicted values with the observed ones leaves out of sight the statistical nature of the analyzed values. Therefore, before estimating the accuracy of the models statistical tests must be applied to rank models by the quality and test the homogeneity of the errors. To solve this problem, nonparametric methods are widely used; in particular, the Wilcoxon signed-rank test. The chapter contains a short survey of nonparametric tests used in business analysis and proposes an alternative nonparametric statistical test that is highly robust, sensitive, and specific. This test is based on comparing the a priori known probability of an event with its observed relative frequency. The known probability is determined according to Hill's assumption, and to compare it with the observed relative frequency, one of the confidence intervals for the binomial proportion of success in the Bernoulli scheme is used. The results of computer modeling and comparison with the Wilcoxon signed-rank text both in theoretical and practical contexts are presented.

Keywords Business analytics · Predictive analytics · Time series · Forecasting · Ranking · Nonparametric test

1 Introduction

The traditional hierarchical scheme of business analytics consists of four levels: (1) descriptive, (2) diagnostic, (3) predictive, and (4) prescriptive analytics.

At first, the lowest level, business data are collected and investigated using descriptive data analytics methods to discover what happened. The input data at this level are, for example, indicators of technological processes in industries or stock exchange data in financial markets. By analyzing such data, we can, for example, identify

D. Klyushin (✉)
Department of Computer Science and Cybernetics, Taras Shevchenko National University of Kyiv, Kyiv 03680, Ukraine

W. Pedrycz et al. (eds.), *Computational Intelligence for Business Analytics*, Studies in Computational Intelligence 953,
https://doi.org/10.1007/978-3-030-73819-8_12

moments of failure of technological equipment and points of a trend change in the stock market.

At the second, diagnostic level, information is analyzed to discover reasons for happened events. At this level, the toolkit becomes more complex because the goal is to find the causes of the observed phenomena, classify them, and discover patterns. Data collected and analyzed at this level allows you to identify internal relationships between data and discover cause-effect relationships between them. For example, it is possible to find the correlation between two time series and find out their relationship.

At the third, predictive level, using data obtained at the first two levels, we forecast events that may occur in the future to determine what can happen. The fundamental complexity of forecasting requires the use of very complex tools of mathematical statistics, machine learning, data mining, and simulation. For example, predictive analytics can be used to predict future stock price in the stock market or determine the optimal time for repairs to prevent the breakdown of process equipment.

At the fourth, the highest level, with the help of prescriptive analytics methods, complex problems are solved that allows finding the optimal decision. Here, besides methods applied at the third level, effective methods of optimization and optimal control, as well as methods of decision-making theory are used. For example, a stock market broker can find the optimal solution for a deal on the stock market and determine the most appropriate moment to make a transaction.

The subject of this chapter is data analysis techniques at the predictive analytics level that do not place any special demands on the data, except for the most natural ones (e.g., do not assume that the data follow a known distribution). Such methods include methods of nonparametric statistics, in particular, methods for testing the hypothesis of samples homogeneity.

Predictive analytics has broad promising applications in commercial, financial, and industrial processes. In the field of trade, the main areas of application of predictive analytics are direct marketing based on customer base segmentation, targeted advertising based on the classification of customers according to their preferences, and customer retention by analyzing the peculiarities of their behavior patterns and recommendation services that suggest products to customers based on their previous views or comments in social networks. In the field of credit and insurance, the current areas of application for predictive analytics are credit scoring and fraud detection by recognizing nonstandard patterns of behavior of clients of banks and insurance companies. In the financial sector, predictive analytics enables investment risk management by assessing investment prospects.

In industry, predictive analytics allows solving problems of analyzing and predicting the quality of manufactured products, predicting equipment failures and planning an optimal repair schedule, predicting production volume and resource consumption, and recognizing non-standard situations in time. In complex high-tech enterprises with a high degree of automation of the technological process, it is necessary to automatically monitor and optimize operational processes. To do this, it is necessary to establish an automated collection and analysis of indicators characterizing the state of equipment and production as a whole. This allows you to optimize the production process by predicting the number of consumed resources in order to avoid unexpected shortages, as well as create an optimal schedule for equipment maintenance and repair. Continuous analysis of big production data avoids disruptions and failures, optimizes costs, and improves product quality.

To solve the problems described above, complex software systems are being intensively developed that implement modern methods of mathematical statistics, data mining, machine learning, and artificial intelligence. All the listed areas of knowledge imply a widespread application of data analysis methods, including those that do not put forward unrealistic data requirements. Such methods, in particular, include nonparametric hypothesis testing methods.

The motivation of this chapter is to survey nonparametric methods used in predictive analytics and describe a new approach to test homogeneity of two samples in the context of estimation of forecasting effectiveness of predictive models.

The chapter is organized in the following way. In Sect. 2, we provide a survey of nonparametric tests for homogeneity. Subsection 2.1 describes purely nonparametric tests, and Subsect. 2.2 is devoted to conditionally nonparametric tests. Section 3 contains a description of the nonparametric tests used in predictive business analytics. Section 4 describes a new test for estimation effectiveness and ranking forecasting models. Section 5 describes the statistical properties of the Klyushin–Petunin test and Wilcoxon signed-rank test. Section 6 contains conclusions and describes possible directions for the development of new tests.

2 Nonparametric Tests for Homogeneity of Samples

Let $x = (x_1, x_2, \ldots, x_n)$ and $y = (y_1, y_2, \ldots, y_m)$ be two samples drawn from general populations G_1 and G_2 following absolutely continuous distribution functions $F_1(x)$ and $F_2(x)$ respectively. Using the information contained in the samples x and y it is necessary to construct nonparametric tests for testing the null hypothesis $F_1(x) = F_2(x)$ against the alternative hypothesis $F_1(x) \neq F_2(x)$.

Criteria for testing such hypotheses can be classified in different ways. In particular, they can be classified according to the principle of verification: permutation tests, rank tests, randomization tests, and distance tests. In addition, these tests are divided into general criteria that are valid against any pair of alternatives, such as Kolmogorov–Smirnov test [1, 2] and tests that are valid against pairs of different alternatives from a certain class [3–8], etc. Given that recently there have been works that apply mixed principles of hypothesis testing (for example, the Dufour test that uses permutations and distances simultaneously), it is appropriate to generalize the classification and divide these criteria into two large groups, which we will call purely nonparametric and conditionally nonparametric criteria.

The first group includes criteria for testing the hypothesis of equivalence of general populations, which are nonparametric in nature, regardless of whether the distribution is continuous or discrete, and the second includes those that, under certain conditions depend on the distribution.

1. Purely nonparametric tests [3–8] etc.
2. Conditionally nonparametric tests.

2.1. Goodness of fit tests: Kolmogorov–Smirnov [1, 2] and Cramer–von Mises [9–14].
2.2. L_1–, L_2– and L_∞–distance tests [15].
2.3. Tests using difference between moments [16–18].
2.4. Test using simulation (bootstrap and Monte Carlo) [19–21]).

2.1 Purely Nonparametric Tests

The nonparametric tests are score tests (Wilcoxon signed-rank test, van der Waerden normal score test, logistic score test, Cauchy score test), and Fisher randomization test. As far as we will compare our proposal with the Wilcoxon signed-rank test, let us consider this test more detail.

Wilcoxon proposed a test for a case when samples have the same size [6]. An extended variant of these tests in which the samples may have unequal sizes was developed by Mann and Whitney [7]. Let $x = (x_1, x_2, \ldots, x_n)$ and $y = (y_1, y_2, \ldots, y_m)$ be two samples drawn from the general populations G_1 and G_2 following absolutely continuous distribution functions $F_1(x)$ and $F_2(x)$ respectively. Introduce a random variable:

$$z_{ij} = \begin{cases} 1 \text{ if } x_i > y_j, \\ 0 \text{ otherwise.} \end{cases} \quad (1)$$

The Wilcoxon statistics is

$$U = \sum_{i=1}^{n} \sum_{j=1}^{m} z_{ij} \quad (2)$$

If null hypothesis does not contradict to the data, then all the values z_{ij} have the same probability, i.e., the average value of $z_{ij}z_{ij}$ equals to 1/2 and

$$U = \frac{1}{2}nm \quad (3)$$

According to the Wilcoxon-Mann–Whitney test (1)–(3), the null hypothesis is rejected if $U \geq U_\beta$ (one-sided case), or $nm - U \geq U_\beta$ (two-sided case), where U_β is chosen in such in such way that the null hypothesis is true only if the number of permutations satisfying the condition $U > U_\beta$ does not exceed $\beta n!$ The Wilcoxon and Mann–Whitney tests are consistent not for every possible alternative. Also, the power of these tests is close to the power of Student test, which is most powerful for normal distributions. If distributions are not normal, the Wilcoxon and Mann–Whitney tests are more powerful [22]. However, the Wilcoxon and Mann–Whitney tests do not require anything from a hypothetical distribution function except being continuous.

Tanizaki [23] shown that the score tests for testing hypotheses on the homogeneity of the distributions based on a shift of location may be described by the following general scheme. $x = (x_1, x_2, \ldots, x_n)$ and $y = (y_1, y_2, \ldots, y_m)$ be two samples drawn from general populations G_1 and G_2 following absolutely continuous distribution functions $F_1(x)$ and $F_2(x)$ respectively. Mix x and y constructing a new sample z having the size n. There are C_{n+m}^n possible variants. All score tests and the Fisher test entails the comparison z from the one side and x and y from the other side. In the score tests, the samples x and y are arranged, and we have two variance series $x = (x_1, x_2, \ldots, x_n)$ and $y = (y_1, y_2, \ldots, y_m)$. A general test statistics is,

$$s_0 = \sum_{i=1}^{n} a\left(x_{(i)}\right) \tag{4}$$

where the function $a(x)$ depends on the particular test.

For every possible permutation among C_{n+m}^n variants, we compute the sum of ranks s_m, $m = 1,\ 2,\ \ldots,\ C_{n+m}^n$. At least one of these statistics equals to s_0 (4), and others may be less or more. Introduce the following notations: L is the number of statistics s_0 that are equal to s_0, N is the number of statistics s_m which are less that s_0, and M is the number of statistics s_m that is more than s_0. Thus, $L + N + M = C_{n+m}^n$. The hypothesis testing is reduced to the estimation of the probabilities

$$P(s < s_0) = \frac{N}{C_{n+m}^n},\ P(s = s_0) = \frac{L}{C_{n+m}^n},\ P(s > s_0) = \frac{M}{C_{n+m}^n} \tag{5}$$

where s is a random value chosen among the statistics s_m.

If the probability $P(s < s_0)$ is small, then the hypothesis that $F_1(x) < F_2(x)$ is plausible. If the probability $P(s > s_0)$ is small, then the hypothesis that $F_1(x) > F_2(x)$ is plausible. Thus, we may formulate the following rule of decision making based on the probabilities (5): let α be the significance level, then the null hypothesis is rejected if $P(s < s_0) \leq \alpha$ or $P(s > s_0) \leq \alpha$.

Using Tanizaki scheme, we may write the Wilcoxon and Mann–Whitney tests as,

$$a\left(x_{(i)}\right) = x_{(i)},\ w_0 = \sum_{i=1}^{n} x_{(i)} \tag{6}$$

As Tanizaki noted, (6) is not only a possible variant. In particular, as the Wilcoxon statistics, we may choose,

$$w_1 = \frac{w_0 - Mw}{\sqrt{Dw}} \tag{7}$$

where w is a random value following normal distribution, $Mw = \frac{n(n+m+1)}{2}$ is its mathematical expectation, and $Dw = \frac{nm(n+m+1)}{12}$ is its variance. The statistics (7) is called the normalized Wilcoxon statistics.

The function $a(x)$ and the test statistics of the normal score test are the following:

$$a(x) = \Phi^{-1}\left(\frac{x}{n+m+1}\right), \; ns_0 = \sum_{i=1}^{n} \Phi^{-1}\left(\frac{x_{(i)}}{n+m+1}\right) \tag{8}$$

where Φ is the standard normal distribution. It was shown [24] that the power of this test for normal distributions is equivalent to the power of the Student t-test, and in other cases, it exceeds the power of the Student t-test.

The function $a(x)$ and the test statistics of the logistic score test are the following:

$$a(x) = F^{-1}(x), \quad ls_0 = \sum_{i=1}^{n} F^{-1}\left(\frac{x_{(i)}}{n+m+1}\right) \tag{9}$$

where $F(x) = \frac{1}{1+e^{-x}}$ is the logistic distribution function.

The function $a(x)$ and the test statistics for the Cauchy score test are the following:

$$a(x) = F^{-1}(x), \quad ls_0 = \sum_{i=1}^{n} F^{-1}\left(\frac{x_{(i)}}{n+m+1}\right) \tag{10}$$

where $F(x) = \frac{1}{2} + \frac{1}{\pi} tg^{-1} x$ is the Cauchy distribution function.

The functions (8)–(10) allow considering the score tests as a variant of a general test and using a general approach to their investigation.

2.2 Conditionally Nonparametric Tests

In this subsection, we consider so called conditionally nonparametric tests, i.e. tests, which imply additional conditions on the underlying distribution function.

The Kolmogorov–Smirnov test uses the statistics proposed in [1, 2]:

$$D_{n,m} = \sup_x \left| \tilde{F}_{1,n}(x) - \tilde{F}_{2,m}(x) \right| \tag{11}$$

where $\tilde{F}_{1,n}(x)$ and $\tilde{F}_{2,m}(x)$ are empirical distribution functions constructed on the samples $x = (x_1, x_2, \ldots, x_n)$ and $y = (y_1, y_2, \ldots, y_m)$. Due to simplicity, this test is very popular, but it has some drawbacks. At first, if the functions $F_1(x)$ and $F_2(x)$ are continuous, the test does not depend on an underlying distribution, but the limit distribution of the statistics (11) is not standard [25, 26]. Second, for discrete general populations, the Smirnov test is not a nonparametric one [27].

The Cramer–von Mises statistics has the following form:

$$W_{n,m}^2 = \frac{mn}{(m+n)^2}\left\{\sum_{i=1}^{n}\left[\tilde{F}_{1,n}(x_i) - \tilde{G}_{1,m}(x_i)\right]^2 + \sum_{j=1}^{m}\left[\tilde{F}_{1,n}(y_j) - \tilde{G}_{1,m}(y_j)\right]^2\right\}. \tag{12}$$

The statistics (12) is the weighted Kolmogorov–Smirnov statistics [27]; thus it has the same drawbacks whish were considered in [14, 28–30].

The tests using distances between kernel estimations of the probability densities in the spaces L_1, L_2 and L_∞ were investigated in [20]. Introduce the probability density functions f_1 and f_2 corresponding to the distribution functions $F_1(x)$ and $F_2(x)$.

Allen considered the following kernel estimations of a probability density:

$$f_{1,n}(x) = \frac{C_X}{n}\sum_{i=1}^{n} K[C_X(x - x_i)],$$
$$f_{1,m}(y) = \frac{C_Y}{m}\sum_{j=1}^{m} K[C_Y(y - y_j)] \tag{13}$$

where $C_X = \frac{n^{1/5}}{2s_X}$, $C_Y = \frac{m^{1/5}}{2s_Y}$, $K(x) = \begin{cases}\frac{1}{2}, \text{ if } |x| \le 1,\\ 0, \text{ if } |x| > 1,\end{cases}$, $s_X = \sqrt{\frac{\sum_{i=1}^{n}(x_i - \bar{x})^2}{n-1}}$.

If $s_X = 0$, then $C_X = 1$, and the function (13) is the relative frequency of x. The estimation of $f_{2,m}(x)$ is constructed similarly.

The tests based on the L_1-, L_2- and $L_\infty-$ distances use the following statistics:

$$\hat{L}_1 = \sum_{i=1}^{n} |f_n(x_i) - g_m(x_i)| + \sum_{j=1}^{m} |f_n(y_j) - g_m(y_j)| \tag{14}$$

$$\hat{L}_2 = \left\{\sum_{i=1}^{n} [f_n(x_i) - g_m(x_i)]^2 + \sum_{j=1}^{m} [f_n(y_j) - g_m(y_j)]^2\right\}^{1/2} \tag{15}$$

$$\hat{L}_\infty = \max_{1\le i\le n,\ 1\le j\le m}\left\{|f_{1,n}(x_i) - f_{2,m}(x_i)|, |f_{1,n}(y_j) - f_{2,m}(y_j)|\right\}. \tag{16}$$

In contrast to the Kolmogorov–Smirnov test and the Cramer–von Mises test, which are nonparametric when the distribution functions $F_1(x)$ and $F_2(x)$ are continuous, the Allen tests (14)–(16) are nonparametric only if the samples are infinite. If the samples x and y are finite, these tests depend on distributions.

The third group of tests uses the difference between the sample average values:

$$\hat{\theta}_1 = \overline{x} - \overline{y}. \tag{17}$$

The tests based on the statistics (17) where developed in [15, 16, 31], and afterwards were extended in [21]. In particular, this work describes the tests based on the statistics

$$\hat{t} = \frac{(\overline{x} - \overline{y}) \Big/ \sqrt{\frac{1}{n} + \frac{1}{m}}}{\left\{ \left[\sum_{i=1}^{n} (x_i - \overline{x})^2 + \sum_{j=1}^{m} (y_j - \overline{y})^2 \right] \Big/ (n + m - 2) \right\}^{1/2}} \tag{18}$$

It was shown [32] that the tests based on the statistics (17) and (18) are equivalent. These arguments allow the development of these tests using a comparison of the moments o f ahigher order.

$$\hat{\theta}_2 = \left| \frac{1}{n-1} \sum_{i=1}^{n} (x_i - \overline{x})^2 - \frac{1}{m-1} \sum_{j=1}^{m} (y_j - \overline{y})^2 \right| \tag{19}$$

$$\hat{\theta}_3 = \left| \frac{1}{n} \sum_{i=1}^{n} \left(\frac{x_i - \overline{x}}{s_X} \right)^3 - \frac{1}{m} \sum_{j=1}^{m} \left(\frac{y_j - \overline{y}}{s_Y} \right)^3 \right| \tag{20}$$

$$\hat{\theta}_4 = \left| \frac{1}{n} \sum_{i=1}^{n} \left(\frac{x_i - \overline{x}}{s_X} \right)^4 - \frac{1}{m} \sum_{j=1}^{m} \left(\frac{y_j - \overline{y}}{s_Y} \right)^4 \right| \tag{21}$$

where $s_X^2 = \frac{1}{n} \sum_{i=1}^{n} (x_i - \overline{x})^2$, $s_{X'}^2 = \frac{1}{m} \sum_{i=1}^{m} (y_i - \overline{y})^2$, and if $s_X = 0$ then $\frac{x_i - \overline{x}}{s_X} = 0$.

If samples x and y are finite, then all the tests based on the statistics (17)–(21), except the Dwass test, depend on the distributions, i.e., they are not nonparametric. To eliminate this dependence from distributions, it was proposed [33–35] to use arbitrary permutations of a joint sample $x_1, x_2, \ldots, x_n, y_1, y_2, \ldots, y_m$. As far as these permutations are equiprobable, the test will reject the null hypothesis basing on the critical value obtained on the conditional distribution under the given order statistics of the joint sample, and it may be considered as nonparametric [36].

The Dufour scheme is organized as follows. Let T be a pivot statistics of the test, i.e., statistics that do not depend on unknown parameters. Denote as T_0 a test statistics computed on an observable sample. The null hypothesis is rejected under the large values of T_0 and the critical region corresponding to the significance level α may be described by the inequality $G(T_0) \leq \alpha$ where $G(x) = P(T \geq x | H_0)$ is an observable confidence level. Let us generate N independent samples $\left(x_1^{(i)}, x_2^{(i)}, \ldots, x_n^{(i)}, y_1^{(i)}, y_2^{(i)}, \ldots, y_m^{(i)} \right)$, $i = 1, 2, \ldots, N$ drawn from the general population with the distribution function F_0 and compute N independent statistics $T^{(i)} = T\left(x_1^{(i)}, x_2^{(i)}, ..., x_n^{(i)}, y_1^{(i)}, y_2^{(i)}, ..., y_m^{(i)}\right)$. Then compute an empirical significance level:

$$\hat{p}_N(x) = \frac{N\hat{G}_N(x) + 1}{N + 1} \tag{22}$$

where

$$\hat{G}_N(x) = \frac{1}{N}\sum_{i=1}^{N} I(x)\left(T^{(i)} - x\right),$$
$$I(x) = \begin{cases} 1 \text{ if } x \geq 0, \\ 0 \text{ otherwise.} \end{cases} \tag{23}$$

Using (22), (23), the critical region corresponding to the Monte Carlo test is defined by the inequality,

$$\hat{p}_N(T_0) \leq \alpha \tag{24}$$

where $\hat{p}_N(T_0)$ is an estimation of $G(T_0)$. It was shown [33, 35] that if the statistics T follows a continuous distribution, then,

$$P\left(\hat{p}_N(T_0)\middle|H_0\right) \leq \frac{\lfloor \alpha(N + 1)\rfloor}{N + 1}, \quad 0 \leq \alpha \leq 1. \tag{25}$$

Thus, if we chose N in (25) such that the number $\alpha(N + 1)$ is integer, then the significance level of the Monte Carlo test does not exceed α.

As we see, purely and conditionally nonparametric tests have some problems: (1) testing of hypotheses often are tied with considerable computational complexities, (2) properties of the test has, as a rule, an asymptotic behavior, and (3) no all tests are consistent under all possible alternatives.

3 Nonparametric Tests Used in Business Analysis

One of the most challenging tasks for the forecasting analyst is choosing the best forecasting method. This choice is influenced by many factors, including both the properties of the initial data and the features of the applied model, in particular, its simplicity and accuracy. Naturally, the overwhelming majority of the main quality criterion of the model is its accuracy.

To assess the accuracy of the model, the standard error (MSE), mean absolute percentage error (MAPE), and mean absolute deviation (MAD) are commonly used. These indicators give an idea of the quality of the model and allow you to make the best choice. However, on the way to the final choice, the analyst must take into account not only the values that characterize the accuracy, but also answer the question of whether there is a statistically significant difference between them. The choice of the optimal model should be made not only on the basis of the accuracy of each of the models but also on their statistical significance. It is quite possible that the models are ranked by accuracy, but there is no statistically significant difference between them. In this case, each of the models can be considered suitable, and the difference between their levels of accuracy can be explained by random fluctuations.

When analyzing the quality of forecasting models, it is necessary to assess the homogeneity of the samples of forecast errors and the corresponding indicators of accuracy (MAPE, MSE, or MAP). The standard assumption regarding model errors is that they are stationary, unbiased, and homogeneous, i.e., belonging to one general population. For example, it is often assumed that the errors are normally distributed. If the assumption on the normality of distribution has no theoretical or practical basis, it is advisable to check whether the error and accuracy samples of different models belong to the same population.

The general scheme of testing equality of forecast accuracy is described in [37]. Suppose, we have forecasting models $M_j, j = 1, ..., m$ generating predictions $x_i^{(j)}$ of the time series $x_i,\ i = 1, \ldots, n$. Let $\varepsilon_i^{(j)},\ i = 1, \ldots, n; j = 1, \ldots, m$ be forecasting errors of the model $\mathrm{M_j}.\mathrm{M_j}$. Introduce a loss function $g\left(\varepsilon_i^{(l)}\right)$, for example, one of the accuracy measures: deviation or mean square error. Null hypothesis about equal accuracy of models M_k and M_l is equivalent to the hypothesis that the mathematical expectation of $d_i^{(k,l)} = g\left(\varepsilon_i^{(k)}\right) - g\left(\varepsilon_i^{(l)}\right)$ is equal to zero, i.e., $E\left(g\left(\varepsilon_i^{(k)}\right)\right) = E\left(g\left(\varepsilon_i^{(l)}\right)\right)$. When as a loss function, we use the deviation $\varepsilon_i^{(k)} - \varepsilon_i^{(l)}$, the problem may be reduced to testing the hypothesis that $E\left(\varepsilon_i^{(k)}\right) = E\left(\varepsilon_i^{(l)}\right)$.

To test this hypothesis, the variety of tests is used. These tests are divided on tests for pairwise ([6, 37, 38] etc.) and tests for multiple comparisons ([39–44] etc.). They are very effective for testing the hypothesis on distribution location shift. An extensive and detail survey may be found in [41, 45, 46]. These methods are widely used for comparing predictions in econometrics (e.g., [38, 47–53]).

Note, that the hypothesis $E\left(\varepsilon_i^{(k)}\right) = E\left(\varepsilon_i^{(l)}\right)$ is an only partial case of the general hypothesis that accuracy measures follow the same distribution. That is why it is reasonable to consider this general hypothesis and to propose tests for its verifying.

We proposed the alternative method that effectively tests not only the hypothesis of distribution location shift but also the hypothesis about the equivalence of distributions when the mathematical expectation is zero, but the standard deviations are different [50]. It tests whether distributions of forecasts generated by two forecasting models are the same. A similar idea but grounded on the Kolmogorov–Smirnov statistics, is described in [38].

As a rule, the Wilcoxon test is used for this. Samples for the Wilcoxon signed-rank test must be of the same size. The null hypothesis is that the median of the differences between the values from the two samples is zero.

Numerical experiments show that the Wilcoxon signed-rank test is effective for testing a hypothesis about a location shift but is inappropriate when samples have the same location and different variances. Thus, there is an acute necessity in a universal test for samples heterogeneity that would be effective both for both shift and scale hypotheses. This test is described in [54].

4 The Klyushin–Petunin Test for Samples Homogeneity

The Hill's assumption $A_{(n)}A_{(n)}$ [55] states that for exchangeable random values $x_1, x_2, \ldots, x_n \in G$ with an absolutely continuous distribution, the following equality holds:

$$P\left(x \in \left(x_{(i)}, x_{(j)}\right)\right) = \frac{j-i}{n+1}, \tag{26}$$

where $j > i$, $x \in G$ is a sample value, and $\left(x_{(i)}, x_{(j)}\right)$ is an interval formed by i-th and j-th order statistics. Using (26), a nonparametric test for detecting the homogeneity of samples without ties was developed [54]. This test estimates the homogeneity of samples $x_1, x_2, ..., x_n$ and $y_1, y_2, ..., y_n$. Suppose that $F_1 = F_2$ and construct the variational series$x_{(1)}, x_{(2)}, ..., x_{(n)}$. Denote as $A_{ij}^{(k)}$ an event $\left\{y_k \in \left(x_{(i)}, x_{(j)}\right)\right\}$. If$j > i$,, then

$$P\left(y_k \in \left(x_{(i)}, x_{(j)}\right)\right) = p_{ij} = \frac{j-i}{n+1}. \tag{27}$$

Let us construct the confidence interval for binomial proportion in Bernoulli trials for an unknown probability of the event $A_{ij}^{(k)}$ (27). For simplicity, we have chosen the Wilson interval, but any confidence interval for binomial proportion in Bernoulli trials may be used here):

$$p_{ij}^{(1)} = \frac{h_{ij}^{(n,k)} n + \frac{g^2}{2} - g\sqrt{h_{ij}^{(n,k)}\left(1 - h_{ij}^{(n,k)}\right)n + \frac{g^2}{4}}}{n + g^2},$$

$$p_{ij}^{(2)} = \frac{h_{ij}^{(n,k)} n + \frac{g^2}{2} + g\sqrt{h_{ij}^{(n,k)}\left(1 - h_{ij}^{(n,k)}\right)n + \frac{g^2}{4}}}{n + g^2}, \tag{28}$$

where $h_{ij}^{(n,k)}$ is the relative frequency of the event $A_{ij}^{(k)}$ in n trials. Then, construct the confidence interval $I_{ij}^{(n)} = \left(p_{ij}^{(1)}, p_{ij}^{(2)}\right)$ with a significance level depended on the parameter g. If g is equal to 3 then the significance level of $I_{ij}^{(n)}$ is less than 0.05 [54]. Let B be an event $\left\{p_{ij} = \frac{j-i}{n+1} \in I_{ij}^{(n)}\right\}$. Put $N = (n-1)n/2$ and find the frequency

L of the event B. The value $h = L/N$ is a homogeneity measure of samples x and y, which we shall call *p-statistics*.

Let us put $h_{ij}^{(n)} = h$, $n = N$ and $g = 3$, and construct the Wilson confidence interval (28) $I_n = (p_1, p_2)$ for $p(B)$. The confidence intervals $I_{ij}^{(n)} = \left(p_{ij}^{(1)}, p_{ij}^{2}\right)$ and $I = (p_1, p_2)$ are called intervals based on the 3 s-rule. The scheme of trials, where the events $A_{ij}^{(k)}$ can arise when the hypothesis that distributions are identical is true, called the generalized Bernoulli scheme [56–58]. If the hypothesis is false, this scheme is called the modified Bernoulli scheme. In the general case, when the null hypothesis can be either true or false, the trial scheme is called MP-scheme (Matveichuk–Petunin scheme) [58]. If $F_1 = F_2$, $\lim\limits_{n\to\infty} \frac{j-i}{n+1} \in (0, 1)$, and $\lim\limits_{n\to\infty} \frac{i}{n+1} \in (0, 1)$, then the asymptotic significance level β of a sequence of confidence intervals $I_{ij}^{(n)}$ based on the 3 s-rule, is less than 0.05 [54].

Let $B_1, B_2, \ldots$ be a sequence of events that may arise in a random experiment E, E, $\lim\limits_{n\to\infty} p(B_k) = p^*$, $h_{n_1}(B_1), h_{n_2}(B_2)$, ... be a sequence of relative frequencies of the events$B_1, B_2, \ldots$, respectively, and $\frac{k}{n_k} \to 0$ as $k \to \infty$. We shall call an experiment E a *strong random experiment* if $h_{n_k}(B_k) \to p^*$ as $k \to \infty$. In a strong random experiment, the asymptotical significance level of the Wilson confidence interval I_n when $g = 3$ is less than 0.05. The test for the null hypothesis $F_1 = F_2$ with a significance level which is less than 0.05 is the following: *if I_n contains 0.95, the null hypothesis is accepted; else, the null hypothesis is rejected.*

The theoretical investigation of the power of this test and the optimal sample size is quite complicated. Using (26), we can say that the desired sample size is $n \geq 39$ because, in this case, the probability (36) is more than 0.95Equation '(26)' is given in the list but not cited in the text. Please provide the respective equation in the list or delete citation.

5 Numerical Experiment and Forecast Model Ranking Using the Klyushin–Petunin and the Wilcoxon Signed-Rank Tests

For comparing the sensitivity and specificity of the tests, we propose empirical evidences obtained in the numerical experiments using samples of different sizes from a normal distribution with various parameters. We have generated 30 pairs of samples containing 10, 20, 30, 40, and 100 random numbers with the same mean and different standard deviations and with different means and the same standard deviation. Then we applied both tests to estimate their sensitivity and specificity empirically. The sensitivity of the Klyushin–Petunin test is the relative frequency of the event when the upper confidence bound for the p-statistics is less than 0.95 for different distributions. The sensitivity of the Wilcoxon signed-rank test is the relative frequency of the event when *p-value* $\leq$0.05 for different distributions. The specificity of the Klyushin–Petunin test is the relative frequency of the event when the upper confidence bound for the p-statistics is more than 0.95 for identical distributions. The specificity of the Wilcoxon signed-rank test is the relative frequency of the event when p-$value \geq 0.05$ for identical distributions. The results are provided in Tables 1, 2 and 3.

As we see, when the samples are drawn from the distributions N(0,1) and N(0.5,1), the sensitivity of the Klyushin–Petunin test and the Wilcoxon signed-rank tests depend on the sample size (see Table 1). Both tests fail for the small sample (n = 10) and have a high sensitivity when the sample size is more than 40, but the sensitivity of the Klyushin–Petunin tests exceeds the sensitivity of the Wilcoxon signed-rank test.

When the compared samples are drawn from the distributions N(0,1) and N(0,2), the Klyushin–Petunin test is effective when the sample size is more than 40 (see Table 2). Meantime, the Wilcoxon signed-rank test fails in all cases. This effect may be explained by the fact that the Wilcoxon signed-rank test verifies the hypothesis about equality of means, but the Klyushin–Petunin test verifies the general hypothesis about the equivalence of the distribution functions.

Table 1 Sensitivity of the Klyushin–Petunin and the Wilcoxon signed-rank tests for the distributions N(0,1) and N(0.5,1)

Size of sample	Klyushin-Petunin	Wilcoxon
10	0.00	0.10
20	0.53	0.50
30	0.73	0.50
40	0.93	0.60
100	1.00	0.90

Table 2 Sensitivity of the Klyushin–Petunin and the Wilcoxon signed-rank tests for the distributions N(0,1) and N(0,2)

Size of sample	Klyushin-Petunin	Wilcoxon
10	0.03	0.100
20	0.30	0.200
30	0.67	0.100
40	0.80	0.100
100	1.00	0.100

Table 3 Specificity of the Klyushin–Petunin and the Wilcoxon signed-rank tests for the distribution N(0,1)

Size of sample	Klyushin-Petunin	Wilcoxon
10	0.97	1.00
20	0.90	1.00
30	0.90	0.97
40	0.90	1.00
100	0.96	1.00

For estimation of specificity of the Klyushin–Petunin test and the Wilcoxon signed-rank test, we again used 30 pairs of samples containing 10, 20, 30, 40, and 100 random numbers from the normal distribution N(0,1) (see Table 3). In contrast to the previous experiments, now the Klyushin–Petunin test demonstrates stable results in every range of the samples sizes, and its specificity is comparable with the specificity of the Wilcoxon signed-rank test in every range of the sample sizes.

Thus, we may conclude that the Klyushin–Petunin test has high sensitivity when $n \geq 40$ and high specificity for all samples sizes and all possible variants of means and standard deviations. The Wilcoxon signed-rank test is valid for testing the hypothesis about distribution location shift but is invalid when different distributions have the same location and different standard deviations (hypothesis of distribution scale).

To illustrate the practical usefulness of the Klyushin–Petunin test, let us consider the dataset from [59]. This paper considers the forecasting accuracy of ARIMA and artificial neural networks model at the example of Dell stock index collected during 23 days from New York Stock Exchange. The authors state the superiority of neural networks model over ARIMA model according to the relative forecast errors (Table 4). Using the Klyushin–Petunin test, we may conclude that these two models produce the errors that are not statistically different because the upper bound of the confidence interval for the p-statistics (0.96) is more than 0.95. Thus these models may be considered as statistically equivalent.

Table 4 Sample results of ANN and ARIMA models for Dell stock index [59]

Forecast error			Forecast error		
Time	ARIMA	ANN	Time	ARIMA	ANN
01.03.2010	0.0302	0.0162	18.03.2010	0.0206	−0.0110
02.03.2010	0.0095	0.0161	19.03.2010	0.0090	−0.0278
03.03.2010	0.0007	0.0110	22.03.2010	−0.0034	−0.0198
04.03.2010	0.0088	0.0000	23.03.2010	0.0019	0.01380
05.03.2010	0.0252	0.0072	24.03.2010	0.0220	−0.0080
08.03.2010	0.0086	0.0093	25.03.2010	−0.006	−0.0229
09.03.2010	0.0183	0.0134	26.03.2010	0.0160	−0.02135
10.03.2010	0.0325	0.0154	29.03.2010	0.0047	−0.0294
11.03.2010	0.0021	0.0007	30.03.2010	−0.003	−0.0354
12.03.2010	−0.0035	−0.0028	31.03.2010	0.0033	−0.0386
15.03.2010	0.0077	−0.0098			
16.03.2010	−0.0133	−0.0147			
17.03.2010	0.0062	−0.0014			

6 Conclusions and Directions for Future Work

Naïve comparison of the predicted values using error measures ignores the stochastic nature of these values. To solve this problem, nonparametric methods are widely used; in particular, the Wilcoxon signed-rank test. We proposed a new test that is effective for comparing forecast models when samples size is more than 40. In a strong random experiment, the asymptotical significance level of this test in less than 0.05. The Klyushin–Petunin test is more universal than the Wilcoxon test, allows arranging the pairs of samples by homogeneity measure, and is easily computed. Application to the practical example demonstrates the usefulness of the proposed test. The future work will be focused on the comparisons of the proposed test with other tests used in this field and on the theoretical properties of the p-statistics for the samples with unequal sizes.

References

1. Smirnov, N.V.: Estimate of difference between empirical distribution curves in two independent samples. Byull. Mosk. Gos. Univ. **2**(2), 3–14 (1939)
2. Smirnov, N.V.: On the deviations of an empirical distribution curve. Mat Sb **6**(1), 3–26 (1939)
3. Dixon, W.G.: A criterion for testing the hypothesis that two samples are from the same population. Ann. Math. Stat. **11**(2), 199–204 (1940). https://doi.org/10.1214/aoms/1177731914
4. Wald, A., Wolfowitz, J.: On a test whether two samples are from the same population. Ann. Math. Stat. **11**(2), 147–162 (1940). https://doi.org/10.1214/aoms/1177731909
5. Mathisen, H.C.: A method of testing the hypothesis that two samples are from the same population. Ann. Math. Stat. **14**(2), 188–194 (1943). https://doi.org/10.1214/aoms/1177731460
6. Wilcoxon, F.: Individual comparisons by ranking methods. Biometrics Bull. **1**(6), 80–83 (1945). https://doi.org/10.2307/3001968
7. Mann, H.B., Whitney, D.R.: On a test of whether one of the random variables is stochastically larger than other. Ann. Math. Stat. **18**(1), 50–60 (1947). https://doi.org/10.1214/aoms/1177730491
8. Wilks, S.S.: A combinatorial test for the problem of two samples from continuous distributions. Proc. Fourth Berkeley Symp. Math. Stat. Prob. **1**, 707–717 (1961)
9. Cramér, H.: On the composition of elementary errors. Scand. Actuar. J. **1**, 13–74 (1928). https://doi.org/10.1080/03461238.1928.10416862
10. von Mises, R.E.: Wahrscheinlichkeit Statistik und Wahrheit. Julius Springer, Wien (1928)
11. Pitman, E.J.G.: Significance tests which may be applied to samples from any populations. Supp. J. R Stat. Soc. B **4**(1), 119–130 (1937). https://doi.org/10.2307/2984124
12. Lehmann, E.L.: Consistency and unbiasedness of certain nonparametric tests. Ann. Math. Stat. **22**(2), 165–179 (1951). https://doi.org/10.1214/aoms/1177729639
13. Rosenblatt, M.: Limit theorems associated with variants of the von Mises statistics. Ann. Math. Stat. **23**(4), 617–623 (1952). https://doi.org/10.1214/aoms/1177729341
14. Anderson, T.W.: On the distribution of the two-sample cramer–von Mises criterion. Ann. Math. Stat. **33**(3), 1148–1159 (1962). https://doi.org/10.1214/aoms/1177704477
15. Dwass, M.: Modified randomization tests for nonparametric hypotheses. Ann. Math. Stat. **28**(1), 181–187 (1957). https://doi.org/10.1214/aoms/1177707045
16. Fisz, M.: On a result be M. Rosenblatt concerning the Mises–Smirnov test. Ann. Math. Stat. **31**(2), 427–429 (1960). https://doi.org/10.1214/aoms/1177705905

17. Barnard, G.A.: Comment on “The spectral analysis of point processes” by M.S. Bartlett. J. R Stat. Soc. B. **25**, 294 (1963)
18. Birnbaum, Z.W.: Computers and unconventional test-statistics. In: Prochan, F., Serfling, R.J. (eds). Reliability and Biometry. SIAM, Philadelphia (1974)
19. Jockel, K.H.: Finite sample properties and asymptotic efficiency of monte carlo tests. Ann. Stat. **14**(1), 336–347 (1986). https://doi.org/10.1214/aos/1176349860
20. Allen, D.L.: Hypothesis testing using L1-distance bootstrap. Am. Stat. **51**(2), 145–150 (1997). https://doi.org/10.1080/00031305.1997.10473949
21. Efron, B., Tibshirani, R.J.: An Introduction to the Bootstrap, of Monographs on Statistics and Applied Probability, vol. 57. Chapman-Hall, New York (1993)
22. Hodges, J.L.J., Lehman, E.L.: The efficiency of some nonparametric competitors of the test. Ann. Math. Stat. **27**(2), 324–335 (1956). https://doi.org/10.1214/aoms/1177728261
23. Tanizaki, H.: Power comparison of nonparametric tests: small-sample properties of Monte Carlo experiments. J. App. Stat. **24**(5), 603–632 (1997). https://doi.org/10.1080/026647697 23576
24. Chernoff, H., Savage, R.I.: Asymptotic normality and efficiency of certain nonpara-metric test statistics. Ann. Math. Stat. **29**(4), 972–994 (1958). https://doi.org/10.1214/aoms/1177706436
25. Massey, F.J.J.: The distribution of the maximum deviation between two sample cumulative step functions. Ann. Math. Stat. **22**(1), 125–128 (1951). https://doi.org/10.1214/aoms/1177729703
26. Gnedenko, B.V., Korolyuk, V.S.: The maximal discrepancy between two empirical distributions. Dokl. Akad. Nauk SSSR **80**(4), 525–528 (1951)
27. Walsh, J.E.: Bounded probability properties for Kolmogorov-Smirnov and similar statistics for discrete data. Ann. Inst. Stat. Math. **15**, 53–158 (1963). https://doi.org/10.1007/bf02865912
28. Burr, E.J.: Distribution of the two-sample Cramer–von Mises criterion for small equal samples. Ann. Math. Stat. **34**(1), 95–101 (1963). https://doi.org/10.1214/aoms/1177704245
29. Burr, E.J.: Small samples distributions of the two-sample CramEr–von Mises’ W2 and Watson’s U2. Ann. Math. Stat. **34**(3), 1091–1098 (1964). https://doi.org/10.1214/aoms/1177703267
30. Darling, D.A.: The Kolmogorov-Smirnov, cramer–von Mises tests. Ann. Math. Stat. **28**(4), 823–838 (1957). https://doi.org/10.1214/aoms/1177706788
31. Fisher, R.A.: The Design of Experiments. Oliver and Boyd, London (1935)
32. Lehman, E.L.: Testing Statistical Hypotheses, 2nd edn. Wiley, New York (1986)
33. Dufour, J.M., Kiviet, J.F.: Exact inference methods for first-order autoregressive distributed lag models. Econometrica **66**(1), 79–104 (1998). https://doi.org/10.2307/2998541
34. Dufour, J.M., Khalaf, L.: Monte Carlo test methods in econometrics. In: Baltagi, B. (eds.) Companion to Theoretical Econometrics, Blackwell, pp. 494–519. Companions to Countemporary Economics, Basil Blackwell, Oxford, U.K. (2001)
35. Dufour, J.M.: Monte Carlo tests with nuisance parameters: a general approach to finite-sample inference and nonstandard asymptotics. J Econom **133**(2), 443–477 (2006). https://doi.org/10.1016/j.jeconom.2005.06.007
36. Dufour, J.-M., Farhat, A.: Exact Nonparametric Two-sample Homogeneity Tests for Possibly Discrete Distributions. Center for Interuniversity Research in Quantitative Economics (CIREQ). Preprint 2001–23. California Press (2001)
37. Diebold, F.X., Mariano, R.S.: Comparing predictive accuracy. J. Bus. Econ. Stat. **20**(1), 134–144 (2002). https://doi.org/10.1198/073500102753410444
38. Hassani, H., Silva, E.S.: A Kolmogorov-Smirnov based test for comparing the predictive accuracy of two sets of forecasts. Econometrics **3**(3), 590–609 (2015). https://doi.org/10.3390/econometrics3030590
39. Friedman, M.: A comparison of alternative tests of significance for the problem of m rankings. Ann. Math. Stat. **11**(1), 86–92 (1940). https://doi.org/10.1214/aoms/1177731944
40. Dunn, O.J.: Multiple comparisons among means. J. Am. Stat. Assoc. **56**(293), 52–64 (1961). https://doi.org/10.1080/01621459.1961.10482090
41. García, S., Fernández, A., Luengo, J., Herrera, F.: Advanced nonparametric tests for multiple comparisons in the design of experiments in computational intelligence and data mining: experimental analysis of power. Inf. Sci. **180**(10), 2044–2064 (2010). https://doi.org/10.1016/j.ins.2009.12.010

42. Holm, S.: A simple sequentially rejective multiple test procedure. Scand. J. Stati. **6**(2), 65–70 (1979)
43. Hochberg, Y.: A sharper Bonferroni procedure for multiple tests of significance. Biometrika **75**(4), 800–802 (1988). https://doi.org/10.2307/2336325
44. Hommel, G.: A stagewise rejective multiple test procedure based on a modified Bonferroni test. Biometrika **75**(2), 383–386 (1988). https://doi.org/10.1093/biomet/75.2.383
45. Carrasco, J., García, S., Rueda, M.M., Das, S., Herrera, F.: Recent trends in the use of statistical tests for comparing swarm and evolutionary computing algorithms: practical guidelines and a critical review. Swarm Evol. Comput. **54**, 100665 (2020). https://doi.org/10.1016/j.swevo.2020.100665
46. Demšar, J.: Statistical comparisons of classifiers over multiple data sets. J. Mach. Learn. Res. **7**, 1–30 (2006)
47. Flores, B.E.: The utilization of the Wilcoxon test to compare forecasting methods: a note. Int. J. Forecast. **5**(4), 529–535 (1989). https://doi.org/10.1016/0169-2070(89)90008-3
48. Pesaran, M.H., Timmermann, A.: A simple nonparametric test of predictive performance. J. Bus. Econ. Statist. **10**(4), 461–465 (1992). https://doi.org/10.2307/1391822
49. Clark, S.D., Grant-Muller, S.M., Chen, H.: Using nonparametric tests to evaluate traffic forecasting performance. J. Transp. Stat. **5**(1), 47–56 (2002)
50. Granger, C.W.J., Newbold, P.: Forecasting Economic Time Series. Academic Press, Orlando (1977)
51. Geurts, M., Kelly, J.P.: Forecasting retail sales using alternative methods. Int. J. Forecast **2**(3), 261–272 (1986). https://doi.org/10.1016/0169-2070(86)90046-4
52. Benavoli, A., Corani, G., Demšar, J., Zaffalon, M.: Time for a change: a tutorial for comparing multiple classifiers through Bayesian analysis. J. Mach. Learn Res. **18**(1), 2653–2688 (2017)
53. Corani, G., Benavoli, A., Demšar, J., et al.: Statistical comparison of classifiers through Bayesian hierarchical modelling. Mach. Learn. **106**, 1817–1837 (2017). https://doi.org/10.1007/s10994-017-5641-9
54. Klyushin, D.A., Petunin, Y.I.: A nonparametric test for the equivalence of populations based on a measure of proximity of samples. Ukrainian Math. J. **55**(2), 181–198 (2003). https://doi.org/10.1023/A:1025495727612
55. Hill, B.M.: Posterior distribution of percentiles: Bayes' theorem for sampling from a population. J. Am. Stat. Assoc. **63**(322), 677–691 (1968). https://doi.org/10.1080/01621459.1968.11009286
56. Matveichuk, S.A., Petunin, Y.I.: Generalization of Bernoulli schemes that arise in order statistics. I. Ukrainian Math. J. **42**(4), 459–466 (1990). https://doi.org/10.1007/BF01071335
57. Matveichuk, S.A., Petunin, Y.I.: Generalization of Bernoulli schemes that arise in order statistics. II. Ukrainian Math. J. **43**(6), 728–734 (1991). https://doi.org/10.1007/BF01058940
58. Johnson, N., Kotz, S.: Some generalizations of Bernoulli and polya-eggenberger contagion models. Statist. Paper **32**, 1–17 (1991). https://doi.org/10.1007/BF02925473
59. Adebiyi, A.A., Adewumi, A.O., Ayo, C.K.: Comparison of ARIMA and artificial neural networks models for stock price prediction. J. App. Math. **2014**, 1–7 (2014). https://doi.org/10.1155/2014/614342

Using an Innovative Model Based on Deep Learning to Determine Reduction of Habitats Associated with Arboreal Birds in Mexico

Alberto Ochoa-Zezzatti, Alberto Hernandez, Luis Alatorre, Luis Bravo-Peña, María Torres-Olave, and José Mejia

Abstract Currently, with a global warming problem, most developing countries such as Mexico or Brazil suffer high levels of deforestation, which directly threatened bird species that have their habitats linked to different tree species. The forests disappear, and hundreds of species disappear along with them. The geography institute of the UNAM estimates that each year more than 500 thousand hectares of forest and rain forest in different parts of the World are lost, placing Mexico in fifth place in global deforestation. That is why it is important to find the definitive factors that influence deforestation; its discovery is key to help promote the prevention of excessive logging, conservation, and continuous reforestation. The use of deep learning and satellite imagery is considered useful to simulate deforestation processes and in the analysis of these factors to determine their reduction. In this research, our objective is to use deep learning to model the effects of deforestation on habitats of bird species with arboreal ecosystems; this will allow us to determine areas in danger of extinction. We plan to simulate an affected area, to be able to visualize over time, through the numerical prediction and the fading of the forest area, and then predict where the next area in danger will be, in order to reduce the effects of this ecological problem. The main problem of deforestation lies not only in the reduction of forests but also in the habitat of different species. Using an innovative artificial intelligence technique such as Deep Learning. We can adequately characterize longitudinal changes in terms of trends in the reduction of this type of locations that require an animal species to survive. There are 10,087 different species of birds that live in diverse habitats; 47.82% have an arboreal habitat, a reason very important and decisive to our study.

Keywords Deforestation · Deep learning · Arboreal habitats associated with tree bird species

A. Ochoa-Zezzatti (✉) · L. Alatorre · L. Bravo-Peña · M. Torres-Olave · J. Mejia
Universidad Autónoma de Ciudad Juárez, 32310 Juárez, México
e-mail: alberto.ochoa@uacj.mx

A. Hernandez
Universidad Autónoma del Estado de Morelos, 62209 Cuernavaca, México

W. Pedrycz et al. (eds.), *Computational Intelligence for Business Analytics*,
Studies in Computational Intelligence 953,
https://doi.org/10.1007/978-3-030-73819-8_13

1 Introduction

The impact of deforestation affects the ecosystem worldwide; by 2015, the total area covered by forests in the world was just 30.825% (data world bank). According to the United Nations, an estimate of 7.3 million hectares is lost every year. The trees and other plants fight against the earth's erosion through their roots, but this situation affects diverse habitats all around. In fact, with the loss of threes and vegetation, the erosion rises, which in some cases causes desert formation? Deforestation also affects the role of the forest as a windbreaker, which in turn helps and nurtures forest fires. Besides, the forests worsening can cause flooding and weather change. Loss of woods contributes between 12 and 17 percent to greenhouse emissions [1, 2]. Mexican territory has 138 million hectares with forests and vegetation, about 70% of the whole country [3]. The mountain forests oversee about 1.7 million hectares, and the main causes of deforestation are land change to turn forests into pastures or farmlands. Another factor that stands against forests is illegal logging, a serious problem because 70% of the national market of wood comes from this activity [1]. However, in every state, you have one or more causes of deforestation.

Chihuahua is a state with 7887 million hectares; the vegetation surface covers 38.17% of the whole territory and the other 61.83% are urban and semi urban areas, areas without vegetation and water bodies. Figure 1 show the most predictive areas with deforestation in Chihuahua. Within the different types of vegetation, the broadleaf is the most popular, and it is mainly composed of evergreen oaks [4].

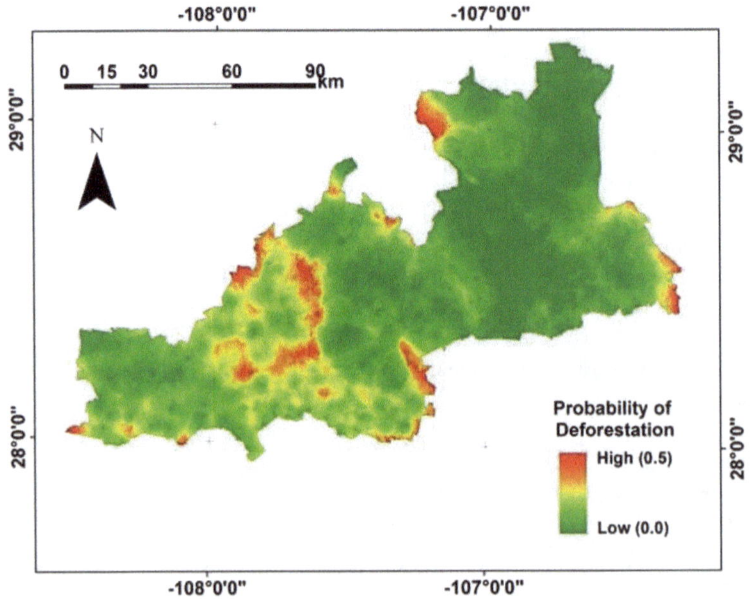

Fig. 1 Model prediction with a forest region with more probability of deforestation [5]

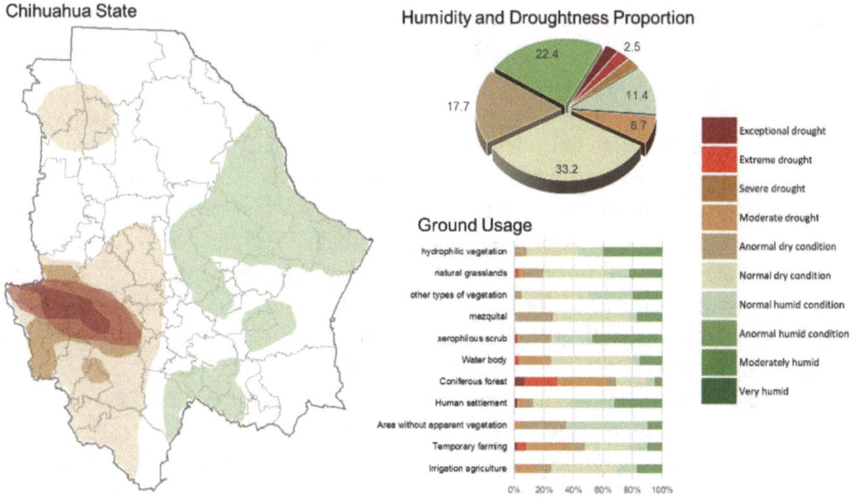

Fig. 2 Frequency of environmental impact in Chihuahua

For oak tree forests, human damage is one of the main causes of the impact of Fig. 2.

To preserve and manage forests in Chihuahua and in the world, it is important to use tools from various disciplines. It is important to note that the transformation of forests to deserts is not necessarily random; there are regions where the rural population is high, and farming lands are scarce, there is poverty and not enough jobs are generated, this causes people to go after the local resources. In these situations, it is frequent that illegal logging and fire provocation are used in order to create farmlands, but these actions only allow a type of nomad agriculture that only causes an irremediable loss of the forests [6]. Deep Learning is a good tool that allows one to consider all the factors mentioned above and is used to model deforestation because of their high level of precision in the area of spatial modeling and they have been used before to support environmental and conservation causes. This article applies a deep learning model to simulate the process of deforestation in the region of the Sierra Madre Oriental.

2 Deep Learning and Deforestation

Deep learning has been used in recent times to model deforestation due to its compatibility with geographic information systems (GIS). The research of land-use simulation has also been used to modeling urban development and, especially, forest fire models. In [2] used a deep learning model to predict deforestation. Rosas [3] used a stochastic model of deep learning to simulate the dynamics of land use in a border threatened by civilization in the Amazonia [2]. Many models of Deep Learning are based on the same structure proposed by cellular automata where the star shows the pixel being analyzed and the blue box its scope of analysis, it is planned to use data mining techniques for the trend analysis of the results each pixel assumes a cell so that an analysis can be carried out under a system of cellular automata [2], whose states depend on the rules that we implement; here values are used within a range that goes from 0 to 1 to establish the final states of the cell. This happens each pixel contains, in the case of deforestation processes, each pixel contains information about its ability to remain as a forest or succumb to urbanism, either by its proximity to roads, populated areas or its proximity to the forest; also applying similar rules for both the growth of cities and for possible areas of logging; this results in a growth model feeds back to itself over time, generating more reliable maps, requiring a smaller number of runs [7], as shown below, see Fig. 3.

Each pixel contains information in the form of indexes that give a reference of the probability in the change of land use, depending on its conditions with its neighbors, these indices can be height above sea level, distance with the roads, distance to the populated areas, distance from crop areas, rainfall, among others, although any number of variables can be included [4, 8]. These indices are recalculated in each interaction, thus making the model more truthful as it advances with time.

The climate change is affecting the small redoubts that connect to different habitats and that allow diversity in the population ecology to invariably be since the space is deforested, the issues of each species are disadvantaged predators because they do not have the minimum necessary forest density, a s can be seen in Fig. 4.

Through the use of a Deep Learning, it has been learned to learn the significant change of the loss of individuals directly in proportion to a species that may be in danger of extinction; in the Chihuahua area there is a species of Great Bustard that in just one generation of twenty-seven years has seen a decrease by almost 47% not only the spaces associated with its habitat, but the population of the species has been reduced to only a quarter of the original population, as can be seen in Fig. 5.

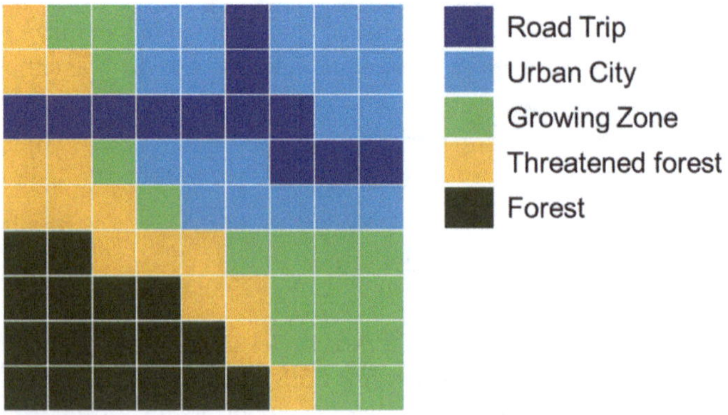

Fig. 3 Representation of scales at a pixel level of each element in a geographic information system associated with a satellite image

Fig. 4 The low deciduous forest is one of the most prone to deforestation given its relevant characteristics of change of season and specifically to lose individuals of some species that provide a habitat and a place of refuge to a huge variety of mammals and birds

Using our intelligent tool developed under the Deep Learning approach, you will find the following characteristics of the 3 different moments in the habitat of a common bustard species in Chihuahua, as is shown in Table 1.

The relevance of our study is to have successfully identified the future trends of deforestation behavior in an area, and how these characteristics can be linked to the environment and finely to the population decline of a species seen from the Population Ecology approach Some species of birds require trees of a certain height to be able to build safe habitats for their chicks and to be able to belong to a flock with certain criteria of survival among them more than 70 members.

3 Experimental Model

The simulation of a forest with deforestation increasing the time is simulated through deep learning is implemented in this case through code written in JavaScript [9] and executed in a development integrated environment (IDE); up next, we describe all the characteristics that make up some deep learning. It's important to notice that this simulation takes into account only the state of its neighbors to evolve.

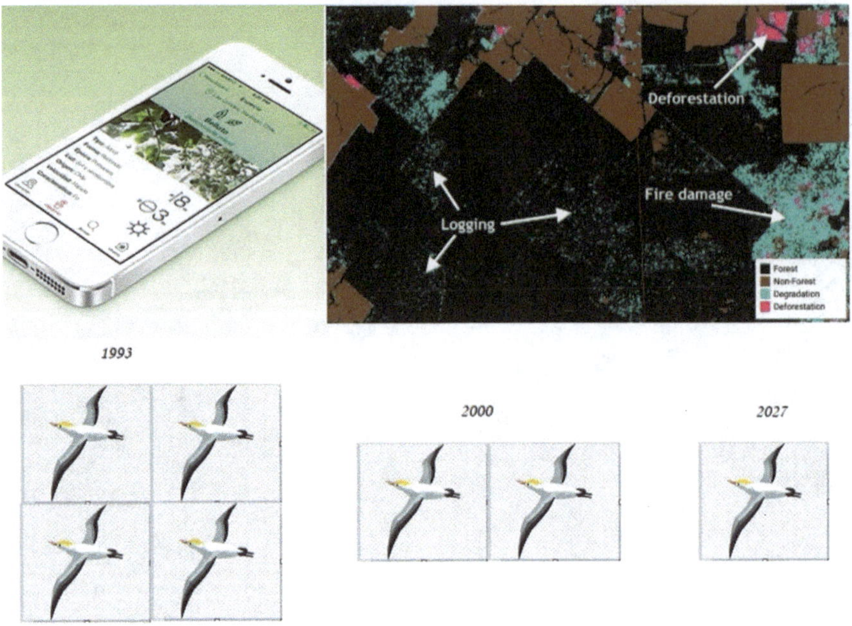

Fig. 5 Predictive model using Deep Learning to determine changes in the arboreal habitats of different bird species in our study area

Table 1 Prediction model of the future situation of a Bustard species in Chihuahua

Year of the longitudinal study	Total, population of the species	Number of arboreal habitats (%)	Number of habitats near an aquifer (%)
1993	7800	73	27
2000	3780	54	11
2027	1470	31	6

Source Ornithology Studies in Chihuahua for the first dates and our Deep Learning model for the numerical prediction of 2027

Factors like ground elevation, distance to roads, urban areas, and water bodies are not considered. It's because of this that further along, we contemplate the possibility of using special software (Dinamica EGO, IDRISI) to develop models that do take in all the parameters that we mentioned above. Using geospatial maps, we were able to adequately identify paradigmatic changes in spaces that have been eroded and with that the species that have habitats in them, to adequately determine emerging actions that must be considered in order to prevent the reduction of said spaces and therefore the end of the species and its inexorable extinction.

4 Study Area in Chihuahua

As part of this research, it's also important to include information obtained from satellite images Landsat/ETM to show the evolution of the vegetation in the area of the Sierra Madre Oriental, Fig. 6, and in this way, complement the simulation made only with deep learning and also take a look into the future of what would this place become if the current rates of deforestation are maintained. The west and southwest region of the state of Chihuahua is a part of the Sierra Madre Oriental, which represents 11.45% of the state's surface. The terrain in this area is abrupt; it has long mountain ranges, canyons, and plains. This area was chosen mainly because it's where most of the ecoregions of the state are concentrated. According to the classifications from INEGI, this area possesses 5 out of the 7 official ecoregions, which are vegetation categories that describe a group by physiognomy, ecology, and floristics [4]. To perform this experiment, we obtained seven satellite maps of the same area during the years 2017 and 2019. The center of these images has coordinates with latitude and longitude (21.219031, −99.471435) and has an area of 185 × 185 km. Once obtained, these images were processed with a technique called semi-automatic classification to create maps of land classification. This process is included in a plug-in called Semi-automatic classification of the software QGIS [10].

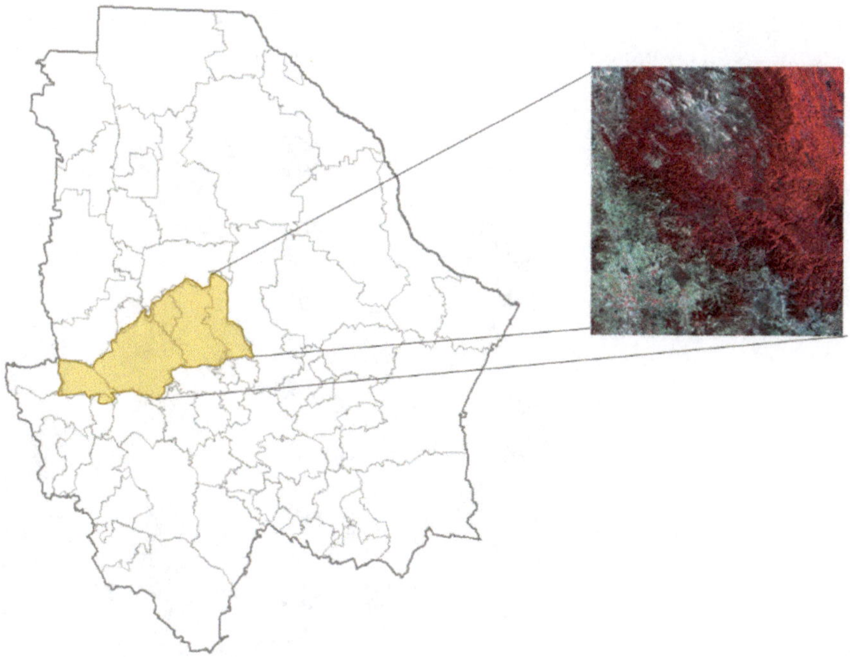

Fig. 6 Study area. *Source* LANDSAT/ETM

The image processing was done through QGIS, freeware and open source software to create, edit, and visualize geospatial information.

The images obtained through Landsat contain different bands, but we only employ the following:

- Band 2: Blue
- Band 3: Green
- Band 4: Red
- Band 5: Near infrared
- Band 6: Short wave infrared 1
- Band 7: Short wave infrared 2.

Once we downloaded bands 2–7, you can see how the different layers of the map are obtained in the following way, Fig. 7.

For the land classification, we then create a layer called "classification" with a single band, as shown in Fig. 8, where we can see the vegetation in red and the urban zones in green light.

The classification layer is then processed to establish 4 classes: Water bodies, Urbanization, Vegetation/Forest, and Uncultivated land. In Fig. 9a, b, both maps are shown with these categories, which represent two different periods of this region of study.

In Fig. 10, we can clearly observe the significant change that has occurred between the two dates given in our study and how it can get worse if there are no public policies associated with the preservation of arboreal habitats for endangered bird species.

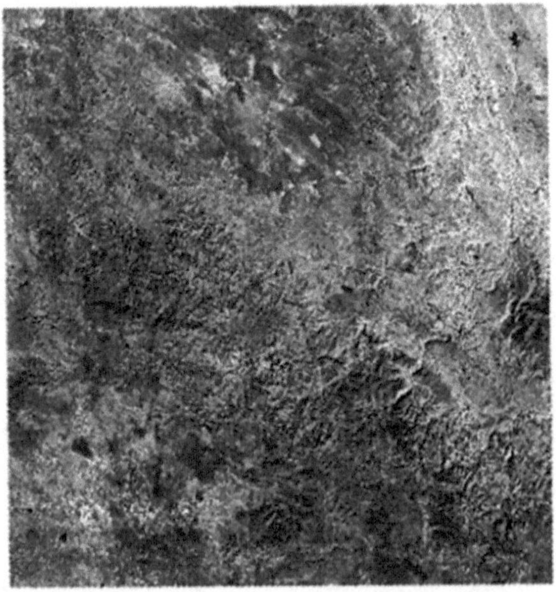

Fig. 7 Band 5 of the satellite image. *Source* LANDSAT/ETM

Fig. 8 Layer classification

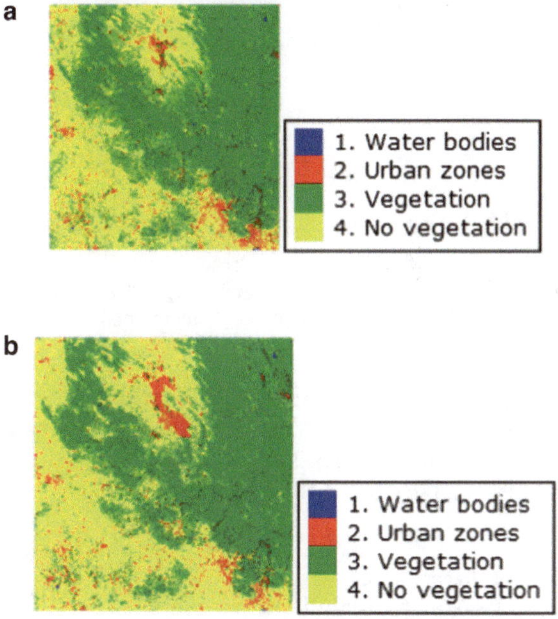

Fig. 9 **a** Maps and legend of the land classification of the area of the Sierra Madre Oriental in 2010. **b** Maps and legend of the land classification of the area of the Sierra Madre Oriental in 2020

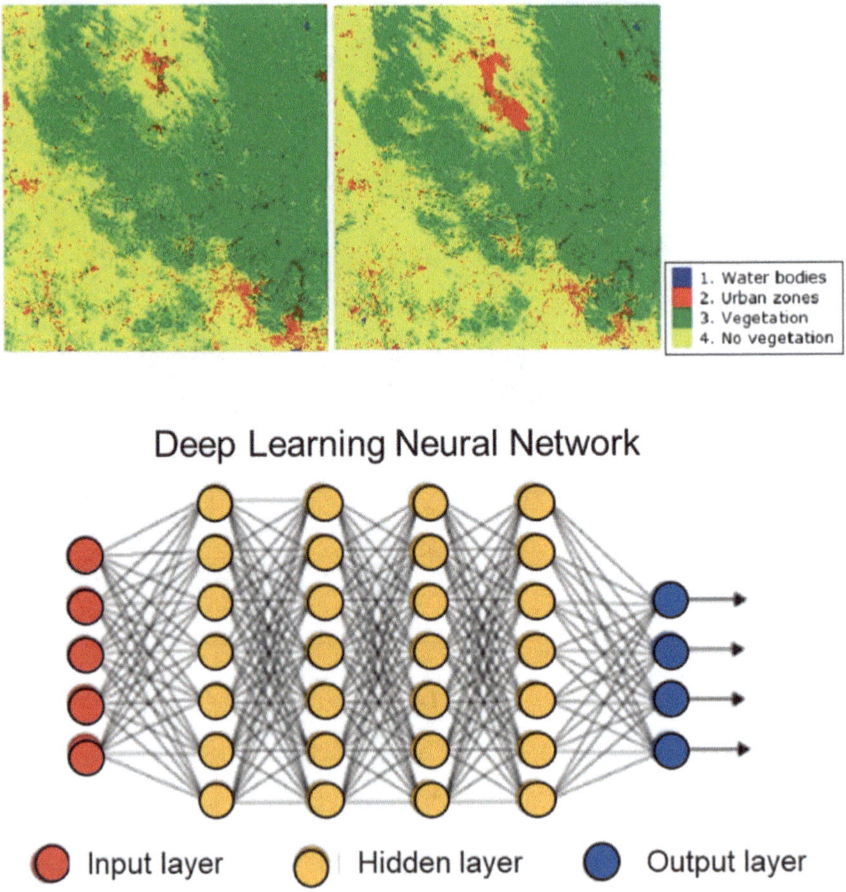

Fig. 10 Comparative map of the land using a classification map based on Deep Learning to the years 2010 and 2019. The orange color in representation of Deep Learning model is associated with the changes in the time

According to the analysis done with the tools before mentioned, we obtained the following tables, Tables 2 and 3, with the number of pixels for each class. We can then obtain the rates of change for each class.

Table 2 Report of the classification of the map 2010

Class	Total pixels	Percentage (%)	Area (m^2)
1.0	53,627	0.1293	48,264,300.0
2.0	1,840,406	4.4400	1,656,365,400.0
3.0	21,030,368	50.7370	18,927,331,200.0
4.0	18,525,317	44.6934	16,672,785,300.0

Table 3 Report of the classification of the map 2019

Class	Total pixels	Percentage (%)	Area (m^2)
1.0	39,051	0.0942	35,145,900.0
2.0	2,081,688	5.0235	1,873,519,200.0
3.0	19,254,367	46.4646	17,328,930,300.0
4.0	20,063,679	48.4178	18,057,311,100.0

5 Result Analysis of This Study

According to the data previously obtained, we can obtain the following rates of change:

- The rate of annual growth of urban zones was of 4.37%
- The rate of annual growth of vegetation zones was of –2.81%
- In the areas with no vegetation, the annual rate of change was of 2.76%.

The increase in the urban areas and decrease of the vegetation areas is a clear indicator of the effect of human presence in the territory of the Sierra Madre Oriental. Nevertheless, it's important to mention that according to the State's inventory [4] these rates of deforestation are negligible because the general forest formations show a high recuperation rate, with a lot of young adult trees, besides the state of the forest was diagnosed as good. This was stated not forgetting the construction of roads and farmlands as the most common causes of deforestation in the whole area. Knowing this, we can add that even if an area is not in extreme danger, the diminishing of the vegetation areas is a reality, and it is growing every year. We consider in Table 4, seven large and common species in these kinds of habitats associated with large and dense forest, with this information consider the future impact of this situation about the reduction of habitats to 2027.

Considering the results of different possible future scenarios, only species 1, 2 and 7 will be able to survive more than 20 years considering not only the reduction of habitat but also the lack of food and the effect that the reduction of diversity will have on the flock, something that directly affects the population ecology index.

6 Conclusions and Future Work

In this research, a small insight of deep learning was used to demonstrate its capabilities to show a true human caused phenomenon such as deforestation. Using a simple model, it was easy to demonstrate the way in which this tool can capture complex behaviors such as a forest fire. Nevertheless, the models can reach a higher level of complexity if decisive factors are taken into account such as road construction in places humans couldn't reach before or a population increase. As shown in the article of the University of Shahid Beheshti in Iran [2], the distance to roads and highways

Table 4 Types of endemic birds on El Bosque de la Primavera forest

#	Species	Socialization[a]	Temp (°C)	Size (cm)	Habitat food[a]	Index population edology[a]	Growing time[a]	Fertility rate[a]	Ecological valuation[b]
1	Suirirí Piquirrojo	4.87	21–31	13	0.72	0.23	0.14	0.72	10
2	Tinamú Canelo	3.98	24–27	9.5	0.91	0.92	0.92	0.77	10
3	Ánade friso	5.76	25.5	10–13	0.43	0.94	0.33	0.98	15
4	Porrón Acollarado	6.97	26.5	8	0.18	0.67	0.79	0.74	15
5	Colín de Moctezuma	5.87	27–31	12.5	0.85	0.52	0.74	0.27	20
6	Achichilique de Clark	3.75	25–31	10–15	0.32	0.47	0.71	0.96	20
7	Tórtola Turca	5.88	22–30	10–13	0.66	0.82	0.36	0.17	15

[a]Lickert Scale

[b]Valuation is given by WWF and uses a scale of 1 to 20 where 1 is less contribution to a specific habitat, and 20 is a very important support to its habitat.

is probably the most important factor in forest destruction; this means that in remote places, the percentage of deforestation is low. The second most important factor is the slope or inclination of the territory, which is also related to the accessibility of the area. We expect to take in account these factors in the continuation of this research [11], as well as the use of a Geographic Information System that involves the deep learning functionality and other mathematical models such as logistic regression to perform a more exact analysis of the future of deforestation in the Sierra Madre Oriental area [12]. Each species of bird in an arboreal habitat will require different approaches to determine if they can survive the next fifty years [13], a correct decision making considering the support of artificial intelligence techniques could improve an ecological process that would make a significant difference [14], although the human factor and time are against. The bird species are a principal key on all ecosystems in all world; lose them affect the rest of our ecosystems; this will be catastrophic to each society [15].

In our future research, it was determined to implement a cluster of drones that can determine the maximum habitat extension of a bird species, with the intention of determining whether it could survive in the long term considering the minimum level for a population ecology model, selecting the platform with multi-criteria analysis [16]; It is a species with at least a thousand issues in the wild, for this reason, is very important to specify an order that the diversity of individuals can ensure the future survival of the entire species; our conceptual diagram is shown in Fig. 11.

Fig. 11 Implementation of an intelligent drone that can determine the size of a flock of a species and its distribution with respect to the entire habitat

A side effect of the reduction in bird habitat is that life support must be provided in the form of artisanal nests so that species have a place where they can fearlessly hatch their eggs and subsequently raise their chicks. This is happening too much on the part of the Z generation in many Smart Cities.

References

1. Greenpeace: La deforestación y sus causas. http://www.greenpeace.org/mexico/es/Campanas/Bosques/La-deforestacion-y-sus-causas/. Accessed 3 Apr 2018
2. Naghdizadegan, M., Behifar, M., Naghdizadegan, B.: Spatial deforestation modelling using deep learning (case study: Central Zagros Forests). ISPRS Int. Arch. Photogrammetry Rem. Sens. Spat. Inf. Sci. **XL-1/W3**, 289–293 (2013). https://doi.org/10.5194/isprsarchives-xl-1-w3-289-2013
3. Rosas, B.: La deforestación en México (2016). http://www.mexicosocial.org/la-deforestacion-en-mexico/. Accedido el 3 abril 2017
4. Secretaria de Medio Ambiente y Recursos Naturales: Colección de Inventarios Estatales Forestales y de Suelos 2013–2014, Chihuahua (2014)
5. Bravo-Peña, L.C., Torres-Olave, M.E., Alatorre, L.C., Wiebe-Quintana, L.C., Moreno-Murrieta, R.L., Granados, A.: Identification of areas in probability of being deforested, through logistic regression. study in Chihuahua (Mexico) for period 2007–2013. In: 2016 IEEE 1er Congreso Nacional de Ciencias Geoespaciales (CNCG), pp. 1–4 (2016). dhttps://doi.org/10.1109/CNCG.2016.7985081
6. Gutierrez, J.; Trejo, O., Camacho, S., Castillo, R., Cruz, S., Castaneda, J.: Distrito Federal: Educación ambiental: caminos ecológicos. Editorial Limusa, S.A. de C.V (1999)
7. Karafyllidis, I., Thanailakis, A.: A model for predicting forest fire spreading using deep learning. Ecol. Model. **99**(1), 87–97 (1997)
8. Mejía, J., Ochoa-Zezzatti A., Contreras-Masse, R., Rivera G.: Intelligent system for the visual support of caloric intake of food in inhabitants of a smart city using a deep learning model. In: Oliva, D., Hinojosa, S. (eds.) Applications of Hybrid Metaheuristic Algorithms for Image Processing. Studies in Computational Intelligence, vol. 890. Springer, Cham (2020). https://doi.org/10.1007/978-3-030-40977-7_19
9. Forest fire: http://rosettacode.org/wiki/Forest_fire. Accessed 3 Apr 2017
10. QGIS Development Team: QGIS Geographic Information System. Open Source Geospatial Foundation Project (2017). http://www.qgis.org/ Accessed 15 May 2017
11. Congedo, L.: Semi-Automatic Classification Plugin Documentation (2016). https://doi.org/10.13140/RG.2.2.29474.02242/1
12. Del-Rio-Ruiz, R., Lopez-Garde, J., Ruiz, J., Legarda, J.: Smart Nests: IoT for Ornithology. GIoTS, 1–6 (2018). https://doi.org/10.1109/GIOTS.2018.8534436
13. Yarwood, M., Weston, M., Garnett, S.: From little things, big things grow; trends and fads in 110 years of Australian ornithology. Scientometrics **98**(3), 2235–2254 (2014). https://doi.org/10.1007/s11192-013-1144-z
14. Koho, M., Hyvönen, E., Lehikoinen, A.: Ornithology Based on Linking Bird Observations with Weather Data. SePublica (2014)
15. Sorokina, D., Caruana, R., Riedewald, M., Hochachka, W., Kelling, S.: Detecting and interpreting variable interactions in observational ornithology data. In: ICDM Workshops, pp. 64–69 (2009). https://doi.org/10.1109/ICDMW.2009.84
16. Contreras-Masse, R., Ochoa-Zezzatti, A., García, V., Pérez-Dominguez, L., Elizondo-Cortés, M.: Implementing a novel use of multicriteria decision analysis to select IIoT platforms for smart manufacturing. Symmetry **12**(3), 368 (2020). https://doi.org/10.3390/sym12030368

Method for Recommending Guardianship to Minors Based on Parental Responsibility Using a Fuzzy Cognitive Map

Hernán Patricio Castillo Villacrés, Mesías Elías Machado Maliza, and Diego Fabricio Tixi Torres

Abstract Under different situations, minors can be left without legal guardianship. When a minor is in distress, he is at the mercy of the decision of the competent entity for the granting of his guardian. However, based on the affective relationships established by the different family members, determining parental responsibility constitutes a highly important decision. This research proposes a solution to the problem described by means of a recommendation system to assign parental responsibility and its incidence in the best interests of minors. The proposed method bases its operation using a Fuzzy Cognitive Map to model uncertainty in causal relationships. A case study is presented to demonstrate the applicability of the proposal.

Keywords Parental responsibility · Recommendations · Fuzzy cognitive map

1 Introduction

The well-being of boys and girls in all their aspects such as health, physical and mental state, home, family, social condition, and education is an international priority. In 1953, the United Nations General Assembly established that UNICEF is a permanent body to later expand its scope to minor issues, being the starting point to create the second Declaration of the Rights of the child [1, 2]. The Declaration of the Rights of the Child indicates that children need special care, establishing adequate legal protection before and after birth [3]. This statement mentions:

- The right to equality, without distinction of race, religion, or nationality.
- The right to have special protection for the child's physical, mental, and social development.
- The right to adequate food, shelter, and medical care.

H. P. Castillo Villacrés (✉) · M. E. Machado Maliza · D. F. Tixi Torres
Universidad Regional Autónoma de los Andes (UNIANDES), Riobamba 060150, Chimborazo, Ecuador
e-mail: ur.c.derecho@uniandes.edu.ec

W. Pedrycz et al. (eds.), *Computational Intelligence for Business Analytics*, Studies in Computational Intelligence 953,
https://doi.org/10.1007/978-3-030-73819-8_14

- The right to education and special treatment for those children who suffer from a mental or physical disability.
- The right to protection against any form of abandonment, cruelty, and exploitation.

In Ecuador, since its existence as a Republic, two legal bodies have been drafted, these are the Juvenile Code with its respective reforms and the Childhood and Adolescence Code [4].

The latter has had to go through several changes, since what is sought is to confer a norm that is appropriate to the comprehensive protection of children and adolescents, and that under the provisions of this norm, they will enjoy equality before the law and not There will be no discrimination in any circumstance, be it because of the sex, religion, social origin, political ideology, affiliation, health, sexual orientation, disability or any other condition, whether of the minor, his parents, representatives or family members. This Childhood and Adolescence Code, which the National Congress described as an "organic law" based on article 142 of the Constitution, as first part establishes the protection that the State, society, and the family must provide to the minor, since the fully enjoy your rights using the effective means that ensure that they are complied with, establishing that this code protects all people from their prenatal state until they turn eighteen [5, 6].

Article 44 of our Magna Carta mentions the principle of the best interest of the child, in which various precepts are enshrined that requires the state and society to respect in all areas the rights of minors, promoting their comprehensive development. Article 45 indicates a list of rights among them, the right to health, education, and family life, in which through this research project, it can be seen that sometimes children are deprived of this right.

The constitution of the Republic of Ecuador is in charge of guaranteeing equality in opportunities and rights to all those who make up the family nucleus since it consecrates the family as the fundamental nucleus of society; therefore, the rights of each person must be protected. One of them promoting a responsible parental relationship giving children care, upbringing, and education, protecting their rights and well-being in the event of being left without legal guardianship [7, 8].

The constitution itself mentions the best interests of the child, where it exalts all the measures that public and private entities must take, always protecting the best interests of the child. Based on this, administrators of justice, especially those in charge of childhood issues, must make their decisions independently of social pressures; when solving a controversy where the minor is involved, the well-being of the child will always be first or girl.

When a minor is in distress, he is at the mercy of the decision of the administrators of justice for the granting of his custody. However, based on the affective relationships established by the different family members, determining parental responsibility constitutes a highly important decision. In our society, we find ourselves with the problems that lie at the moment in which the administrator of justice has the choice of the family nucleus that will assume the care and protection of the minor; in these cases, the law does not have a tacit interpretation, but it is necessary to analyze various aspects that generally end in decision making under uncertainty.

The administrator of justice is in an environment of uncertainty regarding the granting of legal guardianship, because a good is not at stake, but rather a minor, who needs care and attention that is both affective and emotional, satisfying their needs and looking after your interests. A hasty decision will affect both your emotional and psychological state, short and long term.

This research proposes a solution to the problems described, through a method for recommending guardianship to minors based on parental responsibility and its incidence in the best interests of minors.

The research is structured in Introduction, Materials and Methods, Results, and Conclusions. In the end, the bibliographic references are listed. In the introduction, a state of the art is made on the different situations in which the best of age can be left without legal guardianships and the existing legal codes and procedures to assign legal guardianship of a minor. In the Materials and Methods session, a method for recommending guardianship to minors based on parental responsibility is presented, which consists of four basic activities and bases its operation on a fuzzy cognitive map to model uncertainty in causal relationships [9]. The Results show a case study to demonstrate the applicability of the proposal.

2 Materials and Methods

This section describes the operation of the method for recommending child custody based on parental responsibility—the method models the causal relationships between the different concepts [10] using a diffuse cognitive map.

The method supports the following principles:

- Integration of causal knowledge using the Fuzzy Cognitive Map (FCM) for recommending guardianship to minors.
- Identification through the team of experts of causal relationships.
- Orientation of information towards the best welfare of the minor.

The design of the method is structured for the recommendation of guardianship to minors. It has three basic stages: entry, processing, and exit. Figure 1 shows the general outline of the proposed method.

The proposed method is structured to support the management of the inference process for recommending guardianship to minors. Employs a multi-criteria approach as the basis for inference helps experts to nurture the knowledge base [11–13].

The set of evaluative indicators represent one of the inputs of the method that is necessary for the inference activity. The inference activity represents the fundamental nucleus for the reasoning of the method. It bases its processing from the modeling of causal relationships with the use of Fuzzy Cognitive Map [14–16].

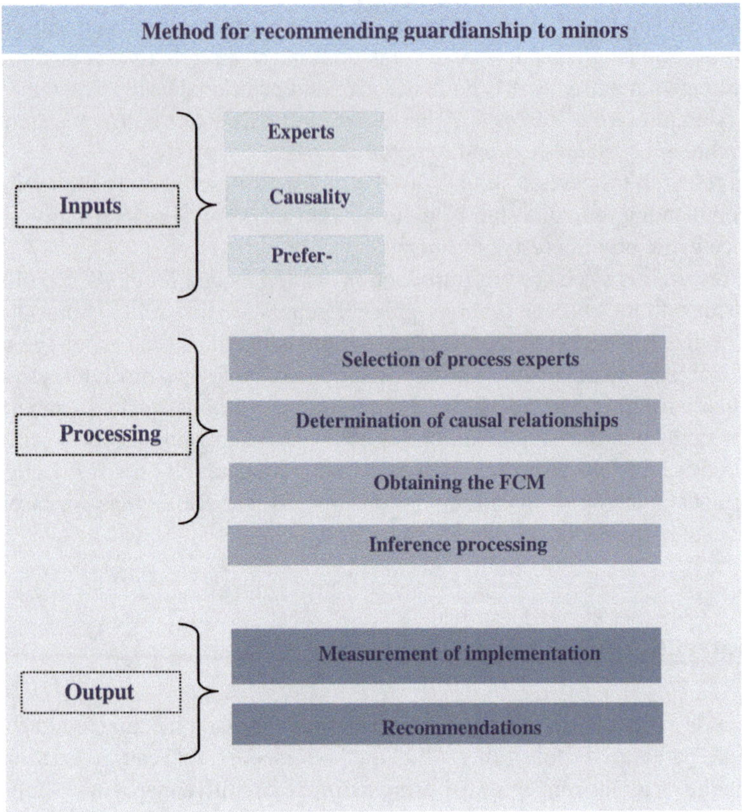

Fig. 1 Structure of the proposed method

2.1 Method Processing Description

This section provides a description of the proposed method. The different activities that guarantee the inference of the processing stage are detailed. The activities are computed by identifying the evaluation criteria, determining the causal relationships, obtaining the FCM resulting from the causal relationships, and inference of the process. Figure 2 shows the flow of the processing stage.

Activity 1: Identification of the evaluation criteria

The activity begins with the identification of the experts involved in the process. From the work of the expert group, the criteria that will be taken into account for the inference of the process are determined.

The activity uses a group work system using a multi-criteria approach. Formally, the problem of recommending guardianship to minors can be defined based on parental responsibility through.

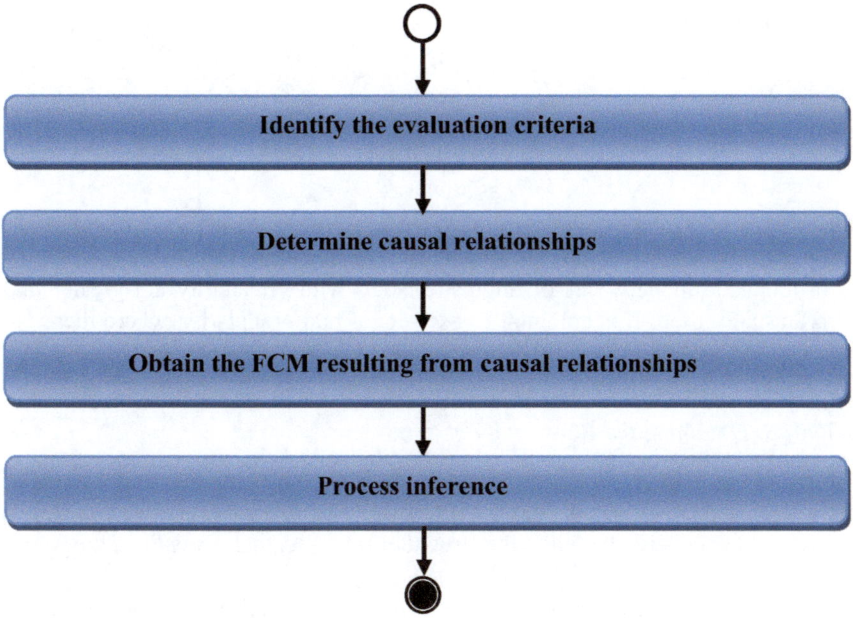

Fig. 2 Processing stage workflow

The number of impact criteria in the guarding process where:

$$I = \{i_1, \ldots, i_n\} \tag{1}$$

$$\forall I_i, 1 \leq i \leq n \tag{2}$$

The number of experts involved in multi-criteria assessment where:

$$E = \{m_1, \ldots, m_n\} \tag{3}$$

$$\forall E_i, \quad 1 \leq i \leq m \tag{4}$$

The result of the activity is obtaining the different impact criteria in the guarding process from the group selection.

Activity 2: Determinations of the causal relationships of the criteria

Once the impact criteria in the guarding process are obtained, the causal relationships are determined [17]. Causal relationships are the expression of causality between the different impact criteria in the guarding process [18].

The determination of the causal relationships consists of establishing from the group work the implication between concepts. The resulting information represents

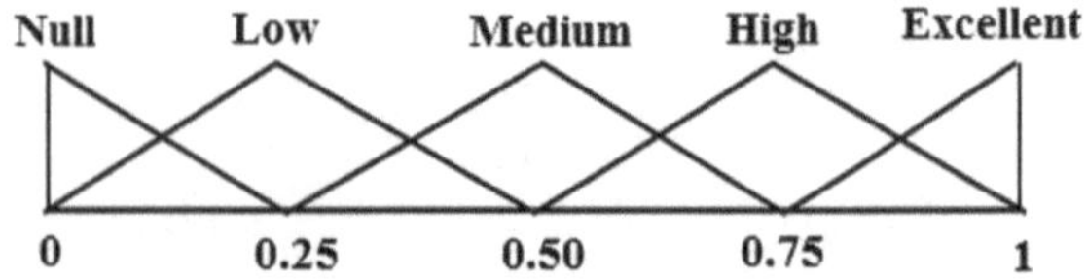

Fig. 3 Linguistic labels set

the primary knowledge to nurture the inference process. Causal relationships are represented by fuzzy variables expressed as linguistic terms [19, 20].

In linguistic models, sets of linguistic labels with granularity not greater than 13 are usually used. It is common to use sets of odd granularity, where there is a central label, and the rest of the labels are symmetrically distributed around it [21, 22]. Figure 3 shows the set of linguistic terms used for the present investigation.

Activity 3: Obtaining the FCM

During the knowledge engineering stage, each expert expresses the relationship between each pair of concepts $C_i y C_j$ of the map. So, for each causal relationship, K rules are obtained with the following structure: If C_i is A then C_j is $B$and the weight W_{ij} is C.

Each node constitutes a causal concept; this characteristic makes the representation flexible to visualize human knowledge. The adjacency matrix is obtained from the values assigned to the arcs. Figure 4 a representation of the FCM and the adjacency matrix [23–25].

The values obtained by the group of experts involved in the process are aggregated, conforming to the general knowledge with the relationships between the criteria. Activity results in the resulting FCM [26–28]. From the obtaining of the causal relationships, the static analysis is performed. The knowledge stored in the adjacency matrix is taken as a reference. For the development of the present method, we work with the degree of output, as shown by Eq. 5 [29–31].

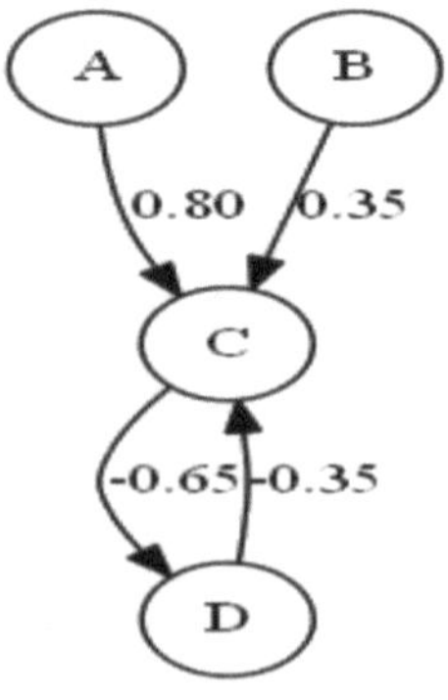

	A	B	C	D
A	0	0	0.8	0
B	0	0	0.35	0
C	0	0	0	-0.65
D	0	0	-0.35	0

Fig. 4 Fuzzy cognitive map and its corresponding adjacency matrix

$$id_i = \sum_{i=1}^{n} \| I_{ji} \| \tag{5}$$

Activity 4: Inference processing

A system modeled by an FCM will evolve over time, where the activation of each neuron will depend on the degree of activation of its antecedents in the previous iteration. This process is normally repeated until the system stabilizes or a maximum number of iterations is reached.

Processing for inference consists of calculating the state vector A over time, for an initial condition A^0 [32].

Analogously to other neural systems, the activation of C_i will depend on the activation of neurons that directly affect the C_i concept and the causal weights associated with the said concept. Equation 6 shows the expression used for processing

$$A_i^{(K+1)} = f\left(A_i^{(K)} \sum_{i=1; j \neq i}^{n} A_i^{(K)} * W_{ji} \right) \tag{6}$$

where,

$A_i^{(K+1)}$: is the value of the concept C_i in the step $k + 1$ of the simulation,

$A_i^{(K)}$: is the value of the concept C_j in the step k of the simulation,

W_{ji}: is the weight of the connection that goes from the concept C_j to the concept C_i y $f(x)$ is the activation function.

Unstable systems can be totally chaotic or cyclical and are frequent in continuous models. In summary, the inference process in an FCM can show one of the following characteristics [17, 33].

Stability states: if $\exists tk \in \mathbb{N} : A_i^{(t+x)} = A_i^{(t)} \forall t > tk$ therefore, after iteration tk the FCM will produce the same state vector. This configuration is ideal, as it represents the encoding of a hidden pattern in causality [34, 35].

Cyclical states: if $\exists tk.P \in \mathbb{N} : A_i^{(t+p)} = A_i^{(t)} \forall t > tk$. The map has a cyclical behavior with period P. In this case, the system will produce the same state vector every P-cycle of the inference process [36, 37].

Chaotic state: The map produces a different state vector in each cycle. Concepts always vary their trigger value [38, 39].

3 Results

This section illustrates the implementation of the proposed method. A case study is described as recommending guardianship to minors based on parental responsibility.

Table 1 Impact criteria in the guarding process

Number	Criterion
1	Acceptance level of the boy or girl
2	Affectivity with the boy or girl
3	Income level
4	Social suitability

The proposal was used as a scenario of implementation of a case as a reference to the canton of Patate. The results of the study are described below:

Activity 1: Identification of the evaluation criteria

For the development of the study, 5 experts who are licensed criminal workers were consulted. The group represents the basis for defining the impact criteria in the guarding process and causal relationships.

From the work carried out by the expert group, the set of criteria were identified. Table 1 shows the result of the identified criteria.

Activity 2: Determination of the causal relationships of the criteria

For the identification of causal relationships, information was obtained from the group of experts participating in the process. As a result, 5 adjacency matrices were identified with the knowledge expressed by each expert. The matrices went through an aggregation process in which a resulting adjacency matrix is generated as a final result. Table 2 shows the adjacency matrix resulting from the process.

Activity 3: Obtaining the FCM

Once the impact criteria in the guarding process and their corresponding causal relationships in Activity 2 have been obtained, the knowledge is represented in the resulting FCM. Figure 5 shows the FCM that represents the process.

Activity 4: Inference processing

The adjacency matrix has the necessary knowledge to determine the weights attributed to each indicator. Equation 5 is used to calculate the weights. Table 3 shows the results of the calculation made.

Once the weights of the indicators have been determined and normalized, the preferences of the relatives' object of analysis of the proposal are determined. For

Table 2 Adjacency matrix of the impact criteria in the guarding process

	C_1	C_2	C_3	C_4
C_1	[0,00]	[1]	[0.25]	[1]
C_2	[1]	[0.00]	[1]	[0,25]
C_3	[0,25]	[0,75]	[0.00]	[1]
C_4	[0,75]	[0.00]	[0.25]	[0.00]

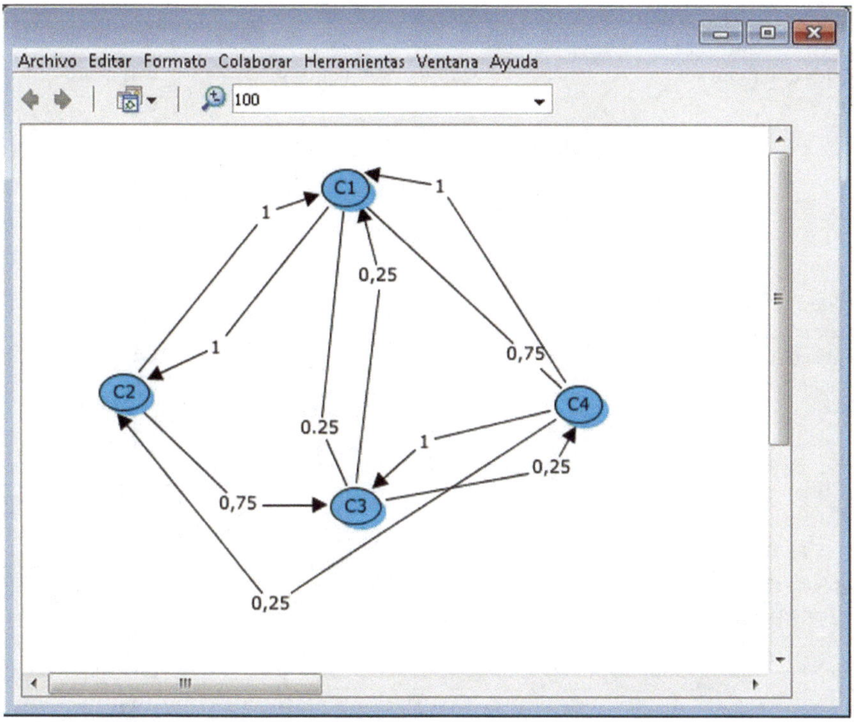

Fig. 5 Resulting fuzzy cognitive map

Table 3 Weight attributed to the indicators

Criteria	Evaluation indicators	Weights	Normalized
C_1	Acceptance level of the boy or girl	0.56	0.25
C_2	Affectivity with the boy or girl	0.56	0.25
C_3	Income level	0.5	0.22
C_4	Social suitability	0.62	0.28

the present case, 3 degrees of the relationship were analyzed (Aunt, Grandmother, and Sister). Tables 4, 5 and 6 show the results of the calculation made for each degree of relationship.

Table 4, presented the processing carried out for the degree of kinship aunt, based on the criteria referred to in Table 1, the degree of kinship preferences is determined; subsequently, the process of information aggregation is carried out as part of the processing of the inference.

Table 4 presented the processing carried out for the degree of kinship grandmother; based on the criteria referred to in Table 1, the degree of kinship preferences is

Table 4 Calculation of preferences attributed to the degree of kinship aunt

Criteria	Weights	Preferences	Aggregation
C_1	0.25	0.85	0.21
C_2	0.25	0.65	0.16
C_3	0.22	1.00	0.22
C_4	0.28	0.65	0.18
Index			0.78

Table 5 Calculation of preferences attributed to the degree of kinship grandmother

Criteria	Weights	Preferences	Aggregation
C_1	0.25	0.75	0.19
C_2	0.25	0.75	0.19
C_3	0.22	1	0.22
C_4	0.28	1	0.28
Index			0.88

Table 6 Calculation of preferences attributed to the degree of kinship sister

Criteria	Weights	Preferences	Aggregation
C_1	0.25	0.65	0.16
C_2	0.25	0.50	0.13
C_3	0.22	1.00	0.22
C_4	0.28	0.65	0.18
Index			0.69

determined; subsequently, the process of information aggregation is carried out as part of the processing of the inference.

Table 4 presented the processing carried out for the degree of kinship sister; based on the criteria referred to in Table 1, the degree of kinship preferences is determined; subsequently, the process of information aggregation is carried out as part of the processing of the inference.

Figure 6 shows a graph with the behavior of the different indicators for each degree of relationship in the proposal.

Once the parental responsibility index has been calculated, the proposed method recommends, in the case of analysis, to hand over the child's custody to the degree of kinship grandmother with a parental responsibility index, $I = 0.88$.

4 Conclusions

From the development of the proposed research, a method is obtained for recommending guardianship to minors based on parental responsibility. The method bases

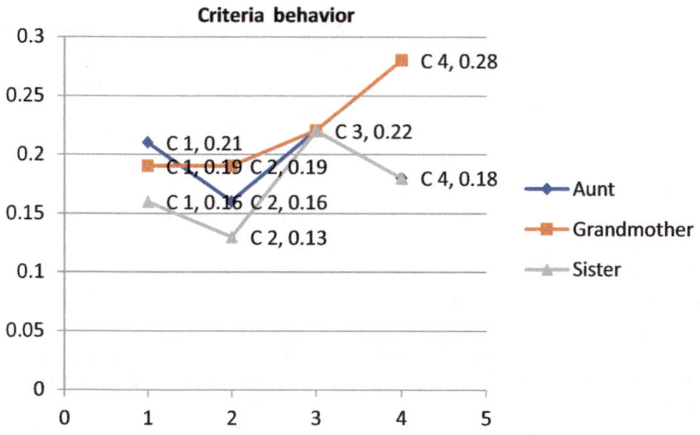

Fig. 6 Behavior of the different criteria

its operation by working in a group of experts to identify criteria with a multi-criteria approach.

With the implementation of the method, the resulting aggregated Fuzzy Cognitive Map is obtained, which expresses the knowledge of the group of experts with the representation of the causal relationships on the evaluation criteria.

The knowledge stored in the Fuzzy Cognitive Map represents the basis for the inference of the operation of the proposed method that guarantees the recommendations on child custody based on parental responsibility.

The application of the method in the case under study makes the applicability to recommend the custody of minors from parental responsibility, taking into account the set of criteria previously defined.

References

1. Zumla, A., Petersen, E.: The historic and unprecedented United Nations general assembly high level meeting on tuberculosis (UNGA-HLM-TB)—'United to End TB: an urgent global response to a global epidemic.' Int. J. Infect. Dis. **75**, 118–120 (2018). https://doi.org/10.1016/j.ijid.2018.09.017
2. Brogan, F.R.: Birds of a feather: exploring the phenomenon of voting cohesion in the United Nations General Assembly. In: Linfield University Student Symposium: A Celebration of Scholarship and Creative Achievement (2017)
3. Mas-Camacho, M.R., Acebo-del Valle, G.M., Gaibor-González, M.I., Chávez-Chacán, P.J., Núñez-Aguiar, F.R., González-Nájera, L.M., Guarnizo-Delgado, J.B., Gruezo-González, C.A.: Domestic violence and its repercussions in children in the Province of Bolivar Ecuador. Revista Colombiana de Psiquiatría (English ed.) **49**(1), 23–28 (2020). https://doi.org/10.1016/j.rcpeng.2018.04.007

4. Friedman, E., Hazlehurst, M.F., Loftus, C., Karr, C., McDonald, K.N., Suarez-Lopez, J.R.: Residential proximity to greenhouse agriculture and neurobehavioral performance in Ecuadorian children. Int. J. Hyg. Environ. Health **223**(1), 220–227 (2020). https://doi.org/10.1016/j.ijheh.2019.08.009
5. Maluf, F., Calaca, I., Freitas, P., Augusto, S.: Nature as subject of rights: a bioethical analysis of the Constitutions of Ecuador and Bolivia. Revista Latinoamericana de Bioética **18**(1), 155–171 (2018). https://doi.org/10.18359/rlbi.3030
6. Galiano-Maritan, G., Tamayo-Santana, G.: Constitutional analysis of personal rights and their relationship with the rights of" Good Living" in the Constitution of Ecuador. Revista de Derecho Privado **34**, 123–156 (2018)
7. Shiraishi Neto, J., Martins, R.L.: Rights of nature: the biocentric spin in the 2008 Constitution of Ecuador. Veredas do Direito **13**(25), 111 (2016)
8. Basantes, A., Naranjo-Toro, M., Zambrano-Vizuete, M., Botto-Tobar, M. (eds.): Internet and legislation on the protection and conservation of cultural heritage in Ecuador. In: International Conference on 'Knowledge Society: Technology, Sustainability and Educational Innovation'. Advances in Intelligent Systems and Computing, vol. 1110. Springer, Cham (2020). https://doi.org/10.1007/978-3-030-37221-7_37
9. González-Ortega, R., Oviedo-Rodríguez, M.D., Leyva-Vázquez, M., Estupiñán-Ricardo, J., Sganderia-Figueiredo, J.A., Smarandache, F.: Pestel analysis based on neutrosophic cognitive maps and neutrosophic numbers for the sinos river basin management. Neutrosophic Sets Syst. **26**(1), 16 (2019)
10. Jadán-Solís, P.Y., Auria-Burgos, B.A., Triana-Palma, M.L., Mackencie-Álvarez, C.Y., Carriel-Paredes, F.R.: Compensatory fuzzy logic model for impact. Neutrosophic Sets and Systems, Book Series, Vol. 26, p. 40: An International Book Series in Information Science and Engineering (2019)
11. Rocchi, L., Paolotti, L., Rosati, A., Boggia, A., Castellini, C.: Assessing the sustainability of different poultry production systems: a multicriteria approach. J. Clean. Prod. **211**, 103–114 (2019). https://doi.org/10.1016/j.jclepro.2018.11.013
12. Moghadas, M., Asadzadeh, A., Vafeidis, A., Fekete, A., Kotter, T.: A multi-criteria approach for assessing urban flood resilience in Tehran, Iran. Int. J. Dis. Risk Reduction **35**, 101069 (2019). https://doi.org/10.1016/j.ijdrr.2019.101069
13. Bagdanavičiūtė, I., Kelpšaitė-Rimkienė, L., Galinienė, J., Soomere, T.: Index based multi-criteria approach to coastal risk assesment. J. Coast. Conserv. **23**(4), 785–800 (2019). https://doi.org/10.1007/s11852-018-0638-5
14. Portilla, I.C.B., Sánchez, I.C.H., Tarquino, I.R.: Diffuse cognitive maps for analysis of vulnerability to climate variability in Andean rural micro-watersheds. Dyna **87**(212), 38–46 (2020). https://doi.org/10.15446/dyna.v87n212.79943
15. Zhang, Y., Qin, J., Shi, P., Kang, Y.: High-order intuitionistic fuzzy cognitive map based on evidential reasoning theory. IEEE Trans. Fuzzy Syst. **27**(1), 16–30 (2018). https://doi.org/10.1109/TFUZZ.2018.2853727
16. Efe, B.: Fuzzy cognitive map based quality function deployment approach for dishwash-er machine selection. Appl. Soft Comput. **83**, 105660 (2019). https://doi.org/10.1016/j.asoc.2019.105660
17. Álvarez-Gómez, L.K., Viteri-Intriago, D.A., Izquierdo-Morán, A.M., Manosalvas-Gómez, L.R., Acurio-Armas, J.A., Mendoza-Alcívar, M.A., Baque-Villanueva, L.K.: Use of neutrosophy for the detection of operational risk in corporate financial management for administrative. In: Neutrosophic Sets and Systems, Book Series, vol. 26: An International Book Series in Information Science and Engineering, pp. 26, 75 (2019). https://doi.org/10.5281/zenodo.3244431
18. Estumpiñan-Ricardo, J., Llumiguano-Poma, M.E., Arguello-Pazmiño, A.M., Albán-Navarro, A.D.: Neutrosophic model to determine the degree of comprehension of higher education students in Ecuador. Neutrosophic Sets Syst. **26**, 55–61 (2019)

19. Ponce-Ruiz, D.V., Albarracín-Matute, J.C., Jalón-Arias, E.J., Albarracín-Zambrano, L.O.: Soft-computing in neutrosophic linguistic modeling for the treatment of uncertainty in information retrieval. In: Neutrosophic Sets Systems, vol. 26 (2019). https://doi.org/10.5281/zenodo.3244320
20. Mar, O., Ching, I., González, J.: Operador por selección para la agregación de información en Mapa Cognitivo Difuso. Revista Cubana de Ciencias Informáticas **14**(1), 20–39 (2020)
21. McCauley, S.M., Christiansen, M.H.: Language learning as language use: a cross-linguistic model of child language development. Psychol. Rev. **126**(1), 1 (2019). https://doi.org/10.1037/rev0000126
22. Wu, Z., Xu, J., Jiang, X., Zhong, L.: Two MAGDM models based on hesitant fuzzy linguistic term sets with possibility distributions: VIKOR and TOPSIS. Inf. Sci. **473**, 101–120 (2019). https://doi.org/10.1016/j.ins.2018.09.038
23. Leyva-Vázquez, M., Pérez-Teruel, K., Febles, A., Gulín-González, J.: Modelo para el análisis de escenarios basado en mapas cog-nitivos difusos: estudio de caso en software biomédico. Ing. Univ. **17**, 375–390 (2013)
24. Papageorgiou, K., Singh, P.K., Papageorgiou, E., Chudasama, H., Bochtis, D., Stamoulis, G.: Fuzzy cognitive map-based sustainable socio-economic development planning for rural communities. Sustainability **12**(1), 1–31 (2019). https://doi.org/10.3390/su12010305
25. Mar-Cornelio, O., Santana, I., Gulín-González, J., Rozhnova, L.: Competency assessment model for a virtual laboratory system at distance using fuzzy cognitive map. Investigación Oper. **38**(2), 169–177 (2017)
26. Mar, O., Gulín, J.: Model for the evaluation of professional skills in a remote laboratory system. Revista científica **3**(33), 332–343 (2018)
27. Anninou, A.P., Groumpos, P.P.: A new mathematical model for fuzzy cognitive maps-application to medical problems. Системная инженерия и информационные технологии **1**(1), 63–66 (2019)
28. Khodadadi, M., Shayanfar, H., Maghooli, K., Mazinan, A.H.: Fuzzy cognitive map based approach for determining the risk of ischemic stroke. IET Syst. Biol. **13**(6), 297–304 (2019). https://doi.org/10.1049/iet-syb.2018.5128
29. White, E., Mazlack, D.: Discerning suicide notes causality using fuzzy cognitive maps. In: IEEE International Conference on Fuzzy Systems (FUZZ-IEEE 2011), pp. 2940–2947. Taipei (2011). https://doi.org/10.1109/FUZZY.2011.6007692
30. Leyva-Vasquez, M.Y., Delgado-Veloz, G.S., Hasan-Saleh, S., Alvarado-Roman, A.M., Alvarado-Flores, R.M.: A model for a cardiac disease diagnosis based on computing with word and competitive fuzzy cognitive maps. Revista de la Facultad de Ciencias Médicas de la Universidad de Guayaquil, **19**(1) (2018)
31. Ladeira, M.J.M., Ferreira, F.A.F., Ferreira, J.J.M., et al.: Exploring the determinants of digital entrepreneurship using fuzzy cognitive maps. Int. Entrepreneurship Manage. J. **15**(4), 1077–1101 (2019). https://doi.org/10.1007/s11365-019-00574-9
32. Giordano, R., Vurro, M.: Fuzzy cognitive map to support conflict analysis in drought management fuzzy cognitive maps. Grecia: Springer-Verlag. Vol. En M. Glykas.,/ 403–425 (2010)
33. Leyva-Vázquez, M., Smarandache, F., Ricardo, J.E.: Artificial intelligence: challenges, perspectives and neutrosophy role. (Master Conference). Dilemas Contemporáneos: Educación, Política y Valore, 6(1, special) (2018)
34. Miao, Y., Liu, Z.-Q., Kheong-Siew, C., Yan-Miao, C.: Dynamical cognitive network-an extension of fuzzy cognitive map. IEEE Trans. Fuzzy Syst. **9**(5), 760–770 (2001). https://doi.org/10.1109/91.963762
35. Amer, M., Jetter, A., Daim, T.: Development of fuzzy cognitive map (FCM)-based scenarios for wind energy. Int. J. Energy Sect. Manage. **5**(4), 21 (2011)
36. Konar, A., Chakraborty, U.K.: Reasoning and unsupervised learning in a fuzzy cognitive map. Inf. Sci. **170**(2–4), 419–441 (2005). https://doi.org/10.1016/j.ins.2004.03.012
37. Felix, G., Nápoles, G., Falcon, R., et al.: A review on methods and software for fuzzy cognitive maps. Artif. Intell. Rev. **5**(3), 1707–1737 (2019). https://doi.org/10.1007/s10462-017-9575-1

38. Alizadeh, S., Ghazanfari, M.: Learning FCM by chaotic simulated annealing. Chaos, Solitons Fractals **41**(3), 1182–1190 (2009). https://doi.org/10.1016/j.chaos.2008.04.058
39. Song, H.J., Miao, C.Y., Shen, Z.Q., Roel, W., Maja, D.H., Francky, C.: Design of fuzzy cognitive maps using neural networks for predicting chaotic time series. Neural Netw. **23**(10), 1264–1275 (2010). https://doi.org/10.1016/j.neunet.2010.08.003

Linguistic Mathematical Morphology w-operators in Fuzzy Color Space

Juan I. Pastore, Virginia L. Ballarin, and Rafael Alejandro Espin-Andrade

Abstract Color is a very important visual feature used in computer vision and image processing. Compared with grayscale images, color images can provide richer information. However, the direct extension of grayscale image algorithms to color is not always straightforward. Usually, Mathematical Morphology (MM) is based on *lattice theory*; therefore, the most elementary requirement to define morphological color operators was thought to establish an ordering of the space of the pixel intensities. Several attempts have been made, and different approaches have been presented in the last years, aiming at building a fuzzy mathematical morphology model. The situation has become more complex when trying to apply fuzzy set theory in color images because of the existence of many different ordering schemes and different definitions for the basic morphological operators. The use of fuzzy set theory is appropriate to manage the imprecision in color description. Moreover, in practical applications, it is usual to work with different color terms, whose number and design depend on the application itself. In this sense, the concept of linguistic color space is useful, among other things, for representing the set of fuzzy colors that are relevant to a certain application. In this article, we propose a novel definition of linguistic w-operators of the mathematical color morphology using diffuse definitions of color spaces, based on the original idea of binary morphology without the need to establish a grid or an order. This innovative proposal allows to reduce the ambiguity in the color description and avoid false colors.

Keywords Mathematical morphology · Fuzzy color space · Image processing

J. I. Pastore · V. L. Ballarin (✉)
Institute of Scientific and Technological Research in Electronics (ICyTE), National University of Mar del Plata-CONICET, 2290 Buenos Aires, Argentina

R. A. Espin-Andrade
Faculty of Administrative Sciences, Autonomous University of Coahuila, 25280 Saltillo, Mexico

W. Pedrycz et al. (eds.), *Computational Intelligence for Business Analytics*,
Studies in Computational Intelligence 953,
https://doi.org/10.1007/978-3-030-73819-8_15

1 Introduction

The Mathematical Morphology (MM) theory was founded by Matheron [1] and Serra [2] and became very popular in the field of non-linear image processing. Originally, the MM had been introduced as a processing technique for binary images, which were regarded as sets; therefore, its elementary operations were based on the set theory [3]. However, the extension to grayscale images, using the idea of anumbra [4, 5], allows a generalization of the basic morphological operations which were subsequently used in many image processing and analysis tasks such as morphological filtering [6], or watershed segmentation [7]. Later studies formalized this methodology of image processing for binary and grayscale images based on the application of the *lattice theory* to spatial structures, establishing an ordering of the space of the pixel intensities [1, 8].

The perception of color is of paramount importance to humans since they routinely use color features to sense the environment, recognize objects, and convey information. That is why it is necessary to use color information for computer vision because in many practical cases, segmentation of scene objects can be obtained only when color information is considered [9]. Humans are able to use a very small amount of colors in comparison with the expressive power of color spaces. A color space is a specification of a coordinate system and a subspace within that system where each color is represented by a single point. The most commonly used color space in practice is RGB because it is the one employed in hardware devices (like monitors and digital cameras). It is based on a cartesian coordinate system, where each color consists of three components corresponding to the primary colors red, green, and blue. Nevertheless, it is well known that RGB is not always the most adequate space for color image analysis. Furthermore, the color components of this space do not have an intuitive interpretation according to the human perception of color. Furthermore, there is no biunivocal link between linguistic terms and colors in a color space, but each linguistic term corresponds to a subset of colors. Unfortunately, the boundaries of such set representations are imprecise, subjective, and depend on the application domain and cultural issues.

In order to reduce the ambiguity in the color description, Soto-Hidalgo et al. [10] introduced a linguistic color space. The use of fuzzy set theory is appropriate to manage the imprecision in color description. Moreover, in practical applications, it is usual to work with different color terms, whose number and design depend on the application itself. In this sense, the concept of linguistic color space is useful, among other things, for representing the set of fuzzy colors that are relevant to a certain application.

Most of the authors propose to extend the morphological operators to color using the concept of lattice as Angulo et al. [11] and Pastore et al. [12]. Other authors model grayscale images and structuring elements as fuzzy sets proposing several different definitions for fuzzy erosion and dilation [13]. In other approaches, the morphological operations are modeled based on a fuzzy notion of distance [14].

However, although these developments solve many problems, they do not provide a unique basis for fuzzy mathematical morphology in color images.

Based on the previous analysis, we cannot assume only a single natural generalization exists, but the solution could be thought of as an obvious extension of the concepts of *background* and *object*, widely used by the binary morphology, in the fuzzy color spaces.

In this chapter, we propose a novel definition of linguistic w-operators of the mathematical color morphology using diffuse definitions of color spaces, based on the original idea of binary morphology without the need to establish a grid or an order. Operators are defined from a local window specification, in which a decision is made. Therefore, we named then w-operators. These new w-operators belong to the family of pseudo-morphologies because they are not based on an ordering of the color space.

This chapter is organized as follows. Section 2 presents the theoretical concepts of the fuzzy color spaces. Section 3 introduces the definition of the Linguistic Mathematical Morphology w-operators in fuzzy color space, and Sect. 4 shows some results. Finally, Sect. 5 discusses the proposed method and experimental its results.

2 Color Spaces

Because of the technological advances, the color information handled by multimedia systems is growing significantly. These have caused color image processing to become essential nowadays. This implies an increasing necessity of techniques for automatic color image analysis and processing. Color images need to be represented computationally in order to be processed and analyzed by computers.

An image is a dataset represented by a matrix of pixels. Each pixel represents color information, considered as a fundamental characteristic of visual content, and it is represented by a vector system. However, it is known that humans use color terms when describing it [13, 14], although there is no direct correspondence between the color representation on a computer and the terms that humans use to identify them, called *semantic gap* [12, 15]. This is one of the reasons why we address the problem of color definition from a linguistic point of view.

Besides this, because the colors imprecision, to find a clear boundary delimiting colors is gradual. In other words, the color boundary between objects is fuzzy. The definition of what is blue or not blue is also subjective, because not all of us define colors in like manner, and they are context-dependent, having different meanings in different areas.

Model color terms are in practice represented by a vector using a color space and color terms represented as linguistic labels. These are models based on a crisp quantization of the space [16–18]. These color spaces are appropriate for mathematical manipulation of color, but they are useless to human description. Some authors use probabilistic models to calculate the probability that a stimulus is assigned to a color appellation [19–22], and fuzzy models which assign a membership degree to a color appellation.

Several authors agree on the fuzzy nature of color as Rosch [23, 24], Kay and McDaniels [25] where linguistic labels are represented by fuzzy subsets of crisp colors. They proposed a fuzzy set that comprises a portion of the color space with fuzzy boundaries, defining a fuzzy partition color space. In that way, each fuzzy subset of colors corresponds to one of the color terms they use.

In this chapter, we propose to use this concept of fuzzy color spaces to define color Mathematical Morphology operators.

In the next section, the notions of fuzzy color and fuzzy color space are introduced as an extension of the classical concepts of color and color space. We review the formal definitions of the notions of fuzzy color and fuzzy color space, the different topologies of spaces, and their properties in order to be able to define linguistic w-operators of the mathematical color morphology in a novelty way.

2.1 *Fuzzy Colors and Fuzzy Color Spaces*

In order to represent the semantic compatibility between crisp colors and linguistic color terms, Soto-Hidalgo et al. [10] introduced the following definitions of fuzzy color and fuzzy color space on a generic crisp color space XYZ where the domain of components being D_X, D_Y, and D_Z:

Definition 1 A fuzzy color $\tilde{C}$ is a linguistic label whose semantics is represented in a color space XYZ by a normalized fuzzy subset of $D_X \times D_Y \times D_Z$.

Definition 2 A fuzzy color space $\widetilde{XYZ}$ is a set of fuzzy colors $\tilde{C}_1, \ldots, \tilde{C}_m$ that define a fuzzy partition of $D_X \times D_Y \times D_Z$, i.e., that satisfies:

1. $\bigcup_{\{1,\ldots,m\}} sup\left(\tilde{C}_i\right) = XYZ$, i.e., the union of the support of the $\tilde{C}_i$ covers the whole space.
2. $ker\left(\tilde{C}_i\right) \cap ker\left(\tilde{C}_j\right) = \emptyset \forall i \neq j$, i.e., the kernels of the $\tilde{C}_i$ and $\tilde{C}_j$ are pairwise disjoint.
3. $\forall i \in \{1, \ldots, m\} \exists\, \boldsymbol{c} \in XYZ$ such that $\tilde{C}_i(\boldsymbol{c}) = 1$, i.e., there is at least one object fully representative of the fuzzy color $\tilde{C}_i$.

Condition 3 is always verified by the definition of fuzzy color. Condition 1$\forall\, \boldsymbol{c} \in XYZ$, $\exists\, i \in \{1, \ldots, m\}$ such that $\tilde{C}_i(\boldsymbol{c}) > 0$. Conditions 2 and 3 imply $\tilde{C}_i \subsetneq \tilde{C}_j\ \forall i \neq j$.

In this work, we use the fuzzy color spaces proposed by Soto-Hidalgo to define the basic linguistic operators of the Mathematical Morphology. These color spaces use the color names provided by well-known ISCC-NBS system [26]. ISCC-NBS provides several color sets in the form of sets of pairs (linguistic term, crisp color). Using the methodology introduced by Soto-Hidalgo, we calculate for each color set a fuzzy color space on the basis of a Voronoi diagram of the crisp color space, calculated using the crisp colors of the set of pairs considered. The Voronoi diagram is a crisp partition corresponding to the 0.5-cut of the fuzzy colors. The kernel and support of each fuzzy color are obtained as scaling with parameters α and β respectively, with $\alpha < 1 < \beta$, and guaranteeing the conditions in definition 2. The membership functions of the fuzzy colors are obtained on the basis of distances in the crisp color space.

In that way, three fuzzy color spaces on the basis of the sets of color names Basic (13 colors), Extended (31 colors), and Complete (267 colors) are obtained in the RGB color space. For instance, the Basic set has color names corresponding to ten basic color terms (pink, red, orange, yellow, brown, olive, green, blue, violet, purple), and 3 achromatic ones (white, gray, and black). The corresponding representative crisp colors are shown in Fig. 1, together with a rough view of the core, the alpha-cuts of level 0.5, and the support of some fuzzy colors in the fuzzy color space obtained from ISCC-NBS Basic. In Fig. 1, some examples of the fuzzy color spaces for the sets extended and complete because of the lack of space are shown.

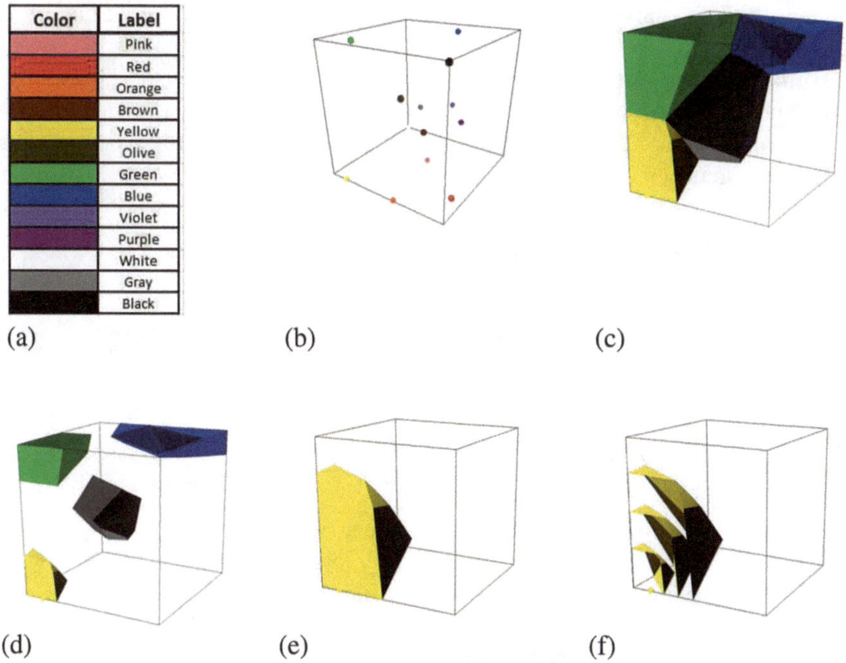

Fig. 1 Part of the RGB fuzzy color space obtained in [10] from the ISCC-NBS Basic set of colors. **a** ISCC-NBS Basic set of colors (representative crisp color and color name). **b** The situation of the representative crisp colors in the RGB color space. **c** Volumes of colors in the 0.5-cut for the fuzzy colors *yellow*, *blue*, *green*, and *gray* obtained from the Voronoi diagram in the RGB cube. **d** Volumes of colors in the kernel of the same fuzzy colors. **e** The volume of colors in support of the fuzzy color *yellow*. **f** Superimposed views of part of the surfaces of the volumes of colors in the kernel (most internal), 0.5-cut (middle), and support (most external) for the fuzzy color *yellow*

3 Linguistic Mathematical Morphology w-operators in Fuzzy Color Space

The principal idea of the Mathematical Morphology (MM) is to enlarge or reduce different objects in an image, comparing them with a small image called Structuring Element (SE). This idea leads naturally to the two basic morphological operators in binary images: *erosion* and *dilation*. The size and shape of the SE are chosen a priori depending on the morphology of the image over which it will interact and according to the shape of the objects to enlarge or reduce. The pixels in the structuring element containing a one, define the neighborhood of the structuring element. Three dimensional, or *non-flat*, structuring elements, extend the structuring element in the x–y plane, adding values to define the third dimension.

In binary images, there are only two kinds of objects: black or white objects. Two dual operators have been defined: erosion and dilation. Erosion enlarges the black objects over the white objects; dilation enlarges the white objects over the black objects. This basic idea about binary morphological operators is used, in this work, to define the linguistic morphological w-operators without the need for the definition of a lattice or order.

Formally, a binary image can be seen as a support set Ω, and X a subset of Ω. Let b be a subset of Ω called the structuring element. We assume that Ω is defined as a translation operation. The erosion $\varepsilon_b(X)$ and the dilation $\delta_b(X)$ of X according to a structuring element b are defined as follows [5, 7]:

$$\varepsilon_b(X) = \cap_{y\in b} X_{-y} = \{p \in \Omega : b_p \subset X\} = \left\{x : \forall p \in \breve{b}, x \in X_p\right\} \tag{1}$$

$$\delta_b(X) = \cup_{y\in b} X_y = \left\{p \in \Omega : X \cap \breve{b}_p \neq \emptyset\right\} = \left\{x : \exists p \in \breve{b}, x \in X_p\right\} \tag{2}$$

where $\breve{b} = \{-p : p \in b\}$ is the transpose of b (or symmetrical set with respect to the origin) and $X_p = \{x + p : x \in X\}$ the translate of X by p. To simplify, we limit the rest of our notation to symmetric structuring elements: $b = \breve{b}$.

In order to extend the binary operators to grey level operators, two concepts were introduced in the bibliography: *umbra* and *top-surface* [23].

Definition 3 Let be a grayscale image defined as a function $f : \mathbb{R}^2 \to \overline{\mathbb{R}}$, where $\overline{\mathbb{R}} = \mathbb{R} \cup \{-\infty, \infty\}$, the umbra of a function (set) f, denoted $U[f]$ is defined:

$$U[f] = \{(x, y, z) \in \mathbb{R}^3 : z \leq f(x, y)\} \tag{3}$$

Definition 4 Let be a grayscale image defined as a function $f : \mathbb{R}^2 \to \overline{\mathbb{R}}$, where $\overline{\mathbb{R}} = \mathbb{R} \cup \{-\infty, \infty\}$, the top surface of the function f (or top of the set A), denoted $T[f]$ is defined (Fig. 2).

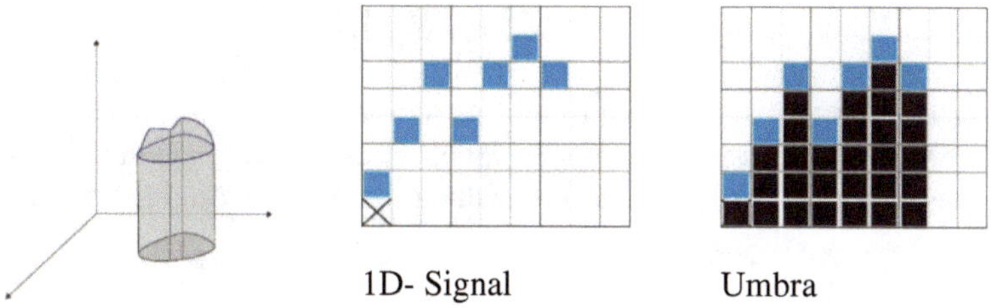

Fig. 2 Examples of umbra and top-surface for grayscale images and 1D-signal

$$T[f] = \left\{(x, f(x)) \in \mathbb{R}^3 : x \in \mathbb{R}^2 \wedge f(y) \in \overline{\mathbb{R}}\right\} \tag{4}$$

Note that $U[f]$ is now a subset of $\mathbb{R}^3$, that is to say, $U[f] \subset \mathbb{R}^3$. Based on these sets, the definitions of dilation and erosion for binary images can be easily extended to grayscale images. Being aware that the structuring element k must be extended from $\mathbb{R}^2$ to $\mathbb{R}^3$. The erosion $\varepsilon_k(f)$ and dilation $\delta_k(f)$ of the gray scale images f by the structuring element k are defined as follows [5, 7] (Fig. 3).

$$\varepsilon_k(f) = T\left[\varepsilon_{U[k]}(U[f])\right] \tag{5}$$

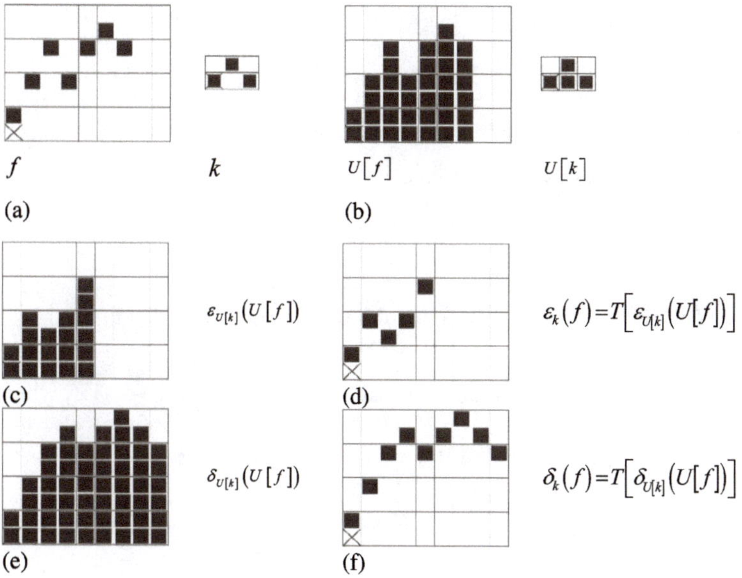

Fig. 3 Examples of erosion and dilation applied to a 1D signal. **a** Original 1D signal *f*. **b** The Umbra *U[f]*. **c** The erosion of *U[f]*. **d** The top surface of the resulted erosion. **e** The dilation of *U[f]*. **f** The top surface of the resulted dilation

$$\delta_k(f) = T\big[\delta_{U[k]}\big(U\big[f\big]\big)\big] \tag{6}$$

Going on with this approach, in the next section, we will define the linguistic w-operators of the mathematical color morphology using diffuse definitions of color spaces, based on the previous concepts, without the need to establish a grid or an order.

3.1 Definition for the Flat Structuring Elements

Definition 5 Let be I a color image defined as a function $I: \mathbb{R}^2 \rightarrow \widetilde{XYZ}$, where $\tilde{\text{XYZ}}$ is a fuzzy color space. The erosion $\varepsilon_{U_x}(I)$ and dilation $\delta_{U_x}(I)$ of the color images I by the structuring element U_x are defined as follows:

$$\varepsilon_{U_x}(I)(x) = \begin{cases} I(x) \; if & \forall z \in U_x, z \in \tilde{C}(x) = 1 \\ I(z) \; if & \exists z \in U_x / z \notin \tilde{C}(x) = 1 \wedge \Psi\big(\tilde{C}(z)\big) \geq \Psi\big(\tilde{C}(w)\big) \forall w \in U_x \end{cases} \tag{7}$$

$$\delta_{U_x}(I)(x) = \begin{cases} I(x) \; if & \forall z \in U_x - \{x\}, z \in \tilde{C}(x) \\ I(z) \; if & \exists z \in U_x - \{x\} / z \notin \tilde{C}(x) = 1 \wedge \Psi\big(\tilde{C}(z)\big) \geq \Psi\big(\tilde{C}(w)\big) \forall w \in U_x - \{x\} \end{cases} \tag{8}$$

where $\Psi: \widetilde{XYZ} \rightarrow \mathbb{R}_{\geq 0}$ is a function that measures the density of the color class $\tilde{C}_i(\boldsymbol{c})$. For this work, Ψ measures the number of pixels in the class $\tilde{C}_i(\boldsymbol{c})$ for the image I.

3.2 Definition for the Not-Flat Structuring Elements

Definition 6 Let be I a color image defined as a function $I: \mathbb{R}^2 \rightarrow \widetilde{XYZ}$, where $\widetilde{XYZ}$ is a fuzzy color space. The erosion $\varepsilon^c_{U_x}(I)$ and dilation $\delta^c_{U_x}(I)$ of the color images I by the not-flat structuring element U_x are defined as follows:

$$\varepsilon^c_{U_x}(I)(x) = \begin{cases} I(x) \; if \; \forall z \in U_x, z \in \tilde{C}(x) = 1 \\ \quad c \quad if \; \exists z \in U_x / z \notin \tilde{C}(c) = 1 \end{cases} \tag{9}$$

$$\delta^c_{U_x}(I)(x) = \begin{cases} I(x) \; if \; \forall z \in U_x, z \notin \tilde{C}(c) \\ \quad c \quad if \; \exists z \in U_x / z \in \tilde{C}(c) = 1 \end{cases} \tag{10}$$

where $\varepsilon^{c}_{U_x}(I)$ reduce the object of color **c** and $\delta^{c}_{U_x}(I)$ extends object of color **c** over their neighbors U_x.

4 Results

Figure 4 shows an example of dilation and erosion, applying the definition in Eqs. 7–8 in a synthetic color image. We also show the representative crisp colors in the RGB color space and the resulting volumes of colors in the 0.5-cut for the fuzzy colors yellow, blue, green, and gray obtained from the Voronoi diagram in the RGB cube

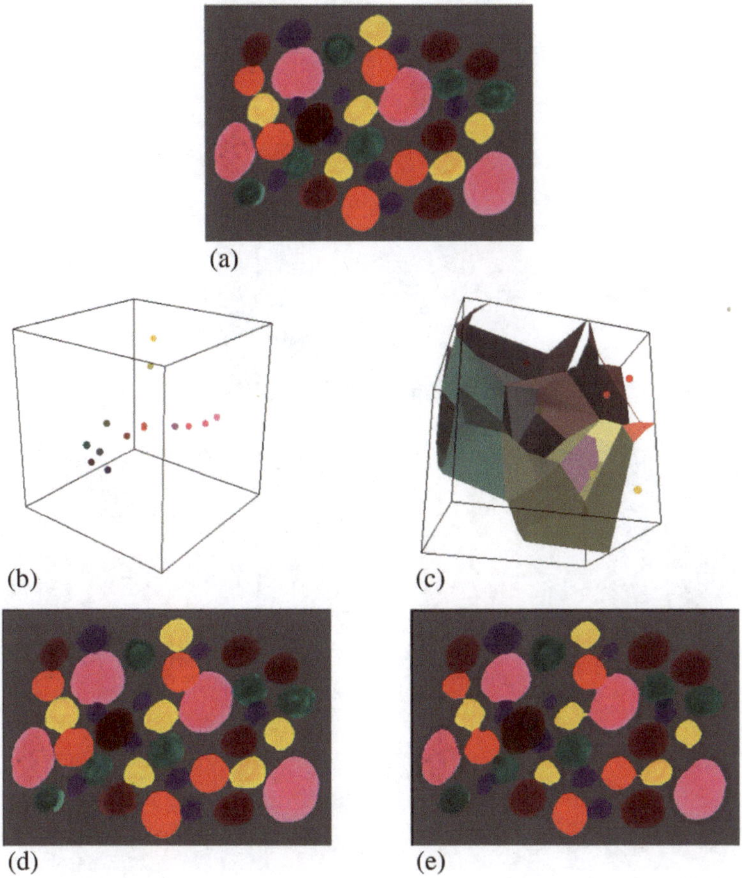

Fig. 4 Example of erosion and dilation applied to a color image. **a** Original color image. **b** The situation of the representative crisp colors in the RGB color space. **c** Volumes of colors in the 0.5-cut for the fuzzy colors yellow, blue, green, and gray obtained from the Voronoi diagram in the RGB cube. **d** Dilation. **e** Erosion

(see Fig. 4b–c). As it can be seen, the dilation (Fig. 4d) as well the erosion (Fig. 4e) works without false colors and discriminating perfectly between two similar colors like red and rose for example.

In Fig. 5, another example of dilation and erosion is shown where the definitions in Eqs. 7–8 were used too. In this case, the definition of dilation (Fig. 5b) as well as the erosion (Fig. 5c) works perfectly, even in the case of transparencies.

Figure 6 shows an example of erosion and dilation applying the definition in Eqs. 9–10 using *non-flat* symmetric elements in synthetic color images. In this figure, the background colors were chosen black and also white in order to show that using these equations, the erosion, and the dilation has the same performance that the binary operators beyond the color of the background, reducing or enlarging the objects.

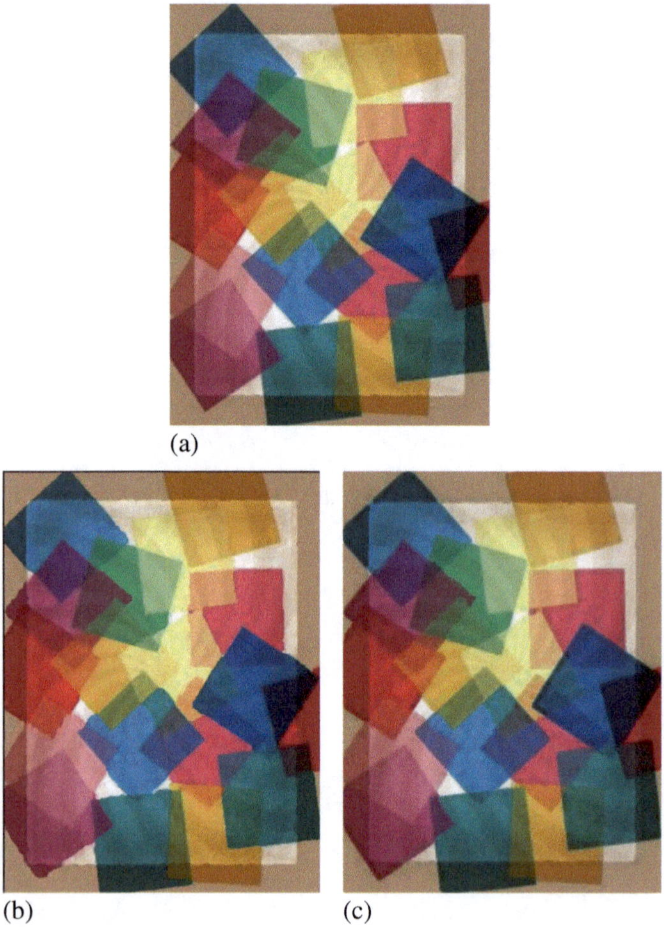

Fig. 5 Example of erosion and dilation applied to a synthetic color image. **a** Original color image. **b** Dilation. **c** Erosion

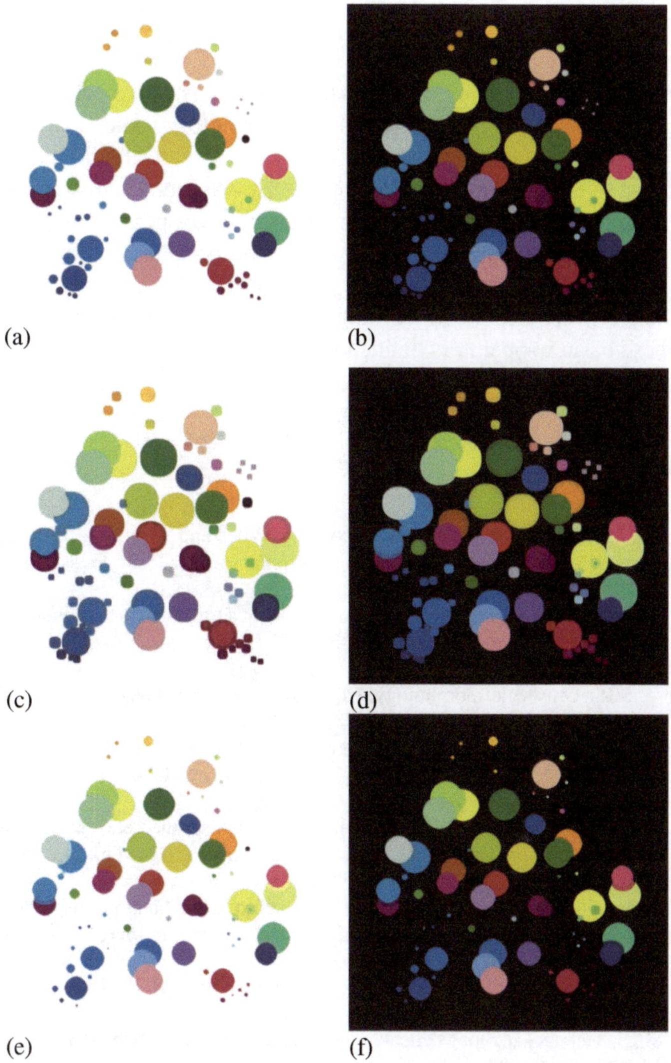

Fig. 6 Example of erosion and dilation applied to a synthetic color image using non-flat structuring elements. **a**, **b** original color images. **b**, **c** Dilation. **e**, **f** Erosion

That is because of the use of the non-flat symmetric element, where the color of the background is involved in the decision mechanism.

5 Conclusions

In this chapter, a new definition of linguistic w-operators of the color mathematical morphology was presented using diffuse definitions of fuzzy color spaces. Operators were defined based on the original idea of binary mathematical morphology without the need to establish a complete lattice in the color space. The results obtained in synthetic images show the expected behavior of these operators, even in images where the boundaries are diffuse. This innovative proposal reduces ambiguity in the description of color and also prevents the appearance of false colors.

As future work, we will go on working on the definition of morphological filters based on these basic w-operators defined in this work.

References

1. Matheron, G.: Random Sets and Integral Geometry. Wiley and Sons, New York (1975)
2. Serra, J.: Image Analysis and Mathematical Morphology. Academic Press, London (1982). https://doi.org/10.1002/cyto.990040213
3. Goutsias, J., Heijmans, H.J.: Fundamenta morphologicae mathematicae. Fundamenta Informaticae Special Issue Math. Morphol. **41**(1–2), 1–31 (2000). https://doi.org/10.3233/FI-2000-411201
4. Sternberg, S.R.: Grayscale morphology. Comput. Vis. Graph. Image Process. **35**, 333–355 (1986). https://doi.org/10.1016/0734-189X(86)90004-6
5. Haralick, R., Sternberg, S., Zhuang, X.: Image analysis using mathematical morphology. IEEE Trans. Pattern Anal. Mach. Intell. **9**(4), 532–550 (1987). https://doi.org/10.1109/TPAMI.1987.4767941
6. Maragos, P., Schafer, R.: Morphological filters–part I: their set-theoretic analysis and relations to linear shift-invariant filters. IEEE Trans. Acoust. Speech Sign. Process. **35**, 1153–1169 (1987). https://doi.org/10.1109/TASSP.1987.1165259
7. Meyer, F., Beucher, S.: The morphological approach of segmentation: the watershed transformation. In: E. Dougherty (eds.). Mathematical Morphology in Image Processing, vol. 12, pp. 43–481. Marcel Dekker, New York (1992)
8. Serra, J.: Image Analysis and Mathematical Morphology, Part II: Theoretical Advances. Academic Press, London (1988)
9. Celebi, M., Schaefer, G.: Color Medical Image Analysis. Springer, Netherlands (2013). https://doi.org/10.1007/978-94-007-5389-1
10. Soto-Hidalgo, J., Chamorro-Martínez, J., Sánchez, D.: A new approach for defining a fuzzy color space. In: IEEE International Conference on Fuzzy Systems, pp. 292–297. IEEE, Barcelona (2010). https://doi.org/10.1109/FUZZY.2010.5584426.
11. Angulo, J.: Geometric algebra colour image representations and derived total orderings for morphological operators—Part I: Colour quaternions. J. Vis. Commun. Image Represent. **21**(1), 33–48 (2010). https://doi.org/10.1016/j.jvcir.2009.10.002
12. Pastore, J., Bouchet, A., Brun, M., Ballarin, V.: New windows based color morphological operators for biomedical image processing. J. Phys: Conf. Ser. **705**(1), 1–11 (2016). https://doi.org/10.1088/1742-6596/705/1/012023
13. Gouras, P.: The Perception of Colour (Perception of Colour, ser.). CRC, Boca Raton, FL (1991)
14. Gage, J.: Color and Meaning: Art, Science, and Symbolism. University of California Press, Berkeley (2000)

15. Smeulders, A., Worring, M., Santini, S., Gupta, A., Jain, R.: Content-based image retrieval at the end of the early years. IEEE Trans. Pattern Anal. Mach. Intell. **22**(12), 1349–1380 (2000). https://doi.org/10.1109/34.895972
16. Lin, W., Lin, J.: Color quantization by preserving color distribution features. Signal Process. **78**(2), 201–214 (1999). https://doi.org/10.1016/S0165-1684(99)00060-2
17. Medeiros, A., Carvalho, P.: Color quantization by pairwise clustering using a reduced graph. Electron. Notes Discr. Math. **7**, 138–141 (2001). https://doi.org/10.1016/S1571-0653(04)00244-6
18. Cheng, S., Yang, C.: A fast and novel technique for color quantization using reduction of color space dimensionality. Pattern Recogn. Lett. **22**(8), 845–856 (2001). https://doi.org/10.1016/S0167-8655(01)00025-3
19. Caetano, T., Barone, D.: A probabilistic model for the human skin color. In: Image Analysis and Processing, pp. 279–283. IEEE, Palermo (2001). https://doi.org/10.1109/ICIAP.2001.957022
20. Andreetto, M., Zelnik-Manor, L., Perona, P.: Non-parametric probabilistic image segmentation. In: Computer Vision, pp. 1–8. IEEE, Rio de Janeiro (2007). https://doi.org/10.1109/ICCV.2007.4408968
21. Chuang, J., Stone, M., Hanrahan, P.: A probabilistic model of the categorical association between colors. In: Color Imaging (2008)
22. Lin, S., Ritchie, D., Fisher, M., Hanrahan, P.: Probabilistic color-bynumbers: suggesting pattern colorizations using factor graphs. ACM Trans. Graph. **32**(4), 1–12 (2013)
23. Rosch, E.: Prototype classification and logical classification: the two systems. In: Scholnick, E. (ed.) New Trends in Cognitive Representation: Challenges to Piaget's Theory, E. Scholnik, pp. 73–86. Hillsdale, NJ (1978)
24. Kay, P., McDaniel, C.: The linguistic significance of the meanings of basic color terms. Language **54**(3), 610–646 (1978). https://doi.org/10.1353/lan.1978.0035
25. Kelly, K., Judd, D.: Color: universal language and dictionary of names. National Bureau of Standards, Washington (1976)
26. Camps, O., Kanungo, T., Haralick, R.: Grayscale structuring element decomposition. IEEE Trans. Image Process. **5**(1), 111–120 (1996). https://doi.org/10.1109/83.481675

Method for Treatment and Its Incidence in the Change of Social Rehabilitation Regime Using Compensatory Fuzzy Logic

José Rodolfo Calle Santander, Eduardo Luciano Hernández Ramos, and Klever Aníbal Guamán Chach

Abstract Throughout life, people can make mistakes that lead to the deprivation of freedom. When people pay off their debt to society, they join a social rehabilitation regime. However, in all cases, complete rehabilitation is not obtained. This research aims to develop a method for treatment and its impact on changing the social rehabilitation regimen. The uncertainty process is modeled using compensatory fuzzy logic. A case study is implemented from which a group of people undergoing rehabilitation is taken to determine their treatment. As a result, the status of compliance with the main treatment axes was obtained for the different cases analyzed.

Keywords Social rehabilitation · Fuzzy logic · Method · Axes of treatment

1 Introduction

Deprivation of liberty constitutes the mechanism used by states to reduce unlawful conduct. Over the years, under a new legal concept, worldwide clear standards regarding the deprivation of liberty were obtained [1]. The fundamental objective of the measure is based on the search for peaceful coexistence among its inhabitants.

In case a person violates the peace and harmony of a state, it will imperatively be brought before the jurisdictional entity [2, 3]. The jurisdictional entity will apply a sanction for undertaking an action classified as a criminal offense.

People who have obtained a sentence of deprivation of liberty have the opportunity to be beneficiaries [4, 5]. The benefits are described through the axes of treatment contemplated in the Comprehensive Organic Penal Code. However, at present, it is not possible to quantify the treatment and its incidence in the change of the rehabilitation regimen [6].

Problems like the one previously exposed have been addressed in the scientific literature with Soft Computing techniques. It represents a methodology widely used

J. R. Calle Santander (✉) · E. L. Hernández Ramos · K. A. Guamán Chach
Universidad Regional Autónoma de los Andes (UNIANDES), Riobamba 060150, Chimborazo, Ecuador
e-mail: ur.josecalle@uniandes.edu.ec

W. Pedrycz et al. (eds.), *Computational Intelligence for Business Analytics*, Studies in Computational Intelligence 953,
https://doi.org/10.1007/978-3-030-73819-8_16

in situations where the data to be considered is not exact but imprecise. These imprecise data are modeled using the fuzzy set theory. From the above analysis, this research aims to: develop a method for treatment and its impact on changing the social rehabilitation regimen.

2 Preliminaries

This section introduces the fundamental elements that facilitate the understanding of the research. The main theoretical references on the social rehabilitation regime and the treatment of the axes of social rehabilitation are proposed.

2.1 *Social Rehabilitation Regime*

The National Social Rehabilitation System is administered by the Ministry of Justice; Human Rights represents the state in the regulation of custody, internal security, and effective rehabilitation of persons deprived of liberty [7].

The Social Rehabilitation System (SRS) comprises a set of principles, regulations, policies, programs, and processes that are fully correlated based on the execution of sentences. The penitentiary system seeks to execute programs that guarantee social rehabilitation before a person deprived of liberty can re-enter their family and social nucleus [8].

The treatment axes establish a set of social indicators that guarantee the social rehabilitation of the individual deprived of liberty. People deprived of liberty have the right to social reintegration, and the state guarantees their fulfillment.

The treatment of persons deprived of liberty, with a view to their rehabilitation and social reintegration, will be based on the axes: labor, education, culture and sport, Health, Family and social ties, Reintegration. The fundamental objective of the axes of rehabilitation is to guarantee comprehensive activities that generate full rehabilitation.

2.2 *Compensatory Fuzzy Logic*

The Compensatory Fuzzy Logic (CFL) represents a logical model used for the simultaneous modeling of deductive and decision-making processes [9, 10]; it represents a logical model used for the simultaneous modeling of deductive and decision-making processes [11, 12].

The CFL uses the Fuzzy Logic scale, which can vary from 0 to 1, to measure the degree of truth or falsity of its propositions. Propositions can be expressed through predicates. A predicate is a function of the universe X in the interval [0;1].

For the processing of the CFL conjunctive operators (∧), disjunction (∨), negation (¬), e implication (→) are defined in a way that restricts the domain [0;1] [13, 14].

An essential property of this logic is the "principle of gradualness", which affirms that a proposition can be both true and false, as long as it is assigned a degree of truth and falsehood. One way to put the principle of gradualness into practice is to define logics where propositions can be expressed by predicates. Precisely the logic of predicates studies the declarative phrases with a degree of detail, considering the internal structure of the propositions.

The different ways of defining operations and their properties determine different multivalent logics that are part of the Fuzzy Logic paradigm [15]. Multivalent logics are generally defined as those that allow intermediate values between the absolute truth and the total falsity of an expression. So 0 and 1 are both associated with certainty and accuracy of what is claimed or denied and 0.5 with maximum vagueness and uncertainty [16, 17].

3 Materials and Methods

For the treatment incidence in the change of social rehabilitation regime, the present method is designed. This method should show whether a person in the social rehabilitation process maintains socially responsible behavior. The method bases its operation through CFL [18, 19].

CFL is based on the geometric mean such that:

$$\begin{aligned} C_1(x_1, x_2, x_n) &= (x_1, x_2, x_n)\frac{1}{n} d_1\left(x_1, x_2, x_i, x_j, x_n\right) \\ &= 1 - [(1 - x_1)(1 - x_2)\ldots(1 - x_n)]\frac{1}{n} o_1[x, y] \\ &= 0.5[c_1(x) - c_1(y)] + 0.5n(x_i) = 1 - x_i \end{aligned} \tag{1}$$

Universal operators are defined at work with the geometric mean in CFL for the discrete domain as [20–22].

The main concepts to be modeled are described below.

Description of second level compound predicates

SRS(x): The Social Rehabilitation Regime is well valued if it adequately complies with the current legal framework and the indicators of the axes of social rehabilitation. If the report of compliance with the legal framework is somewhat unsatisfactory, it must be compensated with very good compliance with the indexes of the rehabilitation axes.

Expression of compound (third level) predicates associated with second level compound predicates.

IL(x): Labor and educational integration.

$IC(x)$: Cultural and sports integration.
$VF(x)$: Family integration.
Expression of second-level predicates in CFL predicates.
From natural or professional language to the LCD predicate, as seen in Eq. 3:

$$SRS_{(x)} = IL_{(x)} \wedge IE_{(x)}{}^{2} \wedge VF_{(x)} \wedge \left(\neg VF_{(x)} \rightarrow \left(IL_{(x)}\right)^{2} \wedge \left(IE_{(x)}\right)^{3}\right) \tag{2}$$

For the present work, a relationship is considered ($SRS(x) \rightarrow$ "Satisfaction") if the truth of the predicate is ≥ 0.9 [23, 24]. From this, the following steps are established:

1. Initial step: Reading the data to perform the discovery.
2. Execution of discovery task.
3. Evaluation of the results considering the sample.
4. Hypothesis approach: Definition of new discovery and evaluation projects under consideration.

Description of third level compound predicates

$IL(x)$: The prison system has adequate labor and educational integration.
$VF(x)$: The prison system has adequate family integration.
Expression of compound (fourth level) and simple predicates associated with third level compound predicates.
Associated Predicates $IL(x)$
$PT(x)$: The system enhances access to decent work.
$PE(x)$: The system promotes inclusion in education.
Associated Simple Predicates $VF(x)$
$IF(x)$: The system promotes adequate family integration.
$IS(x)$: The system promotes adequate social integration.
$RS(x)$: The system promotes adequate social reintegration.

Expression of third-level predicates in predicates of LCD

$$CIL_{(x)} = PT_{(x)} \wedge PE_{(x)} \tag{3}$$

$$CVF_{(x)} = IF_{(x)} \wedge IS_{(x)} \wedge RS_{(x)} \tag{4}$$

Simple Predicates Evaluation Form.
The simple predicates from which the compound predicates will be evaluated will be measured according to the fulfillment of the analyzed values of x that arise from the study of the behavior of the indicators of social rehabilitation.

4 Results and Discussions

The Fuzzy Tree associated with the CFL-based Social Rehabilitation Regime and reflects the relationships between the simple predicates to evaluate, the compound predicates, ssand the final predicate. Figure 1 shows the resulting associated tree.

To obtain the data to be analyzed, the criteria of seven penitentiary institutions was used (Table 1).

Figure 2 shows elements of simple predicates associated with compound predicates in a fuzzy tree view.

The elements assumed for the modeling of the Social Rehabilitation Regime based on the CFL in the proposed predicates are presented in Table 2.

The analysis of the Social Rehabilitation Regime was developed through the modeling of the CFL that allowed evaluating the behavior of compliance with the fundamental axes. Five penitentiary institutions were used as the object of study.

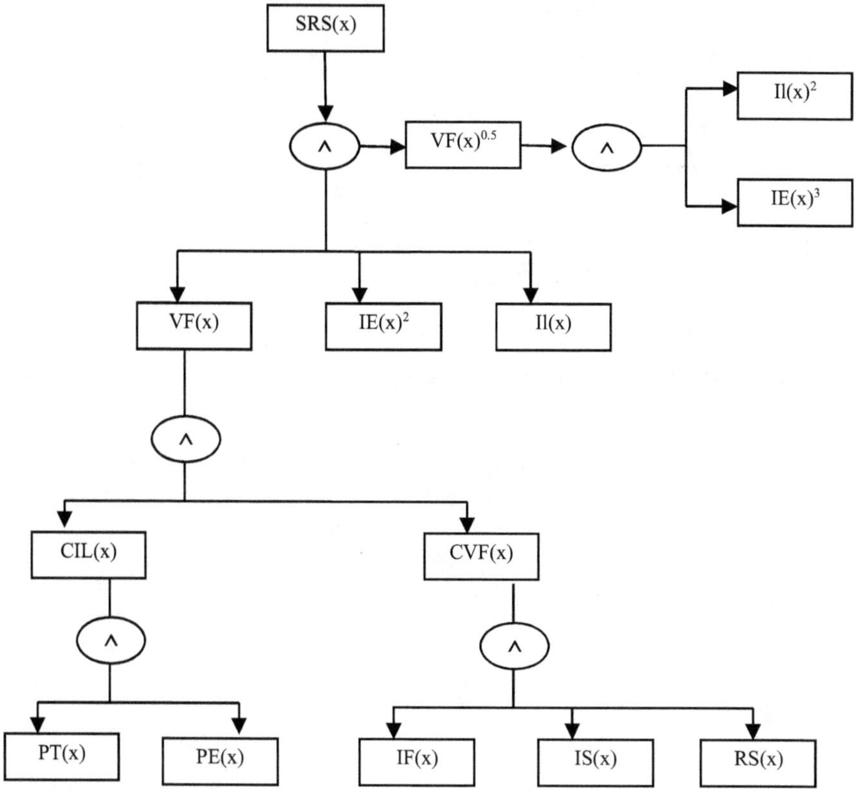

Fig. 1 Fuzzy Tree associated with the FCL-based social rehabilitation regime

Table 1 Results of the predicates in the seven institutions analyzed

Institutions	$PT(x)$	$PE(x)$	$IF(x)$	$IS(x)$	$RS(x)$
I_1	3.00	2.88	2.76	2.92	0.88
I_2	2.74	2.95	2.67	2.63	0.64
I_3	2.45	2.13	2.67	2.89	1.52
I_4	2.96	1.79	1.64	1.85	1.52
I_5	2.60	2.35	2.47	2.29	1.86
I_6	2.75	2.92	3.00	3.00	2.12
I_7	2.70	2.90	2.80	2.60	2.00

I: People from penitentiary institutions

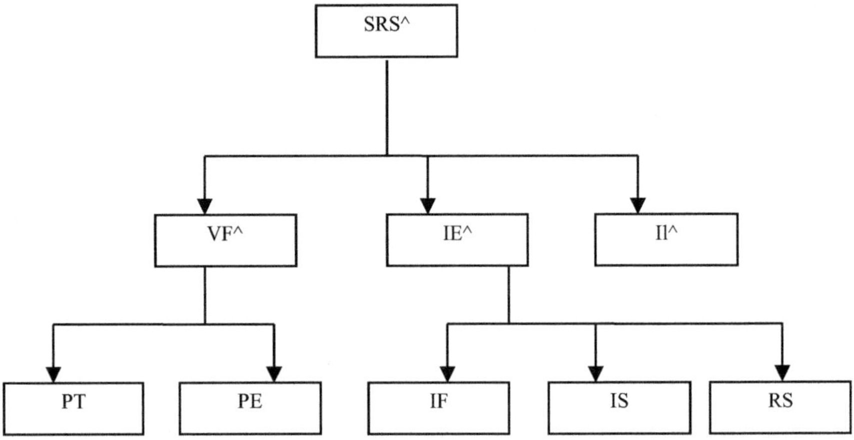

Fig. 2 Fuzzy tree of social rehabilitation based on FCL

Table 2 Social Rehabilitation values obtained through data processing

State	Scale	Incidence of Social rehabilitation	Institutions by state of Social Rehabilitation
1	0–0.2	Inappropriate	14% (1 institution)
2	0.2–0.4	Very low	14% (1 institution)
3	0.4–0.6	Adequate	42.8% (3 institution)
4	0.6–0.8	Well	28.5% (2 institution)
5	0.8–1	Excellent	

The use of the CFL for the analysis of Social Rehabilitation showed that:

- 42.8% representing 3 institutions value the fulfillment of Social Rehabilitation in an appropriate way.

- 28.5% representing 2 institutions value the fulfillment of Social Rehabilitation in a good way.
- However, 14.4% representing 1 institution values compliance with Social Rehabilitation very low.
- 14.4% representing 1 institution value the fulfillment of Social Rehabilitation in an inadequate way.

In the seven institutions taken as an example of the proposed modeling, it can be noted that:

Although the results relation shows how a compound predicate seems to present a good behavior in its internal relations, variations occur.

Variations are conditioned mainly depending on the characteristics of the institution being modeled.

5 Conclusions

The present work is based on the use of information obtained from the social rehabilitation process for the evaluation of the state of the process. It is an interesting way to link the workers of the penitentiary institutions in the evaluation of the social rehabilitation regime.

The application of a mathematical model based on the CFL constitutes an effective instrument for evaluating the treatment of the Social Rehabilitation System.

For future research, the development of knowledge bases on the behavior of treatment in social rehabilitation will be addressed. The deepening of knowledge about the subject in question to conduct further behavioral analysis is another area of research.

References

1. Gil-Betancourt, A.S., Castillo-Núñez, K.T., Cabrera-Granada, J.R., Sánchez-Ramos, H.S.: Derecho a la educación de las personas privadas de libertad en el "Centro de Privación de Libertad" de Santo Domingo. UNIANDES EPISTEME **6**, 952–965 (2020). https://doi.org/10.46377/dilemas.v33i1.2105
2. Argudo-González, E.A., Argudo-González, L.E., Tamayo-Vásquez, F.M.: Derecho penal laboral. La tipificación de infracciones penales en materia laboral en la República del Ecuador. Revista Científica FIPCAEC (Fomento de la investigación y publicación en Ciencias Administrativas, Económicas y Contables). Polo de Capacitación, Investigación y Publicación (POCAIP) **5**(16), 388–407 (2020)
3. Freire, J.M.P.: Responsabilidad Social Universitaria y su Aplicación a la Gestión de Herramientas Administrativas en el Centro de Privación Provisional de Libertad Sector Guayaquil-Ecuador. Una Mirada Reflexiva Humanística. Estudios **36**, 555–571 (2018). https://doi.org/10.15517/RE.V0I36.33514
4. Flores Vallejo, J.C.: La inserción laboral de las personas privadas de libertad con sentencia en el Sistema Penitenciario de Ambato. Pontificia Universidad Católica del Ecuador (2020)

5. Águila, M.R.F., Fuentes, P.E.C.: Los derechos fundamentales de los sancionados a privación de libertad en el Ecuador. Revista Metropolitana de Ciencias Aplicadas **2**(3), 38–47 (2019)
6. Machado-Maliza, M.E., Hernández-Gaibor, E.M., Inga-Jaramillo, M.S., Tixi-Torres, D.F.: Rehabilitación y reinserción social: Una quimera para los privados de libertad. UNIANDES EPISTEME **6**, 857–869 (2020). https://doi.org/10.46377/dilemas.v31i1.1006
7. Velásquez, S.: Prisión preventiva y Constitución del Ecuador 2008. Universidad Santiago de Guayaquil, pp. 283–292 (2016)
8. Gonzalez, J.P.: Los derechos humanos de las personas privadas de libertad. Una reflexión doctrinaria y normativa en contraste con la realidad penitenciaria en Ecuador. Revista Latinoamericana de Derechos Humanos **29**(2) (2018). https://doi.org/10.15359/rldh.29-2.9
9. Espin-Andrade, R.A., Gonzalez, E., Pedrycz, W., Fernandez, E.: An interpretable logical theory: the case of compensatory fuzzy logic. Int. J. Comput. Intell. Syst. **9**(4), 612–626 (2016). https://doi.org/10.1080/18756891.2016.1204111
10. Salas, F.G., del Toro, R.J., Espin, R., Jimenez, J.M.: An approach to knowledge discovery for fault detection by using compensatory fuzzy logic. In: Martínez-Villaseñor, L., Batyrshin, I., Marín-Hernández, A. (eds.) In Mexican International Conference on Artificial Intelligence, vol. 11835. Springer, Cham (2019). https://doi.org/10.1007/978-3-030-33749-0_31
11. Chen, C., Du, H., Lin, S.: Mobile robot wall-following control by improved artificial bee colony algorithm to design a compensatory fuzzy logic controller. In: 2017 14th International Conference on Electrical Engineering/Electronics, Computer, Telecommunications and Information Technology (ECTI-CON), pp. 856–859. IEEE, Phuket (2017). https://doi.org/10.1109/ECTICon.2017.8096373
12. Ortiz-Zezzatti, C.A.O., Rivera, G., Gómez-Santilián, C., Sánchez-Lara, B.: Use of compensatory fuzzy logic for knowledge discovery applied to the warehouse order picking problem for real-time order batching. In: Handbook of Research on Metaheuristics for Order Picking Optimization in Ware-houses to Smart Cities, pp. 62–88. IGI Global (2019)
13. Bělohlávek, R., Dauben, J. W., Klir, G. J.: Fuzzy logic and mathematics: a historical perspective. Oxford University Press (2017)
14. S.J., Singh, H., Bhutani, J., Pandit, S., S. N.N., D. Kumar, S.M.: Congestion aware algorithm using fuzzy logic to find an optimal routing path for IoT Networks. In: 2019 International Conference on Computational Intelligence and Knowledge Economy (ICCIKE), pp. 141–145. IEEE, Dubai (2019). https://doi.org/10.1109/ICCIKE47802.2019.9004351
15. Xiang, X., Yu, C., Lapierre, L., et al.: Survey on fuzzy-logic-based guidance and control of marine surface vehicles and underwater vehicles. Int. J. Fuzzy Syst. **20**(2), 572–586 (2018). https://doi.org/10.1007/s40815-017-0401-3
16. Dhimish, M., Holmes, V., Mehrdadi, B., Dales, M.: Comparing Mamdani Sugeno fuzzy logic and RBF ANN network for PV fault detection. Renew. Energy **117**, 257–274 (2018). https://doi.org/10.1016/j.renene.2017.10.066
17. Nayak, P., Vathasavai, B.: Energy efficient clustering algorithm for multi-hop wireless sensor network using type-2 fuzzy logic. IEEE Sens. J. **17**(14), 4492–4499 (2017). https://doi.org/10.1109/JSEN.2017.2711432
18. Jadán-Solis, P.Y., Auria-Burgos, B.A., Triana-Palma, M.L., Mackencie-Álvarez, C.Y.: Compensatory fuzzy logic model for impact. Neutrosophic Sets Syst. **26**, 40 (2019)
19. Bouchet, A., Pastore, J.I., Brun, M., et al.: Compensatory fuzzy mathematical morphology. SIViP **11**(6), 1065–1072 (2017). https://doi.org/10.1007/s11760-017-1058-y
20. Mellah, R., Guermah, S., Toumi, R.: Adaptive control of bilateral teleoperation system with compensatory neural-fuzzy controllers. Int. J. Control Autom. Syst. **15**(4), 1949–1959 (2017). https://doi.org/10.1007/s12555-015-0309-3
21. Ferdaus, M.M., Anavatti, S.G., Pratama, M., et al.: Towards the use of fuzzy logic systems in rotary wing unmanned aerial vehicle: a review. Artif. Intell. Rev. **53**(1), 257–290 (2020). https://doi.org/10.1007/s10462-018-9653-z
22. Braiki, K., Youssef, H.: Fuzzy-logic-based multi-objective best-fit-decreasing virtual machine reallocation. J. Supercomput. **76**, 1–28 (2020). https://doi.org/10.1007/s11227-019-03029-8

23. Leyva-Vázquez, M., Santos-Baquerizo, E., Sánchez-Delgado, M., Cárdenas-Bolaños, B., Cárdenas-Giler, D.: Performance analysis of researchers using compensatory fuzzy logic. Int. J. Innov. Appl. Stud. **19**(3), 482–486 (2017)
24. Mar-Cornelio, O., Santana, I., Gulín-González, J., Rozhnova, L.: Competency assessment model for a virtual laboratory system at distance using fuzzy cognitive map. Investigación Oper. **38**(2), 169–177 (2017)

A Proposal for Data Breach Detection in Organizations Based on User Behavior

René Palacios and Victor Morales-Rocha

Abstract Data breach has become a big problem for organizations, as the consequences can range from loss of reputation to financial loss. A data breach occurs through outsiders and insiders; however, threats from insiders are the most common and, at the same time, the most difficult to prevent. Data loss detection systems are increasingly implemented in organizations to protect information with techniques like content-based and context-based checking. Machine learning techniques have proven to be useful for data breach detection. In this work, a statistical analysis of data breach incidents is presented. Also, a user behavior characterization is made, mainly based on incidents reported by various organizations. Part of this characterization is used to create a machine learning model with a long short-term memory network with an autoencoder, in order to identify anomalies in user behavior to detect data breaches from insiders.

Keywords Data breach detection · Machine learning · Information security · Information processing · Analytics

1 Introduction

The National Institute of Standards and Technologies (NIST) [1] defines information security as "The protection of information and information systems from unauthorized access, use, disclosure, disruption, modification, or destruction to ensure confidentiality, integrity, and availability". This definition provides three information security objectives confidentiality, integrity, and availability, also known as the CIA triad.

According to the NIST standard "FIPS 199" [2], confidentiality deals with "preserving authorized restrictions on access and disclosure, including means for protecting personal privacy and proprietary information".

R. Palacios · V. Morales-Rocha (✉)
Universidad Autónoma de Ciudad Juárez, 32310 Ciudad Juárez, Chihuahua, México
e-mail: victor.morales@uacj.mx

W. Pedrycz et al. (eds.), *Computational Intelligence for Business Analytics*,
Studies in Computational Intelligence 953,
https://doi.org/10.1007/978-3-030-73819-8_17

Loss of confidentiality occurs when there is a data breach, which is defined as "An incident that involves sensitive, protected, or confidential information being copied, transmitted, viewed, stolen, or used by an individual unauthorized to do so. Exposed information may include credit card numbers, personal health information, customer data, company trade secrets, or matters of national security" [3].

The number of incidents related to data breaches increases every year, directly or indirectly affecting organizations and users around the world. In a report from the Identity Theft Resource Center [4] there were 1244 data breach incidents reported in 2018, exposing a total of 446,515,334 records. The number of exposed records had an increase of 126% compared to the previous year.

The threat of data breach has become a major problem for organizations as the consequences can range from loss of reputation to financial loss. There are two types of costs when a data breach occurs, according to [5], namely, tangible and intangible costs. Intangible costs include, but are not limited to, identity theft, criminal charges against staff members, the increased risk of future attacks on the organization, as well as loss of reputation. A report in [6] shows that when a data breach occurs, 65% of those affected lose their trust in the organization as a result of the incident, and 85% will tell others about their negative experience.

On the other hand, tangible costs refer to the loss of items directly related to the budget. Depending on the nature of the breach, a variety of financial problems can arise. For example, the costs of investigating the causes or vulnerabilities that allowed the incident to occur, the costs of restoring the data if it was deleted, the legal costs of defending against a customer, the cost due to the temporary or permanent loss of availability of the data, loss due to use of the stolen data by a competitor, the costs for paying customers who have suffered some loss or who have been defamed due to disclosure, among others. According to [7], the average cost in 2019 for a data breach was $3.9 million, and since the average of records lost that year was 25,575, the cost per record was approximately $150.

As data breach threats are a source of potential loss, it is important that organizations focus on preventing the loss of sensitive and confidential data as part of a comprehensive business intelligence strategy. A data breach occurs through outsiders and insiders; however, threats from insiders are the most common and, at the same time, the most difficult to prevent.

Data loss prevention has been addressed in different ways. According to the Forrester Wave report in [8], most of the first data loss prevention solutions focused on finding sensitive data by monitoring it at the network level. In the second stage, as removable storage devices matured, data loss prevention solutions began to focus on detecting data breach directly on the devices (workstations, servers, laptops) and providing actions, for example, avoid copying sensitive information to USB devices or CD/DVD, even when the device is not connected to the network. Protection normally begins with the ability to detect potential breach through heuristics, rules, patterns, statistics, classification, and search for anomalies. Prevention occurs as a consequence of detection [9, 10].

Data loss prevention solutions must consider three key objectives, according to [9]:

- Data loss prevention must have the ability to analyze the content and context of confidential data.
- It must be possible to implement data loss prevention to provide protection of confidential data in one or different states, that is, in transit, in use, and at rest.
- They must have the ability to protect data through various corrective actions, such as notification, auditing, blocking, encryption, or quarantine.

Techniques for preventing data breach are based on either content-based checking (analyzing the content of the file or body of text) or context-based checking (analyzing the information beyond the data itself, such as the size of the file, destination, type of file, time of delivery, among others). Machine learning techniques have proven to be useful for data breach prevention and detection. In this work, we propose to analyze users' behavior using long short-term memory network with an autoencoder to prevent a data breach from insiders.

The remainder of this work is organized as follows. Section 2 describes the methodology used in this work, which includes the understanding of the problem, the characterization of the user behavior, and the process of machine learning used to detect anomalies on user behavior. Section 3 presents the conclusions of the work and suggests future directions for research.

2 Methodology

This section describes the methodological approach used in this work. First, the causes that cause data breach in organizations are analyzed. For this purpose, a dataset containing a large number of data breach records was used. Then, we describe the characteristics that we consider to be important to create a user behavior profile, which is later used to create a model that will be approached with a machine learning technique. Finally, using the dataset, the anomalies associated with user behavior are identified.

2.1 The Problem in Numbers

An analysis of data breach has been performed with the dataset in [11]. This dataset contains data breach incidents from 2004 to 2019; each incident has at least more than 30,000 lost records. Each incident is classified according to the breach cause, and a group of incidents was analyzed qualitatively to determine the root cause of the incident. Table 1 describes the fields in the dataset used for the purposes of this work.

Figure 1 shows the number of incidents and records exposed over the years. It shows that the situation has been worsening, as the number of incidents and the number o f ecords affected increases each year.

Table 1 Fields from the dataset

Field	Description
Entity	Affected organization
Records lost	Records reported in the data breach incident
Year	Year in which the incident occurred
Story	Summary of how it happened
Sector	Affected business sector
Method	The method that caused the incident
Source name	The entity that posts the incident
1st source link	Link with the reference
2nd source link	Second link with the reference

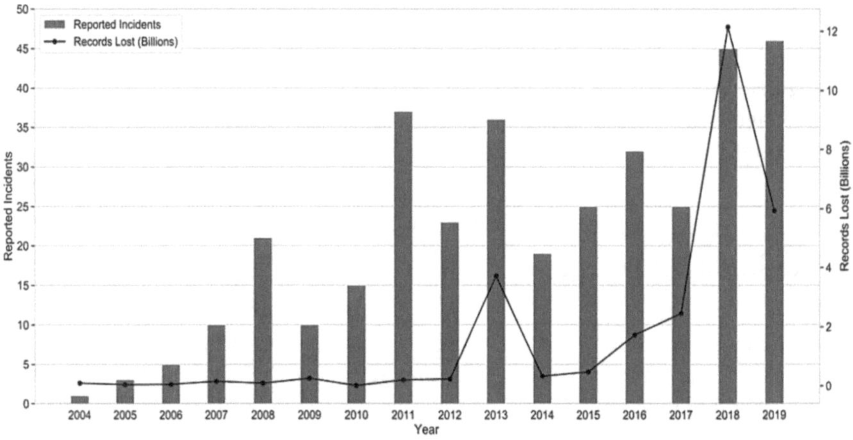

Fig. 1 Number of registered incidents and records compromised per year, from 2004 to 2019

Figure 2 lists the economic sectors most affected by a data breach in terms of incidents and compromised records. It should be clarified that the sector of large web companies, such as Facebook, Apple, Twitter, Dropbox, among others, has been ruled out in this analysis since they are usually specific targets of external intruders and represent a large part of a data breach. The focus of this work will be on organizations where a data breach is most likely due to actions of internal personnel, either accidentally or intentionally. Figure 3 shows the most affected sectors once the Web companies have been discarded.

In Fig. 4 we can see the methods used for a data breach. The hacked method accounts for 8 billions of the 16 billions of total compromised records. By obtaining the top offenders in the percentage of the total records, we can see that the top offender has been "hacked" with 53% and 8.6 billion records compromised, "poor security" with 29% and 4.7 billion records, "oops!" (accident) with 15% and 2.4

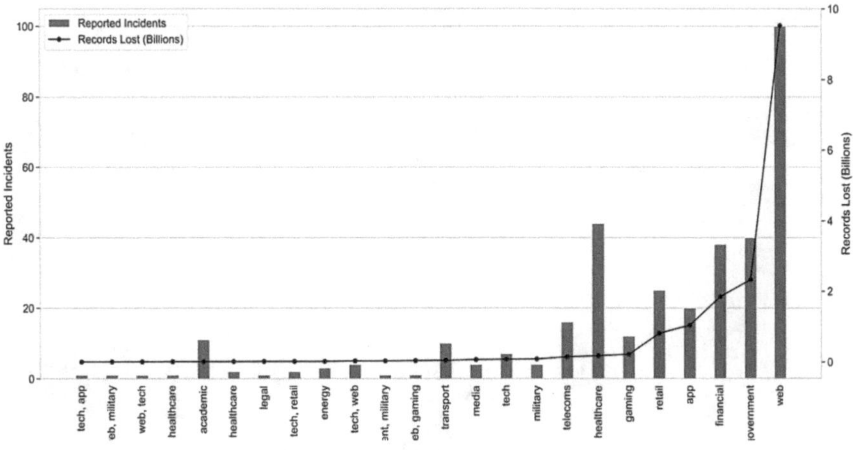

Fig. 2 Incidents and records compromised by economic sector

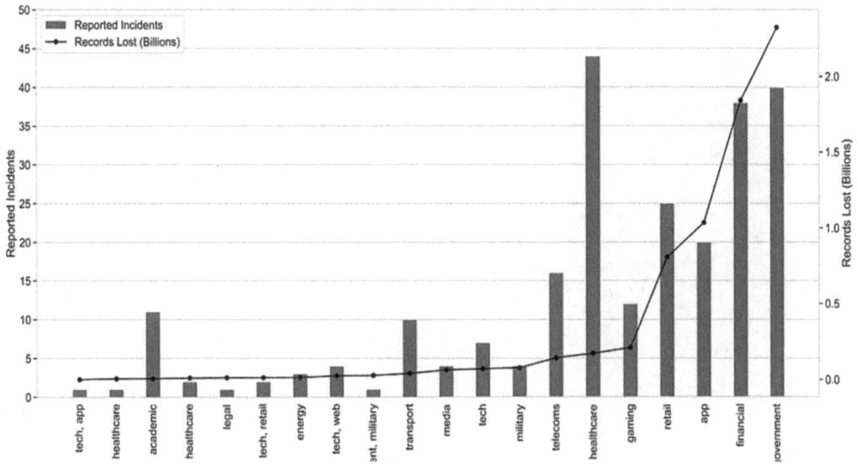

Fig. 3 Incidents and records compromised by the economic sector after removing the web sector

billion records, "inside job" with 2% and 353 million records, and lost device with 1% and 215 million records. This information can be seen in a Pareto chart in Fig. 5.

We grouped the "Oops!", "Inside job" and "lost device" categories into a single category of "insider" that represents 18% of the top offenders. Figure 6 shows the new Pareto after grouping this information.

At this point, it is clear that the "hacked" category represents 53% of data breach problems, "poor security" 29%, and 18% represents the incidents committed by an "insider". An analysis of the "hacked" category was carried out since it is assumed that some of these incidents are due to human oversights.

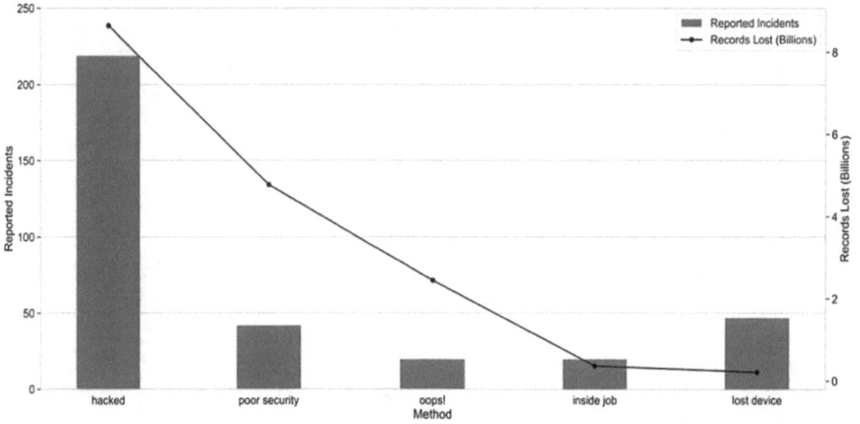

Fig. 4 Reported incidents and total records by the method used that lead to a data breach

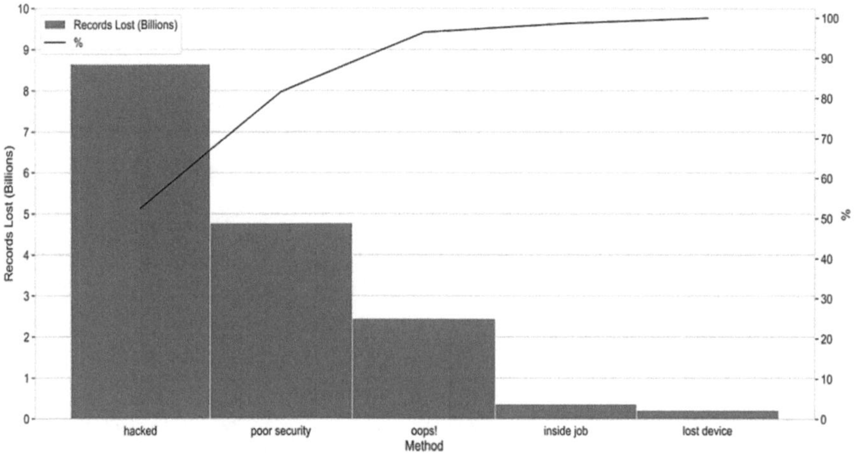

Fig. 5 Pareto chart of total records by the method used that lead to a data breach

By extracting the incidents labeled "hacked" from the previously analyzed dataset, We have a total of 133 such incidents. The calculator in [12] was used to determine a sample of 32 random incidents. These sample of incidents was empirically analyzed, and some subcategories were obtained. Moreover, the root causes that lead to a data breach incident were determined. A summary of subcategories and root causes can be seen in Table 2.

Based on the 32 randomly chosen incidents, 6 incidents were found in misuse accounts, 6 incidents related to improperly secured systems and 4 incidents in phishing attacks were carried out with techniques that did not involve a human factor directly, and 10 (misuse account and phishing attack) were a user was involved that ended up in a data breach.

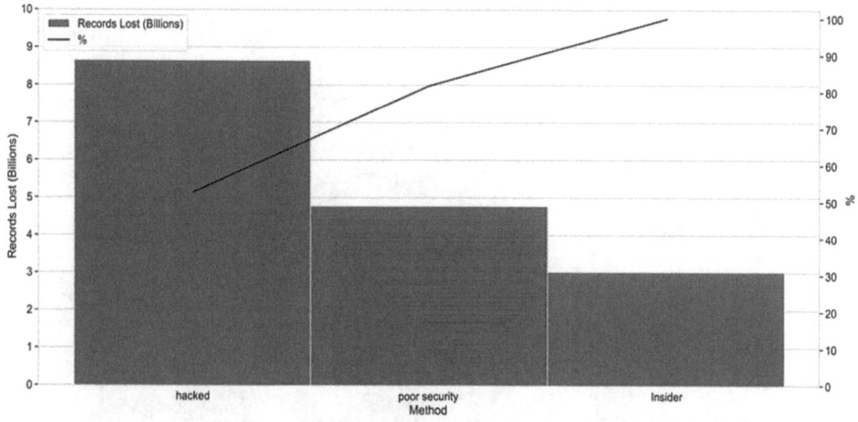

Fig. 6 Pareto chart of total records by the method used that lead to a data breach after grouping "Oops!", "Inside job" and "lost device" categories

Table 2 Description of the 32 randomly chosen incidents of method "hacked"

Subcategory	Root cause	Incidents	Total Records
Hacked	Brute force attack	2	860,083
Hacked	No details	1	270,000
Hacked	Password-guessing attack	1	57,000,000
Hacked	Vulnerability exploitation	12	49,996,000
Insider	Misuse account	6	388,150,000
Insider	Phishing attack	4	14,960,000
Poor security	Improperly secured	6	15,017,000

In Fig. 7, we can see these subcategories from the hacked category.

With the sample of 32 randomly selected incidents, a confidence level of 95% and a confidence interval of ± 20, We can conclude that of the 53% that represents the "hacked" category, 77% have been caused firstly by an "insider". With this analysis, it has been concluded that most of the data breach incidents (around 77%) are caused by an insider. An insider could be a compromised user, a careless user, or a malicious user.

2.2 *User Behavior Characterization*

We propose a user behavior characterization and features selection based on a series of public articles and reports found in the dataset previously analyzed [11].

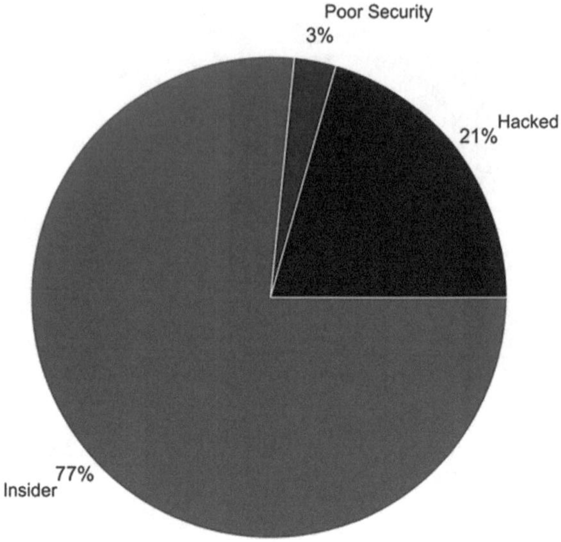

Fig. 7 Percent based pie chart of subcategories of the category hacked

The following unordered list shows examples of compromised users, careless users, and malicious users that lead a data breach in different organizations.

An example of misuse accounts or insiders can be seen in the report in [13]; In this case, around 5.2 million guest records from the Marriot hotels were accessed, apparently with the login credentials of two employees at a franchise property at the end of February 2020. "The company identified that an unexpected amount of guest information might have been accessed", these records included contact details, such as name, mailing address, email address, and phone number.

Desjardins, a financial services company, revealed in June 2019 that "an employee improperly collected information about customers and shared it with a third party outside the financial institution, which is the largest federation of credit unions in North America, with outlets across Quebec and Ontario" [14]. This is a clear example of a data breach inflicted by an insider with access to the information.

Another example of a malicious insider with access to the information occurred in June 2016 [15]. The personal details of 112,000 French police officers "have been uploaded to Google Drive in a security breach … says the details were uploaded by a disgruntled worker … Data includes home addresses."

In 2014, Korea Credit Bureau, a personal credit ratings firm revealed that "an employee has been arrested and accused of stealing the data from customers of three credit card firms while working for them as a temporary consultant" [16]. Certainly, this is another example of how a n insider act.

In 2013, a lawsuit against the Vietnamese identity theft service "contends that the theft of up to 3 million records began in 2010 and was orchestrated by Hieu Minh Ngo. Ngo, posing as a private investigator based in Singapore, gained access to a database of consumer information" [17].

In another case, in 2004 [18], the organization AOL released a statement saying that "A former America Online software engineer stole 92 million screen names and e-mail addresses and sold them to spammers who sent out up to 7 billion unsolicited e-mails."

In August 2007, a job seeker organization called Monster [19] got a trojan by a phishing email. The company said that "A trojan virus stole logins that were used to harvest usernames, e-mail addresses, home addresses, and phone numbers. Soon after, phishing e-mails encouraged users to download a Monster Job Seeker Tool, which was, in fact, a program that encrypted files in their computer and left a ransom note demanding money for their decryption." This is a clear example of a compromised user that led to a data breach.

The Australian National University [20] was a victim of unauthorized access to information. They said, "We believe there was an unauthorized access to significant amounts of personal staff, student and visitor data extending back 19 years ... by a sophisticated operator".

Medical organizations have also suffered from data breaches. In 2014, St. Vincent Medical Group [21] reported: "The investigation has required electronic and manual review of affected emails to determine the scope of the incident. Through the ongoing investigation of this matter, we determined on March 12, 2015, that the employee email account subject to the phishing contained some personal health information for approximately 760 patients".

Another company affected by a phishing email that leads to a data breach was JP Morgan [22], "affecting 76 million households and 7 million small businesses, have apparently originated with spear-phishing campaigns that target a small number of employees who have access to data systems and services housing sensitive customer information".

Based on the reports cited previously, we have identified the potential characteristics that help us to identify possible anomalies in user behavior, for example, login time, active session time, amount of data transfer, accessed directories, among others. It is clear that in many of these scenarios, the users of the organization itself are involved, either through deception, for example, when they are victims of phishing, or by carelessness, for example, users who do not comply with the security policies of their organization. Another possible scenario is when a malicious user, with legitimate access to the organization's resources, intentionally extracts data.

Table 3 contains the features used to characterize users behavior.

2.3 Scope Definition

The dataset CSE-CIC-IDS2018 [23] was used to extract all the user behavior previously defined in Sect. 2.2 with the features available in the evtx and pcap files; one of the principal characteristics of this dataset is that it has user profiles that contain abstract representations of events and behaviors seen on a network. This dataset

Table 3 Selected features of users behavior and their description

Feature	Description
Login time	Time in which users gain access to a computer system by identifying and authenticating themselves
Active session time	Time in seconds a user spends with an active valid session
Amount of usual data transfer	Amount of data a user transfer through the network
Data transfer protocol used by a user	Protocols utilized by the user (i.e., HTPPS, FTP, SSH)
Software used	List of software commonly used
Software recurrency	Recurrency of the used software
Software data amount transfer	Amount of data transferred or downloaded by the software
Web pages used	List of commonly visited web pages used by the user
Web pages data transfer	Amount of data transferred through the website
Web pages recurrency	Recurrency of the web pages visited
Accessed directories	List of commonly network directories accessed
Accessed directories data amount transfer	Amount in GB's transferred or downloaded from the directories to a local media
Accessed directories recurrency	Recurrency of access to directories
External media	List of external media connected
External media data amount transfer	Amount of data transferred or downloaded from external media
External media recurrency	Recurrency of connected media

includes an attacking infrastructure with 50 machines and a victim organization with 5 departments that includes 420 machines and 30 servers.

This dataset has *pcap* files containing packets information of the network and *evtx* files containing the list of events logged by Windows from user profiles.

All the events of the machines are saved individually in the *evtx* files in a proprietary binary format that can only be viewed within the Event Viewer program of Windows.

It is necessary to extract all the features available in the evtx files into a plain text file, specifically into a comma-separated values file, in order to process the data and train a machine learning model. To do that, a script with the capacity to extract all the features from these files and save them in comma-separated values format was created. By doing this, we can extract all the features and log information of all the machines and extract features like: date and time of the event created, time a user logged in, time that user kept an active session, programs used, time lasted with an opened program, among others. On the other hand, we extracted all data streams generated by computers on the network from the *pcap* files.

Anomaly detection is a task of finding rare events [24]. Supervised and unsupervised approaches to anomaly detection have been proposed. Some of these approaches include techniques like Bayesian networks, cluster analysis, support vector machines, and neural networks.

In this work, we used a long short-term memory autoencoder neural network to detect anomalies on user behavior. Long short-term memory networks are a type of recurrent neural network capable of learning order dependence to address sequence prediction problems. Firstly, introduced by Hochreiter and Schmidhuber [25] in 1997, and a non-comprehensive contribution of works by Gers [26], Graves and Schmidhuber [27], Wang and Nyberg [28].

There are other techniques to approach time series data such as Markov chains, multilayer perceptron, convolutional neural networks, among others; however we selected long short-term memory autoencoder neural network as in our experience, it is the easiest way to address our particular problem.

In machine learning problems, it is common to have sets of data; these sets are used to train a model and can be seen as an observation of the problem domain. The order of the observations given to the model is not important [29]. On the other hand, when we have a sequence, the order of the observations given to the model is important [30]. Sequence prediction involves predicting the next value given a sequence; for example, given an input sequence of numbers from 1 to 8 to a sequence prediction model, the expected output is 9.

An autoencoder is a type of artificial neural network used to learn features in an unsupervised way. An autoencoder attempts to learn features by training the network to ignore the noise and to force the model to learn representations of the input to assume useful properties.

In order to detect anomalies in user behavior, the autoencoder was prepared as follows:

- The autoencoder is trained on normal sequential data.
- It will be tested taking a new sequence and trying to reconstruct it using the autoencoder.
- If the error for the new sequence is superior to the defined threshold, the given element is labeled as an anomaly.

All the experiments have been done in *Jupyter Notebook,* and the programming language used is *Python.* The python library *Pandas* was used for data manipulation; the *Python* library *TensorFlow* was used to develop and train the machine learning model; the *Python* library *scikit-learn,* that provides useful algorithms for machine learning was also used; finally, the *Python* library *Matplotlib* has been used for all the visualizations presented in the following sections.

2.4 Data Preparation

Based on the information extracted, two features were selected, date and time lasted on the active session.

The feature "date" is used to express the year, month, and day a user was active, and "actTime" is used to express the active session time in seconds.

Having a look at the selected dataset in Fig. 8, we can see in a linear chart the active time feature for two months.

Before training the model, we need to standardize the dataset. Standardization of a dataset is a common requirement for many machine learning estimators as they might behave poorly or slow down the learning of the model if the individual features do not look like the standard normally distributed data. We were able to accommodate the data with the *scikit-learn* function *StandardScaler*; after that, we had a dataset that looks like Fig. 9.

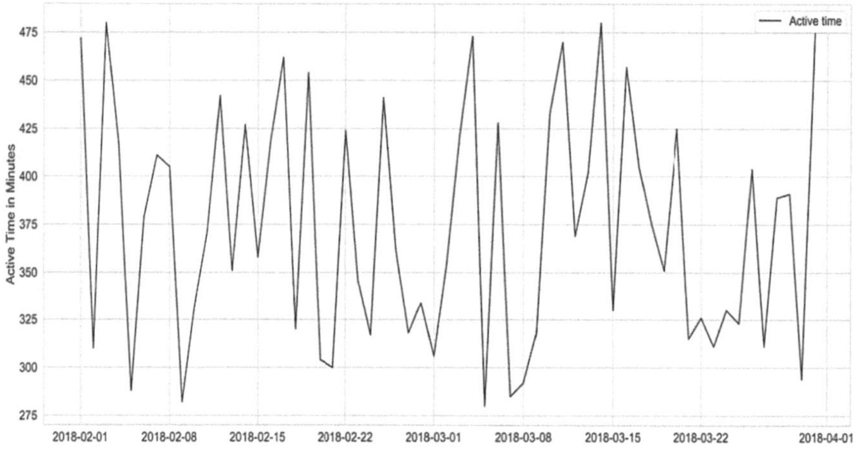

Fig. 8 Lineal representation of the active session time feature

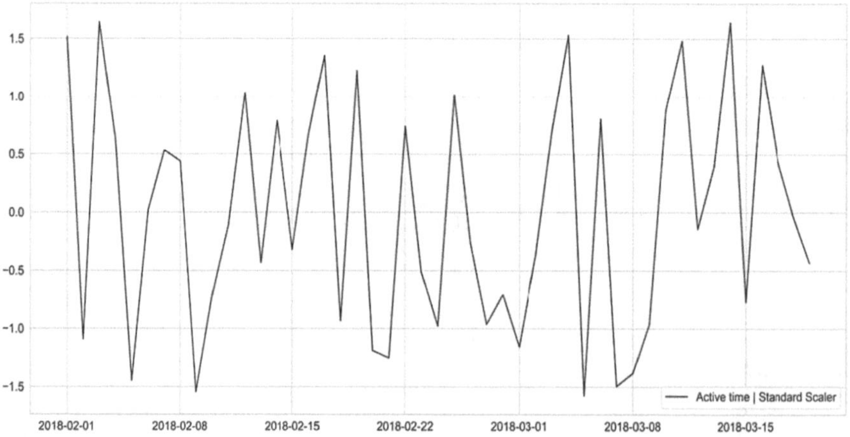

Fig. 9 Lineal representation of the active session in two months period after data rescaling

Table 4 Arguments and values used in the model configuration

Arguments	Value
Dropout rate	0.5
Compile loss	Mean absolute error
Compile optimizer	Adam algorithm

Table 5 Model layer architecture and parameters

Layer (type)	Output shape	Param #
Lstm (LSTM)	(None, 64)	16,896
Dropout (Dropout)	(None, 64)	0
Repeat_vector (RepeatVector)	(None, 2, 64)	0
lstm_1 (LSTM)	(None, 2, 64)	33,024
Dropout_1 (Dropout)	(None, 2, 64)	0
Time_distributed (TimeDistri	(None, 2, 1)	65
Total params: 49,985		
Trainable params: 49,985		
Non-trainable params: 0		

2.5 *Model Configuration*

The first step is to define a neural network in *Keras*; this network is defined as a sequence of layers contained in a Sequential class. To define a model, an instance of Sequential class is created. Layers are added to this class, and in the end, each layer can be connected. Table 4 presents the arguments and the selected values used for this model. The model was defined as follows:

- Dropout rate. Temporarily remove units from the network to prevent overfitting.
- Compile Loss. Used to judge the performance of the model minimized by the optimization algorithm.
- Compile Optimizer. Optimization algorithm to train the network.

After defining the loss function, the optimizer, and the metrics, the function *Compile* of *Keras* is used to be able to train our model. Table 5 shows the description of the layers with the values of the model.

2.6 *Model Training*

Once the model is successfully compiled without errors, it needs to be fitted or adapted according to the weights on the training dataset. To accomplish this, the training data needs to be specified with the input and output patterns (X, y). The model is trained using backpropagation through time algorithm, already defined in *Keras*, optimized

Table 6 Arguments and values used for model training

Arguments	Values
Epoch	10
Batch	32

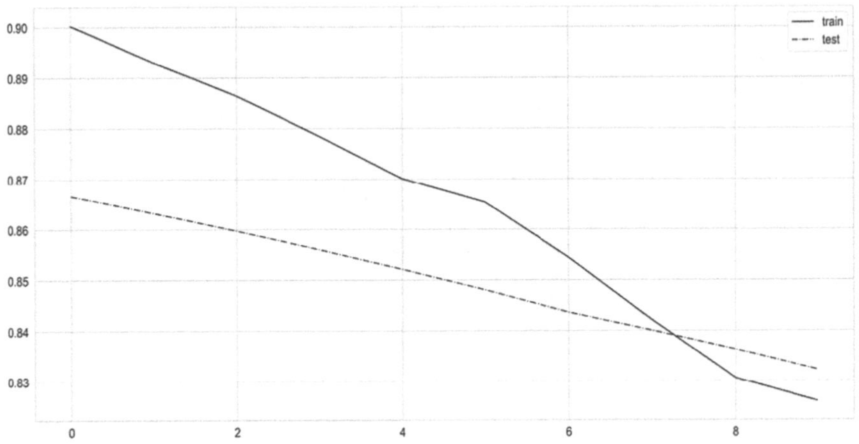

Fig. 10 Performance obtained with 10 epochs

with the Adam algorithm, and for the loss function, the mean absolute error (MAE) was defined in the model configuration. Table 6 presents the arguments used with the selected values. The model was trained with the following parameters:

- Epoch. “Used to separate training into distinct phases, which is useful for logging and periodic evaluation” [31].
- Batch. “Approximates the distribution of the input data better than a single input” [31].

Once fit, an object is returned with the information of the performance during training. We can see the performance returned in Fig. 10.

In Fig. 11, we present the MAE calculated to see the average magnitude of errors in the predictions set on the training data.

A threshold of 0.70 is defined since the loss is not greater than that. If there is an error greater than the established threshold, that element is declared as anomalous behavior. In Fig. 12, we can see the loss and all of the elements above the threshold.

2.7 Model Predictions

Once the model is fit, we can make predictions with the model, simply by calling the *Keras* function that performs a prediction with an array of new input patterns. In

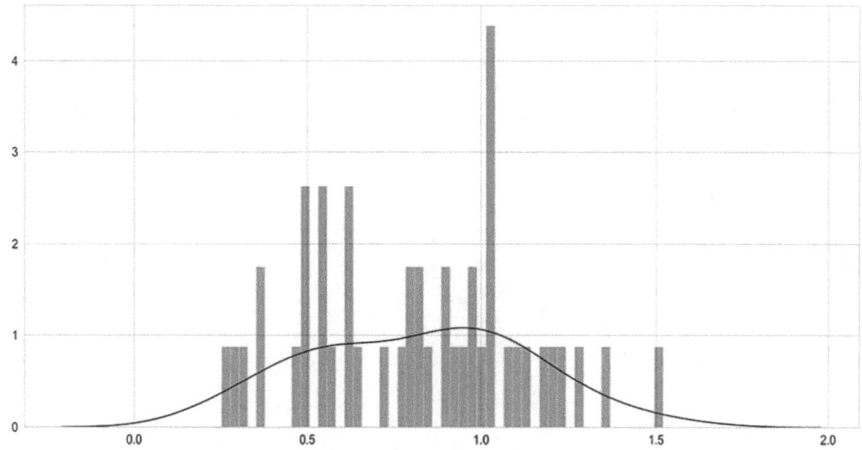

Fig. 11 Mean absolute error of the prediction set

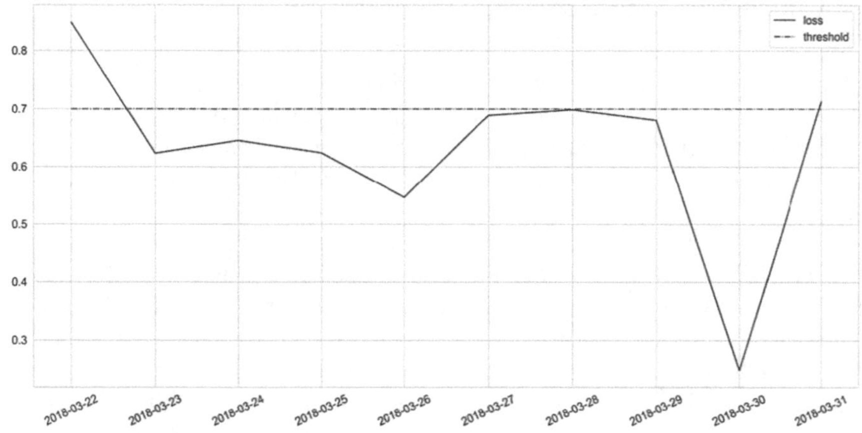

Fig. 12 Threshold and loss from the training dataset

Fig. 13, we can see the anomalies found in the testing data. The dots show the points where there is an abrupt change.

Using two features of the user behavior characterization proposed, we described our data breach anomaly detection. The combination of autoencoders and long short-term memory resulted in a model able to find anomalies on user behavior. The model shows an accuracy of 0.8169, which is considered satisfactory, especially if we take into account that our model was trained without showing a single anomaly.

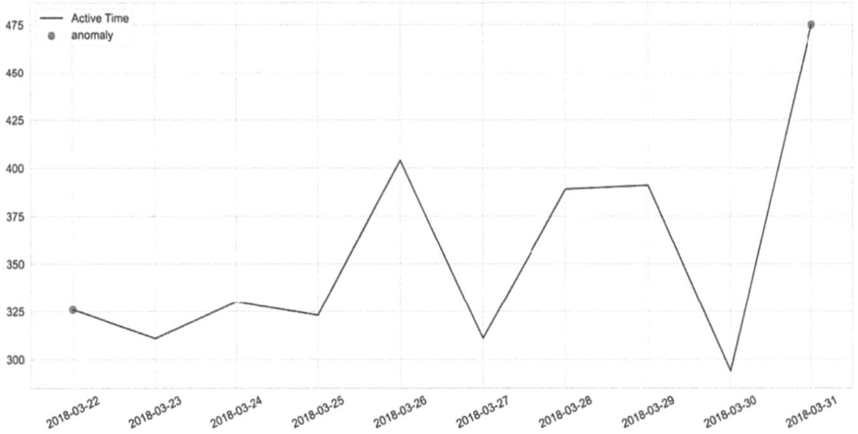

Fig. 13 Test dataset with detected anomalies (dots)

3 Conclusions and Research Directions

There is no doubt that data breach is an ongoing and relevant problem in the information security field as it affects the reputation and finances of organizations. For this reason, organizations must implement systems or mechanisms that allow them to detect and monitor data leakage attempts as part of their business intelligence strategy.

Having carried out an analysis to determine the causes of data breaches in organizations, it is concluded that computer users (insiders) are one of the main causes that lead to a data breach either compromised, careless or malicious. In this sense, characterization of user behavior has been proposed. The proposed user behavior characterization has 16 features, which can be considered as the general characteristics for the majority of users of computer equipment. However, this characterization can be adapted, either reducing or expanding the characteristics according to the needs of each organization.

In this work, a machine learning model and the combination of autoencoders and long short-term memory have been tested. This work has proven that this combination is suitable to detect anomalies in user behavior, based on the characterization proposed. Even though the model was not able to detect all the anomalies, its accuracy was around 80%. However, its accuracy could be improved, either improving the model architecture or diversifying the training data with more parameters and features.

This work can be extended in several ways. For instance, we only used two features from all the proposed characterization. In order to be able to identify a potential data breach, it should be necessary to extend this work using all the features of the characterization. As shown in this work, we can try to tune the model and work with the threshold to get better results.

Another future line of research could be the implementation of other machine learning techniques to the proposed characterization by using a single model or a combination of machine learning models to detect anomalies in user behavior.

Further, additional features can be analyzed to be added to the proposed characterization to understand all the behavior of a computer user that can lead to a data breach by accident or intentionally.

Finally, a combination of different data breach techniques like data content analysis and data context analysis, along with organizational policies such as external devices and external network communications restrictions, as well as procedural measures like user training to identify threats in the form of malicious links or attachments, could be used to have a more complete approach.

In this work, it is estimated that the use of machine learning techniques applied to the detection of a data breach will contribute favorably to the area of information security by exposing an approach to the detection of a data breach through the analysis of user behavior.

References

1. Nieles, M., Dempsey, K., Pillitteri, V.Y.: An introduction to information security. NIST Spec. Publ. **800**(12) (2017). https://doi.org/10.6028/NIST.SP.800-12r1
2. Bement, A.L.: Standards for Security Categorization of Federal In-formation and Information Systems. FIPS, 199 (2004)
3. NIST. csrc.nist.gov/glossary/term/Information_Technology_Laboratory_NIST/ (2019). Accessed 29 Jan 2020
4. The Identity Theft Resource Center. End of Year Data Breach (2019)
5. Layton, R., Watters, P.A.: A methodology for estimating the tangible cost of data breaches. J. Inf. Secur. Appl. **19**(6), 321–330 (2014). https://doi.org/10.1016/j.jisa.2014.10.012
6. The Impact of Data Breaches on Reputation and Share Value. The Ponemon Institute (2017)
7. Cost of a Data Breach Report 2019. Security I., Institute P. (2019)
8. Jaquith, A., Balaouras, S., Crumb, A.: The Forrester WaveTM: Data Leak Prevention Suites, Q4 2010 (2010)
9. Alneyadi, S., Sithirasenan, E., Muthukkumarasamy, V.: A survey on data leakage prevention systems. J. Netw. Comput. Appl. **62**, 137–152 (2016). https://doi.org/10.1016/j.jnca.2016.01.008
10. Petkovic, M., Popovic, M., Basicevic, I., Saric, D.: A host based method for data leak protection by tracking sensitive data flow. In: Proceedings—2012 IEEE 19th International Conference and Workshops on Engineering of Computer-Based Systems, ECBS 2012, pp. 267–274. IEEE, Serbia (2012). https://doi.org/10.1109/ECBS.2012.5
11. McCandless, D., Evans, T., Barton, P., et al.: World's Biggest Data Breaches. www.informationisbeautiful.net/visualizations/worlds-biggest-data-breaches-hacks/ (2020). Accessed 31 Jul 2020
12. Creative Research Systems. Sample Size Calculator. www.surveysystem.com/sscalc.htm (2012). Accessed 31 Jul 2020
13. Marriott International. Marriott International Notifies Guests of Property System Incident. news.marriott.com/news/2020/03/31/marriott-international-notifies-guests-of-property-system-incident (2020). Accessed 19 Aug 2020
14. MacFarlane, J.: 4.2 million Desjardins members affected by data breach, credit un-ion now says. www.cbc.ca/news/canada/montreal/desjardins-data-breach-1.5344216 (2019). Accessed 19 Aug 2020

15. BBC News. French police hit by security breach as data put online. www.bbc.com/news/world-europe-36645519 (2016). Accessed 12 Aug 2020
16. The Straits Times. 20 million people in South Korea fall victim to latest data leak. www.straitstimes.com/asia/20-million-people-in-south-korea-fall-victim-to-latest-data-leak (2014). Accessed 19 Aug 2020
17. Krebs, B.: Experian Sold Consumer Data to ID Theft Service. krebsonsecurity.com/2013/10/experian-sold-consumer-data-to-id-theft-service/ (2013). Accessed 19 Aug 2020
18. Wired. AOL Worker Sells 92 Million Names. www.wired.com/2004/06/aol-worker-sells-92-million-names/ (2004). Accessed 12 Aug 2020
19. BBC News. Monster attack steals user data. news.bbc.co.uk/2/hi/6956349.stm (2007). Accessed 19 Aug 2020
20. The Guardian. Australian National University hit by huge data breach. www.theguardian.com/australia-news/2019/jun/04/australian-national-university-hit-by-huge-data-breach (2019). Accessed 19 Aug 2020
21. Doe, D.: IN: St. Vincent Medical Group notifies patients after successful phishing attempt compromises PHI. www.databreaches.net/in-st-vincent-medical-group-notifies-patients-after-successful-phishing-attempt-compromises-phi/ (2015). Accessed 19 Aug 2020
22. Roman, J.: Chase Breach: Prosecutors Demand Details. www.bankinfosecurity.com/chase-breach-prosecutors-demand-details-a-7798 (2015). Accessed 19 Aug 2020
23. University of New Brunswick, Canadian Institute for Cybersecurity. A Realistic Cyber Defense Dataset (CSE-CIC-IDS2018). registry.opendata.aws/cse-cic-ids2018/ (2019). Accessed 31 Jan 2020
24. Chandola, V., Banerjee, A., Kumar, V.: Anomaly detection: a survey. ACM Comput. Surv. **41**(3) (2009). https://doi.org/10.1145/1541880.1541882
25. Hochreiter, S., Schmidhuber, J.: Long short-term memory. Neural Comput. **9**(8), 1735–1780 (1997). https://doi.org/10.1162/neco.1997.9.8.1735
26. Gers, F.A., Schmidhuber, J., Cummins, F.: Learning to forget: continual prediction with LSTM. Neural Comput. **12**(10), 2451–2471 (2000). https://doi.org/10.1162/089976600300015015
27. Graves, A., Schmidhuber, J.: Framewise phoneme classification with bidirectional LSTM and other neural network architectures. Neural Netw. **18**(5–6), 602–610 (2005). https://doi.org/10.1016/j.neunet.2005.06.042
28. Wang, D., Nyberg, E.: A long short-term memory model for answer sentence selection in question answering. In: ACL-IJCNLP 2015—53rd Annual Meeting of the Association for Computational Linguistics and the 7th International Joint Conference on Natural Language Processing of the Asian Federation of Natural Language Processing, Proceedings of the Conference, pp. 707–712. Association for Computational Linguistics (ACL), Beijing (2015). https://doi.org/10.3115/v1/P15-2116
29. Burkov, A.: The Hundred-Page Machine Learning, illustrate. Andriy Burkov (2019)
30. Brownlee, J.: Long short-term memory networks with python. Mach. Learn. Mastery Python **1**, 228 (2017)
31. Keras.: Keras FAQ. keras.io/getting_started/faq/#what-do-sample-batch-epoch-mean (2020). Accessed 2 Feb 2020

Descriptive Analytics

Estimation of the Yield Curve for Costa Rica Using Combinatorial Optimization Metaheuristics Applied to Nonlinear Regression

Andrés Quirós-Granados and Javier Trejos-Zelaya

Abstract The term structure of interest rates or yield curve is a function relating to the interest rate with its own term. Nonlinear regression models of Nelson-Siegel and Svensson were used to estimate the yield curve using a sample of historical data supplied by the National Stock Exchange of Costa Rica. The optimization problem involved in the estimation process of model parameters is addressed by the use of four well known combinatorial optimization metaheuristics: Ant colony optimization, Genetic algorithm, Particle swarm optimization, and Simulated annealing. The aim of the study is to improve the local minima obtained by a classical quasi-Newton optimization method using a descent direction. Good results with at least two metaheuristics are achieved, Particle swarm optimization and Simulated annealing.

Keywords Yield curve · Nonlinear regression · Nelson-siegel model · Svensson model · Ant colony · Genetic algorithm · Particle swarm · Simulated annealing

1 Introduction

The interest rate is essential in the modern economy; it refers to the payment of money from a debtor to a creditor by use of capital [1]. There are many factors that determine the level of interest rates: inflation risk, uncertainty, quality of information, random fluctuations, and the period of investment, among others. Remaining constant all factors affecting the level of interest rates, except the period of investment is called term structure of rates interest [1].

In a technical document authored by Bank for International Settlements [2] presented methodologies and models used by 13 nations in the estimation of the yield curve, which highlights the parametric models of Nelson-Siegel and Svensson.

A. Quirós-Granados
School of Mathematics, University of Costa Rica, San José 11501, Costa Rica

J. Trejos-Zelaya (✉)
CIMPA, University of Costa Rica, San José 11501, Costa Rica
e-mail: javier.trejos@ucr.ac.cr

W. Pedrycz et al. (eds.), *Computational Intelligence for Business Analytics*,
Studies in Computational Intelligence 953,
https://doi.org/10.1007/978-3-030-73819-8_18

One of the papers, which is an important reference, is presented by the Central Bank of Canada [3]. The paper introduces the parametric models of Nelson-Siegel and Svensson for estimating the yield curve in the Central Bank of Canada. The optimization problem was faced with two methods called partial-estimation algorithm and full-estimation algorithm. It was concluded that the optimization process could be improved. Moreover, given the large size of the search space, genetic algorithms were suggested as a method that can improve the estimation.

The stock market in Costa Rica is small; therefore, many of the existing methods are not feasible to implement. The paper of Barboza et al. [4] mentions that after reviewing existing models to estimate the curve, the most suitable for the Costa Rica market is the Svensson model. It also proposes a modification to the objective function, in order to consider topics such as historical data and volatility.

The optimization problem in the area of the yield curve for Costa Rica was worked by Piza et al. [5]. In that study, numerical methods such as Gauss-Newton, gradient descent, and Marquardt were used [6]. It was concluded that a successful optimization depends on the initial values and, also, indicated that only local minima were obtained. It is recommended the use of Metaheuristics to address the problem of finding the global minimum.

In the present paper, Metaheuristics are implemented to improve local minima that are achieved using methods that work with descent direction in the problem of estimating the parameters of the nonlinear regression models of Nelson-Siegel and Svensson.

The article is divided as follows: In Sect. 2, we present the data, and it's particularities. Section 3 describes the yield curve models used, Nelson-Siegel and Svensson, and the optimization criterion to be minimized. In Sect. 4 are presented the heuristics we have used and the characteristics of their implementation. Results in the real Costarican data are contained in Sect. 5, and we conclude in Sect. 6, mainly that Simulated annealing and Particle swarms achieved better results.

2 Data

Historical data were provided by the Bolsa Nacional de Valores (BNV, Costa Rican National Stock Exchange). These are bonds and zero-coupon bonds issued by the Central Bank and the Treasury Costa Rica, which are called *tp0*, *tp*, *bem0* and *bem*.

Data are for the period of February 23, 2015 to March 12, 2015; only emissions in colones, the Costa Rican national currency, were considered, and there is not any restriction on the amounts of transactions.

Data were provided by the BNV and contain the following information:

Description and acronym of the bond issuer.

Classification of the instrument.

Identification of the bond.

Date of issuing.

Date of expiration.

Date of next coupon.

ISIN code (for international identification).

Currency.

Periodicity: interval of times for coupon payments.

Net rate: rate paid by the coupon issuer.

Rate type: fixed, variable or without rate.

From the *book of closed operations*, we obtain the following information:

Type of operation: in the primary or secondary market.

Date of operation: day of the operation.

Nominal yield: net yield obtained by the financial instrument in operation.

Price: the price paid in the transaction, as well as

Value of transaction in colones.

From the *book of buy and sell offers,* we obtain the following information:

Offer: quotation identifier.

Facial amount: quotation amount.

Yield: quote yield, net of tax.

Price: proposed price in the quotation.

Position: indicated whether it is a buy or a sell.

In the case that a financial instrument is present several times, we keep only the last appearance in the book of closed operations. From the books of buy and sell offers, we calculate the average *bid-ask spread*; in some cases, this spread cannot be calculated since there are only buy offers or only sell offers, or there was no offer at all, in these cases the observation is not used.

Prices in these books are *clean prices*; for our estimation, we use *dirty prices*, that is, the clean price added by cumulated interests.

The database with 32 entries was reduced to 25 entries, after the elimination of observations that concentrated too much weight.

3 Yield Curve Estimation

The yield curve relates interest rates with its own term [1, 7]; this rate is called spot interest rate.

The forward interest rate is an interest that is negotiated today for a transaction that will occur in the future [8]. The forward rate is an expectation of what the spot rate will be in the future [7].

If there are continuous rates δ_t and δ_s for terms t and s, $(s < t)$, it is defined as the forward continuous rate as [2, 4]:

$$f_{t,s} = \frac{t\delta_t - s\delta_s}{t - s}.$$

The instantaneous forward rate is obtained as a limit [4, 8, 9]:

$$f_t = \lim_{s \to t} f_{t,s}.$$

The Nelson-Siegel model [10], from 1987, proposes a continuous function to describe the shape of the instantaneous forward rate depending on the term t,

$$f_t = \beta_0 + \beta_1 e^{-\lambda t} + \beta_2 \lambda t e^{-\lambda t}. \tag{1}$$

From Eq. 1 a continuous function is obtained for the spot rate,

$$\delta_t = \beta_0 + \beta_1\left(\frac{1 - e^{-\lambda t}}{\lambda t}\right) + \beta_2\left(\frac{1 - e^{\lambda t}}{\lambda t} - e^{-\lambda t}\right).$$

The Svensson model [11] extends the Nelson-Siegel model by incorporating two parameters more: $\beta_3 y \lambda_2$. Thus, the continuous function for forward rate is,

$$f_t = \beta_0 + \beta_1 e^{-\lambda_1 t} + \beta_2 \lambda_1 t e^{-\lambda_1 t} + \beta_3 \lambda_2 t e^{-\lambda_2 t} \tag{2}$$

and from (2) the function for the spot rate is

$$\delta_t = \beta_0 + \beta_1\left(\frac{1 - e^{-\lambda_1 t}}{\lambda_1 t}\right) + \beta_2\left(\frac{1 - e^{-\lambda_1 t}}{\lambda_1 t} - e^{\lambda_1 t}\right) + \beta_3\left(\frac{1 - e^{-\lambda_2 t}}{\lambda_2 t} - e^{\lambda_2 t}\right).$$

If the spot rates for different maturities are available, the price of a bond can be calculated as [7].

$$Pr = \sum_{k=1}^{t} c e^{-\delta_k k} + F e^{-\delta_t t} \tag{3}$$

where c is the coupon, and F is the face amount of the bond.

On the other hand, with a sample of bonds price the parameters of the Nelson-Siegel and Svensson models can be estimated. The estimation is obtained by minimizing the objective function (4) with respect to $\theta = (\beta_0, \beta_1, \beta_2, \lambda)$ parameters of Nelson-Siegel model or $\theta = (\beta_0, \beta_1, \beta_2, \beta_3, \lambda_1, \lambda_2)$ parameters of Svensson model.

The objective function is given by the least square criterion with weighting factors proposed in [4]. These weighting factors allow using historical observations, and it also reduces volatility through a stock measure:

$$\sum_{k=1}^{n} \frac{\left(Pr_k - \widetilde{Pr_k}\right)^2}{H_k(1 + ND_k)} \tag{4}$$

where Pr_k is the observed price for the bond k, $\widetilde{Pr_k}$ is the estimated price for the bond k obtained by (3) as a function of θ, ND_k is the number of days from the bond k was traded, and H_k is the bid-ask spread:

$$H_k = \left| \frac{\sum_{i=1}^{m_s} OS_{k,i} fS_{k,i}}{fS_{k,total}} - \frac{\sum_{i=1}^{m_b} OB_{k,i} fB_{k,i}}{fB_{k,total}} \right|$$

where $OS_{k,i}$ (respectively $OB_{k,i}$) is the i-th sell (resp. buy) offer, $fS_{k,i}$ (resp. $fB_{k,i}$) is the facial amount of the i-th sell (resp. buy) offer, and $fS_{k,total}$ (resp. $fB_{k,total}$) is the total amount of facials sell (resp. buy) offers.

A set of constraints, similar to those used in [3], have been implemented with two goals, results economically feasible and speed in the optimization process.

The following constraints are for the Nelson-Siegel model:

$$0\% < \beta_0, \beta_2 < 25\%; -20\% < \beta_1 < 20\%; 1/300 < \lambda < 12; 0 < \beta_0 + \beta_1;$$

$$1/300 < \lambda < 12. \tag{5}$$

For the Svensson model constraints are:

$$0\% < \beta_0, \beta_2, \beta_3 < 25\%; -20\% < \beta_1 < 0\%; 1/300 < \lambda_1, \lambda_2 < 12$$

$$0 < \beta_0 + \beta_1. \tag{6}$$

4 Optimization Methods

In order to minimize (4), it is frequently used nonlinear regression methods based on Gauss-Newton, gradient descent, or Marquardt iterative procedures [6]. However, it is well known that these procedures are suboptimal since they are based on local search; thus, they usually find a local minimum of the objective function. In order to avoid this suboptimality problem, in this article, the following metaheuristics were used: Genetic algorithm [12], Ant colony [13], Particle swarm [14] and Simulated annealing [15]. These metaheuristics were programmed in R [16].

The results obtained with the metaheuristics were compared with the results of the Quasi-Newton algorithm BFGS [17] applied through an adaptive barrier method [18]. To implement these methods, the built-in R [16] functions *constrOptim* and *optim* were used.

4.1 Genetic Algorithm

An algorithm based on ideas of genetic evolution and biology [19, 20]. It starts with a population of solutions chosen randomly; in each iteration, a new population is obtained from the previous one by pairing, mating, and mutation. In our implementation, we use a population of $M = 100$ chromosomes, with a chromosomic representation based on a numerical vector of nonlinear regression parameters (4 parameters for Nelson-Siegel model, 6 parameters for Svensson model). Initial chromosomes are chosen at random, satisfying the parameter constraints.

In this work, fitness is the inverse of the cost function (4). The population matrix is ranked from best to worst. The best 50% are automatically kept as an elitist selection, and the rest are replaced with the offspring generated by pairing, crossover, and mutation.

For pairing, two chromosomes, the mother $\theta^{mo} = \left(\theta_1^{mo}, \theta_2^{mo}, \ldots, \theta_p^{mo}\right)$ and the father $\theta^{fa} = \left(\theta_1^{fa}, \theta_2^{fa}, \ldots, \theta_p^{fa}\right)$ are selected with probability:

$$p_i = \frac{M/2 - i + 1}{\sum_{m=1}^{M/2} m}.$$

Crossover is as follows: a crossing point is selected as the integer part of up plus one, with $u \sim U(0, 1)$ and p the number of parameters or variables in the regression model. Position k of children is defined as

$$\theta_k^{ch1} = \theta_k^{mo} - \alpha\left(\theta_k^{mo} - \theta_k^{fa}\right), \theta_k^{ch2} = \theta_k^{fa} + \alpha\left(\theta_k^{mo} - \theta_k^{fa}\right), \alpha \sim U(0, 1).$$

Children are defined by the exchange of variables a t he right side of k:

$$child_1 = \left(\theta_1^{mo}, \ldots, \theta_{k-1}^{mo}, \theta_k^{ch1}, \theta_{k+1}^{fa}, \ldots, \theta_p^{fa}\right)$$

$$child_2 = \left(\theta_1^{fa}, \ldots, \theta_{k-1}^{fa}, \theta_k^{ch1}, \theta_{k+1}^{mo}, \ldots, \theta_p^{mo}\right).$$

If $k = p$, then all positions at left are exchanged.

A mutation operator is performed over 1% of $(M - 1) \times p$ positions in the population, excluding the best chromosome. The selected variable is replaced by a continuous random number (with uniform distribution) in the domain.

The algorithm stops if the standard deviation of fitness in the population is less than 0.5 or if the maximum number of iterations (10,000) is attained.

4.2 Ant Colony

Ant colony optimization (ACO) is a metaheuristic that takes its ideas from the way ants get food [13, 21–23]. Usual ACO is usually designed for combinatorial optimization problems. In this study, it is used the version for continuous domains presented in [22] since our case is rather continuous. The pheromones are used by means of an array that stores a number of solutions, and new solutions are built sequentially using the information of the array.

ACO will construct a solution sequentially using a Gaussian kernel,

$$G^i(x) = \sum_{l=1}^{q} w_l g_l^i(x) = \sum_{l=1}^{q} w_l \frac{e^{-(x-\mu_l^i)^2/2(\sigma_l^i)^2}}{\sigma_l^i\sqrt{2\pi}}$$

where parameters are

$$\mu^i = (\mu_1^i, \ldots, \mu_q^i) = (\theta_1^i, \ldots, \theta_q^i), \sigma_l^i = \xi \sum_{h=1}^{q} \frac{|\theta_h^i - \theta_l^i|}{k-1},$$

ξ being the evaporation rate of pheromone in ACO.

The weights are defined by:

$$w_l = \frac{e^{-(l-i)^2/2v^2q^2}}{vq\sqrt{2\pi}}$$

where l is the order of the l-th solution in decreasing order, and v is a user-defined parameter for speeding the convergence.

For constructing a solution g_l^i is chosen with probability $p_l = w_l / \sum w_{l'}$. We take a random sample with distribution g_l^i for completing a solution.

Best q solutions are stored in $P = \left(S_1, \ldots, S_q\right)$. I n our implementation we used the following parameters: 2 ants, $q = 50$, $\xi = 0.4$ locality of the search and $v = 1.1$ speed of convergence. As in the genetic algorithm, the procedure stops if the standard deviation of fitness in the population is less than 0.5, or if the maximum number of iterations (10,000) is attained.

4.3 Particle Swarm

Based on the social behavior of some groups of animals [24, 25]. The performance of an individual is influenced by its best historical performance and the best overall performance of the group up to the present iteration.

In our implementation, each particle is a vector θ in 4 dimensions (for the Nelson-Siegel model) or in 6 dimensions (for the Svensson model). We use a population $(\theta_1, \ldots \theta_M)$ of $M = 47$ particles.

Let $\theta^*(t)$ be the overall best particle and $\theta_m^*(t)$ the best value for particle m up to iteration t. Then next position of particle m in iteration $t + 1$ is:

$$\theta_m(t+1) = \theta_m(t) + v_m(t+1)$$

where $\theta_m(t)$ is its position in iteration t and

$$v_m(t+1) = w(t)v_m(t) + \lambda_1 r_1\left[\theta^*(t) - \theta_m(t)\right] + \lambda_2 r_2\left[\theta_m^*(t) - \theta_m(t)\right]$$

is the velocity vector, that defines the direction of the particle in the new iteration, with:

$$w(t) = w_{\max} - (w_{\max} - w_{\min})\frac{t}{T_{\max}}.$$

Here, λ_1 is a cognitive parameter and λ_2 is a social parameter; $r_1, r_2 \sim U[0, 1]$ are random numbers. We suppose that velocity is bounded $\left|v_{mj}(t)\right| \leq v_{\max}$ so the particles do not diverge, $w_{\max}$ and $w_{\min}$ are bounding parameters and $T_{\max}$ is the maximum number of iterations. We iterate until $\theta_m^*(t)$ does not change or iterations reach $T_{\max}$.

Taking into account the recommendations made by [26], $w = -0.1832$, $\lambda_1 = 0.5287$ as cognitive parameter and $\lambda_2 = 3.1913$ as social parameter. Stop criterion is the same as in the Genetic algorithm and Ant colonies: once the standard deviation of fitness in the population is less than 0.5 or 10,000 iterations are made.

4.4 Simulated Annealing

Based on the physical process named annealing, which takes a solid to a high temperature and then let it cool very slowly in order to get a more resistant and pure state of the solid [15, 27, 28]. Also, it uses the Metropolis criterion of acceptation whose purpose is to get out of local minimum zone [27–29]; this criterion accepts better states of the problem, but may also accept a worse state with a certain probability, that decreases as the temperature cools down.

It is well known that, from a Markov chain modeling, simulated annealing converges asymptotically to the global optimum under some conditions [15]. The basic conditions of the Markov chains are reversibility, connectedness, and length of the chains.

In this paper, it is used the version named very fast simulated reannealing [30], which allows to work with restrictions.

Let θ be a state of the problem, that is, a set of 4 or 6 nonlinear regression parameters, depending on dealing with the Nelson-Siegel or the Svensson model, respectively. A new state θ' will be defined by components generated as

$$\theta_i' = \theta_i + \lambda_i\left(\theta_{\underset{i}{\max}} - \theta_{\underset{i}{\min}}\right)$$

where $\lambda_i \in [-1, 1]$ and $\theta_{\underset{i}{\max}}, \theta_{\underset{i}{\min}}$ are bounds of the i-th parameter. Let T be the simulated annealing temperature, we use λ_i distributed with

$$g_T(\lambda_i) = \frac{1}{2(|\lambda_i| + T)\ln(1 + 1/T)}$$

where λ_i is generated as $\lambda_i + \text{sgn}(u - 0.5)T[(1 + 1/T)^{|2u-1|} - 1]$, and $u \sim U[0, 1]$.

The size of the Markov chain was established in 100, and the temperature is updated with the factor 0.95, that is $T_{k+1} = 0.95T_k$.

For estimating the initial temperature, we follow [1]. Given a value $\chi_0 \approx 0.95$ that represents the fact that in the beginning, almost 95% of new states that worsen the objective function F in Eq. (4) will be accepted in the Metropolis rule. Then, is we make 1000 blank iterations let m_1 be the number of times that F decreases and m_2 the number of times that F increases; if $\overline{\Delta F}^+$ is the average in F differences for those blank iterations that increase the value of f, then T_0 is estimated with

$$T_0 = \overline{\Delta F}^+ / \ln\left(\frac{m_2}{m_2\chi_0 - m_1(1 - \chi_0)}\right).$$

Metropolis rule works as follows: a new state is accepted if f decreases, or it is accepted with probability

$$\exp(-\Delta F/T),$$

$$\text{where } \Delta F = F(\theta') - F(\theta).$$

The iterations stop when $T \approx 0$, or a maximum number of iterations is reached, or after one complete Markov chain, there are no improvements in the cost function.

4.5 An Adaptive Barrier with a BFGS Quasi-Newton Algorithm

BFGS algorithm is a local search method [17] where the search is given by a modified Newton direction.

Let $\theta(t)$ be the current state of the problem; the new state is defined by a vector direction $p(t)$ as in several descent methods:

$$\theta(t+1) = \theta(t) + \alpha_t p(t)$$

such that $F(\theta(t+1)) \leq F(\theta(t))$, where $\alpha_t \in \mathbb{R}$, $\alpha_t = \underset{(\alpha>0)}{\arg\min} F(\theta(t) + \alpha p(t))$.

With a second order Taylor approximation for $F(\theta(t) + p(t))$ it is obtained

$$p(t) = -\left(\nabla^2 F(\theta(t))\right)^{-1} \nabla F(\theta(t))$$

supposing $\nabla^2 F(\theta(t))$ is positive definite. In the BFGS algorithm [20], Hessian is replaced by an approximation calculated in each iteration:

$$H_{t+1} = \left(I - \frac{z(t)y'(t)}{y'(t)z(t)}\right) H_t \left(I - \frac{y(t)z'(t)}{y'(t)z(t)}\right) + \frac{z(t)z'(t)}{y'(t)z(t)}$$

$$z(t) = \theta(t+1) - \theta(t), y(t) = \nabla F(\theta(t+1)) - \nabla F(\theta(t)).$$

In order to satisfy the constraints in the Nelson-Siegel and Svensson models, we have used an adaptive barrier method, that transforms the minimization problem

$$\min_{\theta} F(\theta) \textit{ subject to } L_j(\theta) = u_j^{'}\theta - c_j \geq 0, j = 1, \ldots, p$$

into

$$\min_{\theta} F(\theta) - \mu \sum_{i=1}^{p} \left[L_j(\vartheta_k) \ln L_j(\theta) - u_j^{'}\theta \right],$$

where $F(\theta)$ has been added with a so-called logarithmic barrier that considers the regression constraints, ϑ_k is an interior point of the feasible region. Parameter μ tends to 0 with the goal to neglect more and more the barrier [18].

The objective function depends on the vector θ which has to satisfy (5) or (6) so that the constrained optimization problem is changed into an unconstrained problem, and an adaptive barrier method is used [31]. In this case, a logarithmic barrier is added to the objective function in order to handle the constraints (5) or (6). If the minimum lies on the boundary, the barrier will not allow to reach it; to deal with this, the logarithmic barrier has a component that changes in each iteration [18].

In the minimization of the barrier method, the BFGS procedure is used.

5 Results

For each method, a multistart strategy [25] of size 2000 was made. The way of comparison is as follows: the best objective function value for the metaheuristics is the expected value from their multistart; in the case of the adaptive barrier, the best objective function value is the minimum value that was achieved from its multistart.

Tables 1 and 2 contain a summary of the results for the Nelson-Siegel and Svensson models, respectively. The following values are reported: the objective function value, the coefficient of variation information taken from the multistart, the goodness of fit, and the average time of running the R-code measured in seconds.

Table 1 Summary metaheuristics performance in estimating the Nelson-Siegel model

Algorithm	Objective function value	Coefficient of variation (%)	Goodness of fit (%)	Average time (m)
Particle swarm	441.5018	<1	0.003345	00:22
Simulated ann	441.5034	<1	0.003353	00:28
Adaptive barr	441.5243	240	0.003354	–
Genetic alg	1206.0571	18	0.066502	01:16
Ant colony	1207.6136	36	0.066541	00:18

Table 2 Summary metaheuristics performance in estimating the Svensson model

Algorithm	Objective function value	Coefficient of variation (%)	Goodness of fit (%)	Average time (m)
Particle swarm	251.5805	<1	0.012147	00:43
Simulated ann	251.6899	<1	0.012550	00:46
Ant colony	254.6444	84	0.012345	01:32
Adaptive barr	441.6267	317	0.003164	–
Genetic alg	1138.3852	12	0.052407	00:13

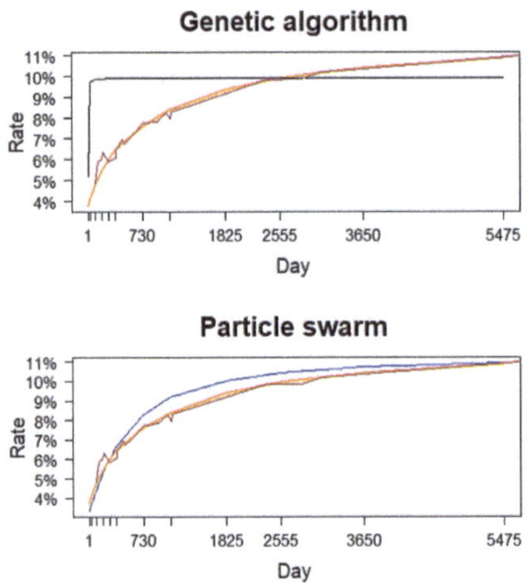

Fig. 1 Yield curve and estimated yield curve with Nelson-Siegel model for March 17, 2015

The results for the Nelson-Siegel model are shown in Table 1. Two metaheuristics got better results than the adaptive barrier method, namely, Particle swarm and Simulated annealing. Their coefficients of variation are almost zero indicating that the same results are obtained almost every time the functions are run. The average time is approximately 20 s.

Figure 1 shows graphically the yield curves obtained with the four metaheuristics for Nelson-Siegel model.

In the case of the Svensson model (see Table 2), three metaheuristics had better performance than the adaptive barrier: Particle swarm, Simulated annealing, and Ant colony. But we highlight Particle swarm and Simulated annealing, which have a coefficient of variation almost zero and an average time of 40 s.

In Fig. 2 are shown graphically the yield curves obtained for the Svensson model.

6 Concluding Remarks and Further Research

Two metaheuristics were better in both models, Particle swarm and Simulated annealing. These metaheuristics, besides having the best results, their algorithms are easy to implement, the execution time is acceptable, and the outcomes are very stable.

Therefore, Particle swarm and Simulated annealing are recommended for getting the parameters of the Nelson-Siegel and Svensson models.

For future research, it is suggested to repeat this study with other sets of sample data so as to confirm the results obtained so far. For another financial data set, similar restrictions for (5) or (6), which are adjusted to the Costa Rican market characteristics, should be determined. Moreover, parameters tuning can be improved with a factorial

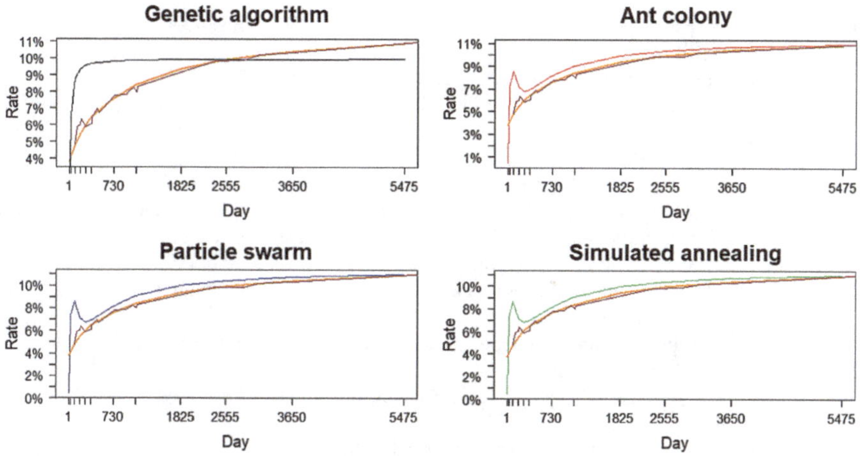

Fig. 2 Yield curve and estimated yield curve with Svensson model for March 17, 2015

design that could suggest better choices. Finally, a review of the configuration used in Genetic algorithm and Ant colony could also be made in order to obtain satisfactory parameters that may make compete for these metaheuristics with the better ones. Also, we will perform further studies with simulated data and controlled parameters, and the use of benchmark data will also be considered.

The implementation of nonlinear regression with some other metaheuristics, such as artificial bee colony [32], bat algorithm [33], and differential evolution [34] are under study. Once the implementation is fine, the application to financial data such as the estimation of the yield curve will be performed.

References

1. Kellison, S.: The Theory of Interest, 3rd edn., Irwin McGraw-Hill. Massachusetts (2008)
2. Bank for International Settlements: Zero-coupon yield curves: technical documentation. BIS, Monetary and Economic Department (2005). https://doi.org/10.2139/ssrn.1188514
3. Bolder, D., Streliski, D.: Yield curve modeling at the Bank of Canada. Technical Report No. 84, Bank of Canada (1999). https://doi.org/10.2139/ssrn.1082845
4. Barboza, L., Ramírez, A., Víquez J.J.: Zero-Coupon Yield Curve for Costarican Market. Preprint, Risk Section, National Bank of Costa Rica, June, San José (2005)
5. Piza, E., Trejos, J., Bermúdez, E.: Optimization of yield curves in Costa Rican market. In: Proceedings of the II International Conference on Optimization and Control, Ulaan Batar (2008)
6. Draper, N.R., Smith, H.: Applied Regression Analysis. Wiley, New York (1969)
7. Hull, J.: Options, Futures, and Other Derivatives, 7th edn. Pearson Prentice Hall, New Jersey (2009)
8. Pereda, J.: Estimation of zero-coupon yield curve for Peru and its use for money analysis. Economía **33**(65), 103–132 (2010)
9. Kladívko, K.: The Czech Treasury yield curve from 1999 to the present. Czech J. Econ. Finan. **60**(4), 307–335 (2010)

10. Nelson, C., Siegel, A.: Parsimonious modeling of yield curves. J. Bus. **60**(4), 473–489 (1987). https://doi.org/10.1086/296409
11. Svensson, L.: Estimating and interpreting forward interest rates: Sweden 1992–1994. NBER Working Paper Series 4871, pp. 1–49 (1994). https://doi.org/10.5089/9781451853759.001
12. Goldberg, D.E.: Genetic Algorithms in Search, Optimization and Machine Learning. Addison-Wesley, Boston (1989)
13. Dorigo, M., Stützle, T.: Ant Colony Optimization. MIT Press, Cambridge (2004)
14. Kennedy, J., Eberhart, R.C.: Intelligent Swarm Systems. Academic Press, New York (2000)
15. Aarts, E., Korst, J.: Simulated Annealing and Boltzmann Machine. A Stochastic Approach to Combinatorial Optimization and Neural Computing. Wiley, Chichester (1989)
16. R Core Team: A language and environment for statistical computing. R Foundation for Statistical Computing, Vienna (2013)
17. Nocedal, J., Wright, S.: Numerical Optimization, 2nd edn. Springer, New York (2006). https://doi.org/10.1007/978-0-387-40065-5
18. Lange, K.: Numerical Analysis for Statisticians. Springer, New York (1999). https://doi.org/10.1007/b98850
19. Haupt, R., Haupt, S.: Practical Genetic Algorithms, 2nd edn. Wiley, New Jersey (2004)
20. Mitchell, M.: An Introduction to Genetic Algorithms. MIT Press, Massachusetts (1996)
21. Socha, K., Dorigo, M.: Ant colony optimization for continuous domains. Eur. J. Oper. Res. **185**, 1155–1173 (2008). https://doi.org/10.1016/j.ejor.2006.06.046
22. Socha, K.: Ant colony optimization for continuous and mixed-variable domains. Unpublished Doctoral Dissertation, Université Libre de Bruxelles, Brussels (2008)
23. Gómez-Santillán, C., Cruz-Reyes, L., Schaeffer, E., Meza, E., Rivera-Zarate, G.: Adaptive ant-colony algorithm for semantic query routing. J. Autom. Mobile Robot. Intel. Syst. **5**, 85–94 (2011)
24. Parsopoulos, K., Vrahatis, M.: Particle Swarm Optimization and Intelligence Advances and Applications. Information Science Reference, New York (2010)
25. Yang, X.: Engineering Optimization: An Introduction with Metaheuristic Applications. Wiley, New Jersey (2010)
26. Hvass, M.: Good parameters for particle swarm optimization. Technical Report No. HL1001, Hvass Laboratories (2010)
27. Lee, K., El-Sharkawi, M.: Modern Heuristic Optimization Techniques Theory and Applications to Power Systems. Wiley, New York (2008)
28. Trejos, J., Villalobos, M.: Simulated annealing optimization in nonlinear regression: algorithm and software. Invest. Operacional **21**(3), 236–246. (2000) (La Habana, Cuba)
29. Spall, J.: Introduction to Stochastic Search and Optimization: Estimation, Simulation, and Control. Wiley, New Jersey (2003). https://doi.org/10.1002/0471722138
30. Jang, J., Sun, C., Mizutani, E.: Neuro-Fuzzy and Soft Computing. Prentice Hall, New Jersey (1997)
31. Luenberger, D., Ye, Y.: Linear and Nonlinear Programming, 3rd edn. Springer, New York (2008). https://doi.org/10.1007/978-0-387-74503-9
32. Karaboga, D. Basturk, B.: Artificial bee colony (ABC) optimization algorithm for solving constrained optimization problems. In: Melin, P. et al. (eds.) Foundations of Fuzzy Logic and Soft Computing Lecture Notes on Artificial Intelligence, 4529, pp. 789–798. Springer, Heidelberg (2007). https://doi.org/10.1007/978-3-540-72950-1_77
33. Yang, X.-S.: Bat algorithm for multi-objective optimization. Int. J. Bio-Inspired Comput. **3**(5), 267–274 (2012). https://doi.org/10.1504/IJBIC.2011.042259
34. Storn, R, Price, K.: Differential evolution—a simple and efficient heuristic for global optimization over continuous spaces. J. Global Optim. **11**, 341–359 (1997). https://doi.org/10.1023/A:1008202821328

Kernel-Based Clustering Driven by Density Index

Edwin Aldana-Bobadilla, Ivan Lopez-Arevalo, Ivan Mendez-Alvarez, Alejandro Molina-Villegas, and Hiram Galeana-Zapien

Abstract We propose a clustering method that can deal with non-linearly separable groups by means of a search process in which we look for the best values for parameters of a set of kernel functions that induce, on the objects to be clustered, the best partition relative to quality in terms of a density-based criterion. In summary, our proposal is an iterative adaptation of a set of kernel parameters, guided by a density-based criterion, which is able to yield clustering solutions that improve those results obtained from the separate application of kernel and density-based methods from the state-of-the-art.

Keywords Clustering · Kernel based clustering · Density index · Non-linearly clusters · Genetic algorithm

1 Introduction

Clustering is a fundamental task in data analysis; its purpose is to divide a set of objects into a set of groups (named clusters), in which objects belonging to a cluster share several properties. It has been addressed in various fields and disciplines such as pattern recognition [1], information retrieval [2, 3], image processing [4–6], computer security [7, 8], etc. Typically, these objects are represented as numerical d-tuples of the form $(x_1, x_2, \ldots, x_d)$. A set of N *d-tuples* is known as a dataset, denoted in what follows as X. Clustering is that process that allows us to find a partition Π on X consisting of k disjoint subsets of X, wherein their elements satisfy a similarity

E. Aldana-Bobadilla (✉)
Conacyt - Cinvestav Tamaulipas, 87130 Victoria, Tamaulipas, Mexico
e-mail: edwyn.aldana@cinvestav.mx

I. Lopez-Arevalo · I. Mendez-Alvarez · H. Galeana-Zapien
Cinvestav Tamaulipas, 87130 Victoria, Tamaulipas, Mexico

A. Molina-Villegas
Conacyt - Centro de Investigación en Ciencias de Información Geoespacial, 97302 Mérida, Yucatán, Mexico

W. Pedrycz et al. (eds.), *Computational Intelligence for Business Analytics*,
Studies in Computational Intelligence 953,
https://doi.org/10.1007/978-3-030-73819-8_19

criterion. The way in which the partition Π is found defines a clustering algorithm. Most algorithms begin by defining the number of clusters in the desired partition. Subsequently, a search process defines a set of k centroids (one per each cluster) and associates iteratively each object in X to the nearest centroid based on a similarity measure, until an optimality criterion is satisfied. In this regard, we can find methods as K-means [9, 10], fuzzy C-means [11, 12], partitioning around medoids (PAM) [13], among others. Other methods not following the previous approach are:

- Hierarchy-based. These methods create clusters by recursively partitioning the dataset [14]. The result is a tree structure known as dendrogram, where the nodes are possible clusters. The root node represents the whole dataset. The nodes at the same level represent a partition Π. The resulting clustering can be obtained by cutting the dendrogram at different levels. Some algorithms with this approach are: Single, Complete, and Ward [15].
- Density-based. The principle of these methods is to create clusters in such a way that objects found in dense regions belong to the same cluster. Some algorithms under this approach are: DBSCAN [16], Denclue [17] and OPTICS [18].
- Probability-based. This approach proposes to model clustering through a process that seeks a partition Π based on a probabilistic model. The objects in X are assumed to be generated based on several probability distributions. Each distribution determines the probability that an object belongs to a cluster. Thus, the objects have a certain probability of belonging to a cluster [19, 20].
- Graph-theory-based. In this approach, objects to be clustered are represented as nodes in a weighted graph. Then, the edges connecting the nodes are weighted by a similarity measure between them. The graph is recursively partitioned into disjoint subgraphs based on several criteria, such as minimum cut [21, 22].

Our main motivation for proposing a new clustering algorithm is that in some real problems related to natural language data, we have to deal with big and very different datasets. Natural language processing (NLP) is a cutting-edge field of research constantly evolving, which nowadays has the need for reliable clustering algorithms. In order to provide an idea of the modern NLP datasets that are currently used, we can mention the Wikipedia Links, containing approximately 13 million documents where each Wikipedia page is treated as an entity, while the anchor text of the link represents a mention of that entity. Another currently used data is Amazon reviews containing around 35 million reviews from Amazon, which includes product and user information, ratings, and the plaintext review. The business firms utilize NLP methods to learn about the customer's opinions about their product and services from online reviews.

In real world scenarios, the clustering approach to apply depends on several premises; an important one is the separability of clusters, which can be linear or non-linear (see Fig. 1). The similarity measure (Euclidean, Mahalanobis, Manhattan, Minkowsky, Cosine, etc.) to use is crucial because it imposes several constraints in the shape and boundaries of clusters in Π, that could not encompass the objects adequately in X (see Fig. 2). Attempting to overcome this problem, other methods have arisen. For instance, Support Vector Clustering (SVC [23]) includes data space

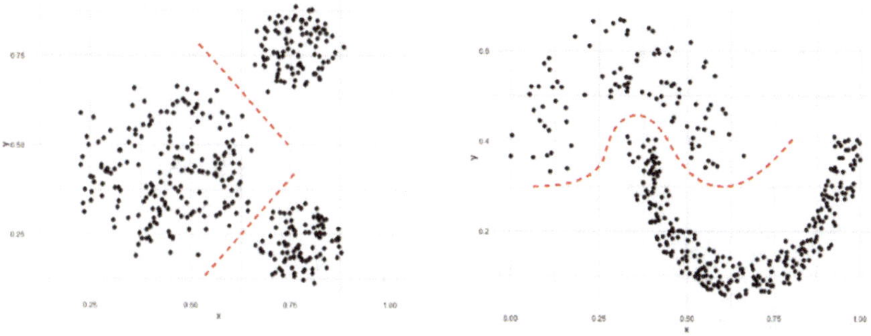

(a) Linear separability (b) Non-linear separability

Fig. 1 Separability of clusters

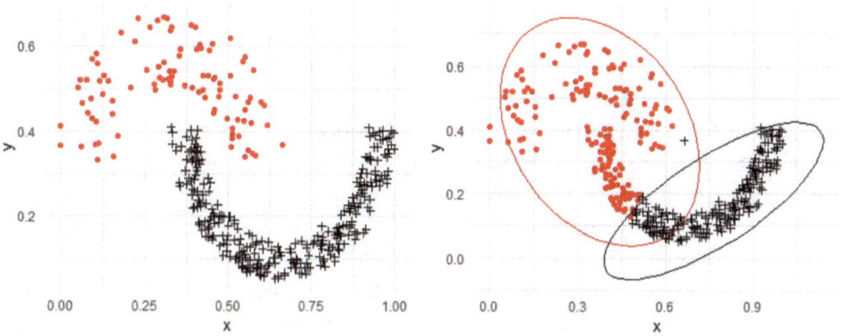

(a) Expected partition (b) Spurious partition

Fig. 2 Constraints imposed by distance measures

transformation techniques, based on parametric functions called kernels, which induce a separability between possible clusters in the dataset (based on the popular supervised method known as Support Vector Machine SVM [24]). As illustrated in Fig. 3, a kernel is essentially a mapping function that transforms a given space into a higher dimensional space where a linear separation between clusters is possible. Other clustering algorithms using kernel functions are Kernel K-means [25], Kernel fuzzy C-means [26, 27], Kernel SOM [28], etc. Other approaches include criteria beyond distance, such as the notion of density, which allows facing the constraints associated with distance measurements. Density-based algorithms locate high-density regions separated by low-density regions.

As mentioned, DBSCAN, Denclue, and OPTICS belong to this category. The DBSCAN algorithm bases its idea on the concept of core objects; these are in areas of high density. Therefore, a cluster is a set of core objects, each close to each other, and a set of non-core objects that are close to a core object. Core objects are those

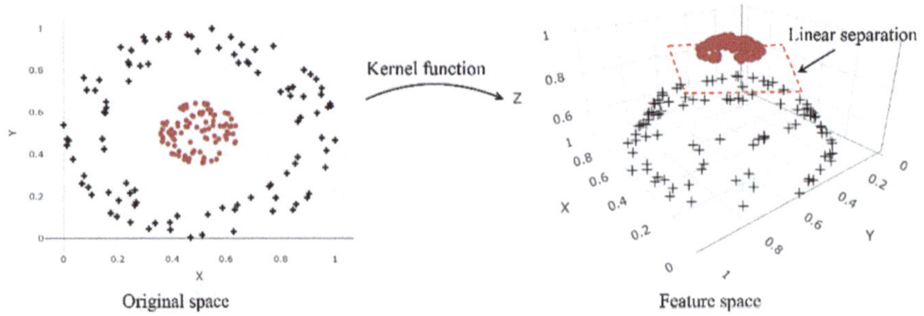

Fig. 3 Example of 2-dimensional space mapped to 3-dimensional space

whose radius neighborhood (*eps*) has a number of objects greater than or equal to a defined threshold (*minPts*). Denclue searches for clusters with local maxima of the estimated density function; data objects that go to the same local maxima are placed in the same cluster. On the other hand, OPTICS is an algorithm that uses similar concepts to DBSCAN but addresses one of the main weaknesses of DBSCAN: the problem of detecting significant clusters in variable density data. DBSCAN is still the most widely used density-based algorithm.

In general, kernel-based and density-based algorithms have proved to be effective in clustering problems when datasets have non-linear separations [29, 30]. Given this effectiveness, it is expected that the combination of these approaches may result in a broader clustering method. In this regard, we propose a clustering approach based on the following idea: we can map a dataset into some high dimensional space through some non-linear kernel function. In this new space, a labeling process is performed, inducing a non-linear partition on the original dataset. This partition can be evaluated in terms of a density-based criterion. The kernel parameters can be adapted iteratively in order to obtain better partitions in subsequent iterations. This process can be repeated until the best partition (relative to the density-based criterion) is found, or a stop criterion is reached.

The discussion and description of our proposal have been organized as follows: Sect. 2 shows what we consider the main concepts associated with our discussion. In Sect. 3, we describe in detail the main body of our proposal. Then, in Sect. 4, experimental results on benchmark datasets are presented. Finally, in Sect. 5, we present the most important conclusions and outlines future research work.

2 Background

Our proposal is based on three important elements: (1) kernel functions, (2) iterative adaptation parameters, and (3) clustering evaluation. In this regard, we started this section by describing the concept of the kernel function. Then, we show that the adaptation of the kernel parameters involves a large search space that requires heuristic-based search approaches to be explored efficiently. In this sense, we present several heuristic-based search approaches and select the one that we consider the most suitable option for our purposes. Finally, we present the most popular metrics of clustering evaluation, known as cluster validity indices. We focus on an index defined in terms of density, which has an important role in guiding the search of optimal values of the kernel parameters and, in consequence, the best partition of the dataset.

2.1 Kernel Functions

Kernel functions are parametric functions that map a dataset from a d-dimensional real space $\mathbb{R}^d$ into a real space with more (even infinite) dimensions $\mathbb{R}^{d+m}$ where it is expected that changes in the parameters create linear separations from the dataset inducing clusters. This situation is illustrated in Fig. 3. The $\mathbb{R}^{d+m}$ space is known as the induced feature space. In general, kernel functions take two vectors as input, $\overrightarrow{x}$ and $\overrightarrow{y}$ to map them to a real value corresponding to the dot product in the feature space. Given a set $X = \{\overrightarrow{x} \in \mathbb{R}^d\}$ its vectors can be mapped by the kernel function ϕ (Eq. 1):

$$\Phi : \mathrm{\mathbb{R}^d} \mapsto \mathrm{\mathbb{R}^{d+m}} \tag{1}$$

such that the dot product of $\overrightarrow{x}$ and $\overrightarrow{y}$ is $\phi(\overrightarrow{x}) \cdot \phi(\overrightarrow{y})$. A kernel is a function k corresponding to this dot product (Eq. 2).

$$\mathrm{K}(\vec{\mathrm{x}}, \vec{\mathrm{y}}) = \phi(\vec{\mathrm{x}}) \cdot \phi(\vec{\mathrm{y}}) \tag{2}$$

There are a variety of kernel functions, the most used in learning tasks are:

$$Polynomial\ Kernel K(\vec{x}, \vec{y}) = (\vec{x} \cdot \vec{y} + 1)^p \tag{3}$$

$$Radial\ basis function\ Kernel K(\vec{x}, \vec{y}) = e^{-\|\vec{x}-\vec{y}\|^2/2\sigma^2} \tag{4}$$

$$Hyperbolic\ tangent\ Kernel K(\vec{x}, \vec{y}) = than(\kappa \vec{x} \cdot \vec{y} - \delta) \tag{5}$$

Kernel functions have been used to different clustering methods to find similarity relationships beyond standard metrics (e.g., Euclidian metric). For instance, there is a *kernelized* version of K-means, known as Kernel K-means [25]. There are also proposals as Kernel SOM [28], Support Vector Clustering [23], and Kernel Fuzzy C-means [26]. What is common to most clustering methods, based on kernel functions, is that they apply the actual clustering on the feature space allowing to find linear partitions that induce non-linear partitions on the original space of the data.

However, as it has been pointed out, kernel functions have parameters that could affect the clustering effectiveness. For this reason, tuning of good parameter values must be applied considering the specific problem to solve. Such parameter tuning could be considered a limitation for this approach.

We consider very important to include a mechanism to estimate the optimal parameters automatically. These values induce a search process that cannot be explored via traditional methods based on iterating along the direction of the negative steepest slope of an objective function. In this direction, some heuristic-based search methods have already been proposed obtaining high accuracy [31–33].

2.2 Heuristic-Based Search Methods

The clustering problem is considered a NP-hard problem [34] wherein the use of an exhaustive method is impractical. Traditional clustering approaches exhibit successful approximations via iterative refinements of a feasible solution based on an optimality criterion in terms of a proximity measure, in which case, it is frequently possible to iterate along the direction of the steepest slope of such a criterion, and, in consequence, to exhibit a complexity of polynomial order. Since we want to incorporate intricate optimality criteria that are beyond a proximity measure (a validity index and kernel parameters), such a complexity order could be unreachable, in which case the use of a heuristic-based search method becomes necessary.

Heuristic-based search is a set of methods (also known just as *heuristics methods*) that attempt to find an approximate solution to a complex optimization problem in a reasonable time. Some of the most popular heuristics are: tabu search [35], simulated annealing [36], ant colony optimization [37], particle swarm optimization [38], and evolutionary computation [39]. Within evolutionary computing, there are some variations, such as: evolutionary strategies [40], evolutionary programming [41], and genetic algorithms (Gas [42, 43]). A problem with most heuristic methods is that it does not guarantee to find the optimal solution; however, Rudolf [44] and Hruschka et al. [45] proved that a GA always converges to the optimal solution under *full elitism* conditions. This convergence is not limited in time, and the choice of the GA variation with the best dynamic behavior should be considered. In this regard, we rely on the conclusions of previous analyses [46, 47], which showed that a breed of GA, called the *eclectic genetic algorithm* (EGA [48]), achieves the best relative performance. We have selected an EGA as the heuristic method in the proposed clustering approach. Such an algorithm incorporates the following:

1. Full elitism over a set of size n of the last population. Given that, by generation t, the number of individual tested is nt, the population in such a generation consists of the best n individuals.
2. Deterministic selection as opposed to the traditional proportional selection operator. Such a scheme emphasizes genetic variety by imposing a strategy

that enforces the crossover of predefined individuals. After sorting the individual's fitness from better to worse, the i-th individual is combined with the $(n - i + 1)$-th individual.

3. Annular (two-point) crossover.
4. Random mutation of a percentage of bits of the population.

A detailed description of EGA can be found in Appendix A.

2.3 Validity Index

A validity index is a measure that allows determining the adequateness of the results of a clustering process obtained via a clustering method. These can be mainly classified into two types: internal and external. The external validation uses some external information in the validation process like class labels, and it is mostly used in supervised learning. On the other hand, internal validation is usually employed in unsupervised learning to evaluate the goodness of clustering without using any external information [49]. Many cluster validity indices (CVIs) have been developed, including classic indices such as the Calinski-Harabasz index (CH [50]), Davies-Bouldin index (DB [51]), Dunn and Dunn index (DD [52]), Silhouette (SILH) [53], Scattering and Dispersion (SD [54]), and density-based indices like Cluster Validity index based on Density-involved Distance (CVDD [55]).

However, most of them are only effective when applied to a dataset with a spherical cluster structure or well separated clusters, and their effectiveness is reduced when the dataset has a complex structure, such as arbitrarily shaped clusters or non-linearly separated clusters. For this work, we used CVDD [55] since it proves that considering the density in the validation increases the efficiency of the clustering evaluation. CVDD employs the minimum pairwise distance between clusters to represent the separation between clusters. According to Eq. 6, a larger value of CVDD indicates a better quality of the partition Π.

$$CVDD(\Pi) = \frac{\sum_{i=1}^{k} sep(C_i)}{\sum_{i=1}^{k} com(C_i)} \tag{6}$$

where $sep(C_i)$ is the separation between the cluster C_i and all other clusters and $com(C_i)$ is the compactness of C_i. A detailed explanation of this index can be found in [55].

3 Proposal

The proposed clustering method finds a partition with non-linear separations induced by kernel functions. The parameter values associated with a kernel function have a high impact on finding the correct partition. These parameters are adapted iteratively until their values induce the best partition Π^*, relative to a quality criterion Q. This can be expressed as a general optimization problem of the form (Eq. 7):

$$\begin{aligned} &\text{Optimize} : Q(\Pi(\vec{s})) \\ &st : g_1(\Pi) \leq \epsilon_1 \\ &g_2(\Pi) \leq \epsilon_2 \\ &\vdots \\ &g_n(\Pi) \leq \epsilon_n \end{aligned} \tag{7}$$

where the partition Π depends on kernel parameters denoted as $\vec{s}$ and the functions $g_i(\cdot)$ represent a set of possible constraints that must be satisfied by all instances of Π. In our case, Q is defined in terms of a density-based validity index (see Eq. 6), in which case, Eq. 7 can be expressed as Eq. 8:

$$\begin{aligned} &\text{Maximize} : Q(\Pi(\vec{s})) = CVDD(\Pi(\vec{s})) \\ &st : |C_i| > 1 \\ &i = 1, \ldots, k \end{aligned} \tag{8}$$

where k is the number of clusters in Π. We have included a constraint that forces more than one object in each cluster (C_i) to avoid sparse clusters.

The above problem represents a large solution space whose cardinality can be expressed in terms of the number of possible partitions for N objects into k clusters, which is given by the Stirling number of the second kind [56] in Eq. 9:

$$S(N, k) = \frac{1}{k!} \cdot \sum_{i=1}^{k} (-1)^{k-i} \cdot \binom{k}{i} \cdot i^N \tag{9}$$

where $N = |X|$ and k is the number of clusters. The above shows why finding the optimal partition involves exploring a large solution space (NP-hard problem) wherein the use of an exhaustive search method is impractical. Relying on the previous discussion, we have chosen EGA as a suitable method to explore this space efficiently.

3.1 Encoding of Kernel Parameters

Let $\overrightarrow{s}_i \in \mathbb{R}^{m+1}$ be a vector representing m kernel function parameters plus one additional parameter corresponding to the cut threshold in a hierarchical clustering process (explained later in Sect. 3.3). The $\overrightarrow{s}_i$ encodes a transformation of the original space, which in turn induces a candidate clustering solution Π as illustrated in Fig. 4. Now, let $\mathbb{S}$ be a set of $\overrightarrow{s}_i$, which are adapted iteratively via the genetic operators of EGA until the best partition Π^* is found. Thus, elements in $\mathbb{S}$ are considered possible solutions prone to mutation, selection, and crossover. The algorithm iteratively will enhance the population of individuals. Finally, when the stop criterion is reached, $\mathbb{S}$ will contain the parameters that induce Π^*.

The adaptation process of $\mathbb{S}$ is driven by the evaluation of each induced partition via the objective function $Q(\Pi)$, as it is illustrated in Fig. 5.

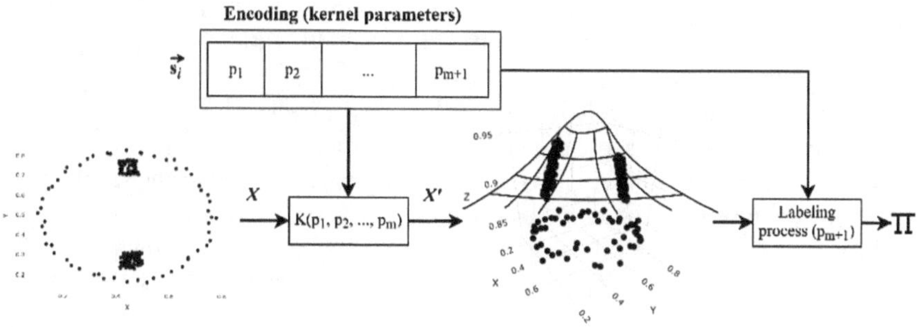

Fig. 4 Proposed encoding

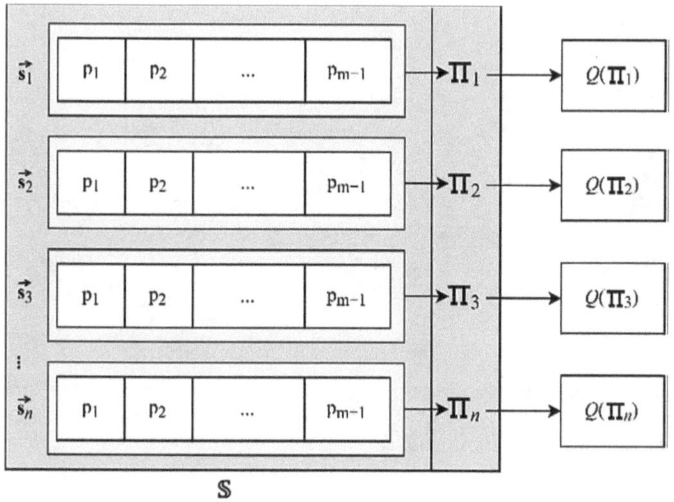

Fig. 5 Set of candidate partitions

3.2 Building Π from $\vec{s}_i$

The proposed clustering method maps the dataset by means of a kernel function to find linear separations in the mapped space. The kernel is essentially a mapping function, one that transforms a given space into some other space. The kernel corresponds to a dot product in a high-dimensional feature space. For this work, the radial basis kernel function was used (Eq. 10). The main property of this kernel is that it generates Gaussian envelopes, with $\vec{\mu}$ being the mean of the curve (highest point) and σ the width of the curve (Eq. 10).

$$K(\vec{x}, \vec{\mu}, \sigma) = e^{-\|\vec{x}-\vec{\mu}\|^2/2\sigma^2} \tag{10}$$

If the dataset is mapped with a specific $\vec{\mu}$ value, then all those objects in the mapped space will be placed close to $\vec{\mu}$. These parameters play an important role in kernel performance and must be adjusted for each dataset to find the best partition. Figure 6 assumes an infinite number of possible partitions depending on the parameters of the kernel function.

Each individual in the population establishes the parameters for a kernel. Given an instance (individual in the population of EGA) of kernel parameters $\vec{s}_i$, the mapping process is carried out as follows: a kernel function is applied over the dataset resulting in a single dimension vector $\vec{z}$ as shown in Eq. 11. Then, we add the resulting $\vec{z}$ to the dataset X in order to build an augmented matrix $X' \in \mathbb{R}^{d+1}$ as shown in Eq. 12, and Fig. 4. Finally, a labeling process next described is performed in the new space X'.

$$\forall \vec{x} \in X \mapsto^{K(\vec{x},\vec{s}_i)} \vec{z} \in \mathbb{R} \tag{11}$$

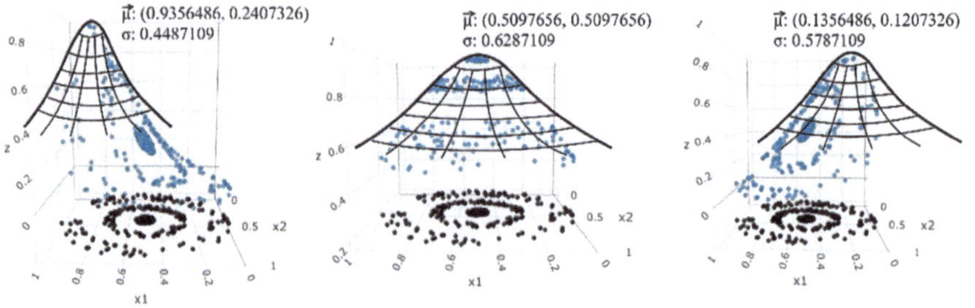

Fig. 6 Examples of mapping a dataset through a radial-based kernel function with different values for $\vec{\mu}$ and σ. The data in the original space is shown in *black* and the mapping of the objects after applying the kernel function is shown in *blue*

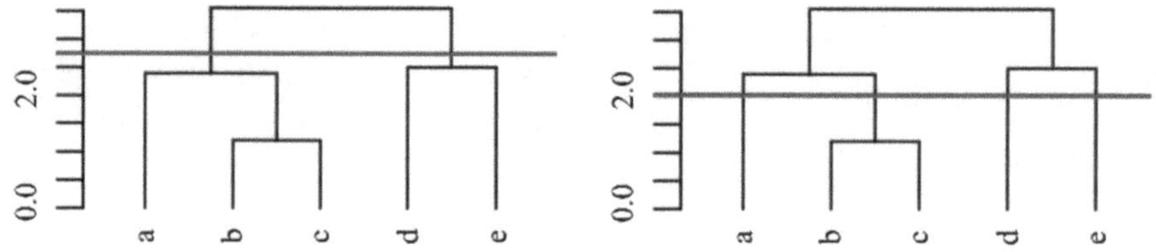

Fig. 7 Example of dendrograms with 5 objects and cut-off value marked with a *red line*

$$X' = \left[\begin{array}{cccc|c} x_{1,1} & x_{1,2} & \cdots & x_{1,d} & z_1 \\ x_{2,1} & x_{2,2} & \cdots & x_{2,d} & z_2 \\ & & \vdots & & \vdots \\ x_{n,1} & x_{n,2} & \cdots & x_{n,d} & z_n \end{array}\right] \tag{12}$$

3.3 *Labeling of Instances*

After the mapping process, each object in X will have a unique association with one in X'; this allows to assign the objects to a cluster in one space and obtain its equivalent in the other space. To assign the objects to a cluster, a labeling process is performed. We analyzed different ways to perform this process and observed that by computing the distance matrix of X' it becomes feasible to perform the labeling as is done in the hierarchical clustering by first getting individual clusters and joining them according to the closest distances between those clusters, with this we get a dendrogram as shown in Fig. 7. Generally, a cut is made at a certain height of the dendrogram to obtain the desired clusters; all elements joined under the cut belong to the same cluster. For example, in Fig. 7 (left) the cut-off value is 2.75 resulting in two subsets $C_1 = \{a, b, c\}$ and $C_2 = \{d, e\}$, while in Fig. 7 (right) the cut-off value is 2.00 resulting in four subsets $C_1 = \{a\}$, $C_2 = \{b, c\}$, $C_3 = \{d\}$ and $C_4 = \{e\}$. This value has a high impact on the induced partition Π. Therefore, we include this cut-off value within $\vec{s}_i$ so that EGA determines an appropriate value. It is important to note that by including the cut-off value as part of the parameters to be estimated, the proposed clustering method does not specify the number of clusters as an input since it is a hyperparameter optimized by the method.

3.4 *Clustering Validation*

The resulting partition Π_i is validated to know its quality, this is carried out by means of the objective function Q, defined in terms of CVDD. As a result of this validation, we can determine the fitness exhibited by Π for a set of kernel parameters $\vec{s}_i$.

CVDD can deal with clusters of spherical and non-spherical forms. This is based on the idea of the DBSCAN algorithm using two key concepts: core objects and density connectivity. The first concept is useful to recognize outliers, and the second one is useful to differentiate clusters separated by density. This process is executed for all $\vec{s}_i \in \mathbb{S}$.

3.5 Adaptation Process

Each individual (solution) in the population is validated to ensure that it is (or their descendants) appropriate to keep in the population. As part of the evolutionary algorithm, each individual must be adapted to obtain better individuals. This is done by following these steps:

1. For all $\vec{s}_i \in \mathbb{S}$ a partition Π_i is obtained and validated.
2. The set $\mathbb{S}$ is ordered (in descending order) according to the fitness of the partition induced by each $\vec{s}_i$.
3. The values of each $\vec{s}_i$ are modified based on the EGA operators (selection, crossover, and mutation).
4. Steps 1–3 are repeated until the number of iterations, or other stop criteria is reached.

We consider the stop criteria the number of iterations or generations Q. The heuristic selected (EGA), additionally requires setting the parameters P_c (crossover probability), P_m (mutation probability) and population size Θ. In Sect. 4.3, the values of these parameters are provided.

4 Experiments and Results

Based on the proposed method, we implemented a functional prototype in *R* language version 3.5.3. All the experiments were run on a computer with Intel® Core™ i7 processor and 8 GB of RAM. For testing the prototype, we used 14 numerical datasets standardized to range [0, 1]. Some of these were taken from the *Fundamental Clustering Problems Suite* (FCPS [57]), others from the works of Jain and Law [58], Zelnik-Manor and Perona [59], Chang and Yeung [60], Veenman and Reinders [61], and others were generated in a synthetic way by means of tools to generate datasets with normal distribution [62–64]. The datasets were intentionally selected to comply with having clusters with linear and non-linear separations and to be able to visualize them in two or three dimensions. They also have known labels a priori for evaluation purposes. Detailed information about these datasets, such as the number of clusters, the number of dimensions, and the number of instances, is shown in Table 1. Figures 8 and 9 show the labeled clusters of the datasets.

Table 1 Dataset description

Dataset	Clusters (k)	Dimensions (d)	Instances (N)	Separation	Source
(1) Noisy_circles1	2	2	200	Non-linear	[63]
(2) Noisy_circles2	3	2	300	Non-linear	[63]
(3) Atom	2	3	800	Non-linear	[57]
(4) Noisy_moons	2	2	373	Non-linear	[58]
(5) Zelnik	3	2	238	Non-linear	[59]
(6) Spiral	3	2	312	Non-linear	[60]
(7) r15	15	2	600	Non-linear	[61]
(8) Varied	3	2	1500	Non-linear	[62]
(9) blobs	3	2	1500	Linear	[62]
(10) synthetic1	5	2	500	Linear	[64]
(11) Synthetic2	5	3	500	Linear	[64]
(12) Synthetic3	10	3	1000	Linear	[64]
(13) Mouse	3	2	490	Linear	[64]
(14) Aniso	3	2	1500	Linear	[62]

4.1 Methodology to Gauge the Effectiveness

The way to evaluate the performance of the clustering methods in the experiment was by getting an error ratio between the CVI value of a pre-labeled dataset versus the CVI value obtained with a clustering method. For this purpose, each evaluated dataset has cluster labels as additional information. Figure 10 shows the procedure to obtain the CVI values. Once we have the true CVI value Q' and the resulting CVI Q from a clustering method, we can measure the absolute error $AE(Q)$ as indicated by Eq. 13, where a lower absolute error value means higher performance.

$$AE(Q) = \left| \frac{Q - Q'}{Q'} \right| \tag{13}$$

4.2 Experimental Design

The experimental design was carried out as follows:

1. A clustering solution Π_i is obtained by a clustering method for each dataset X.
2. For each solution Π_i the value of the CVI Q is calculated and denoted as Q_i.
3. The absolute error is measured by the proximity of the Q_i value obtained to the value of the true CVI Q' indicated by Eq. 13.

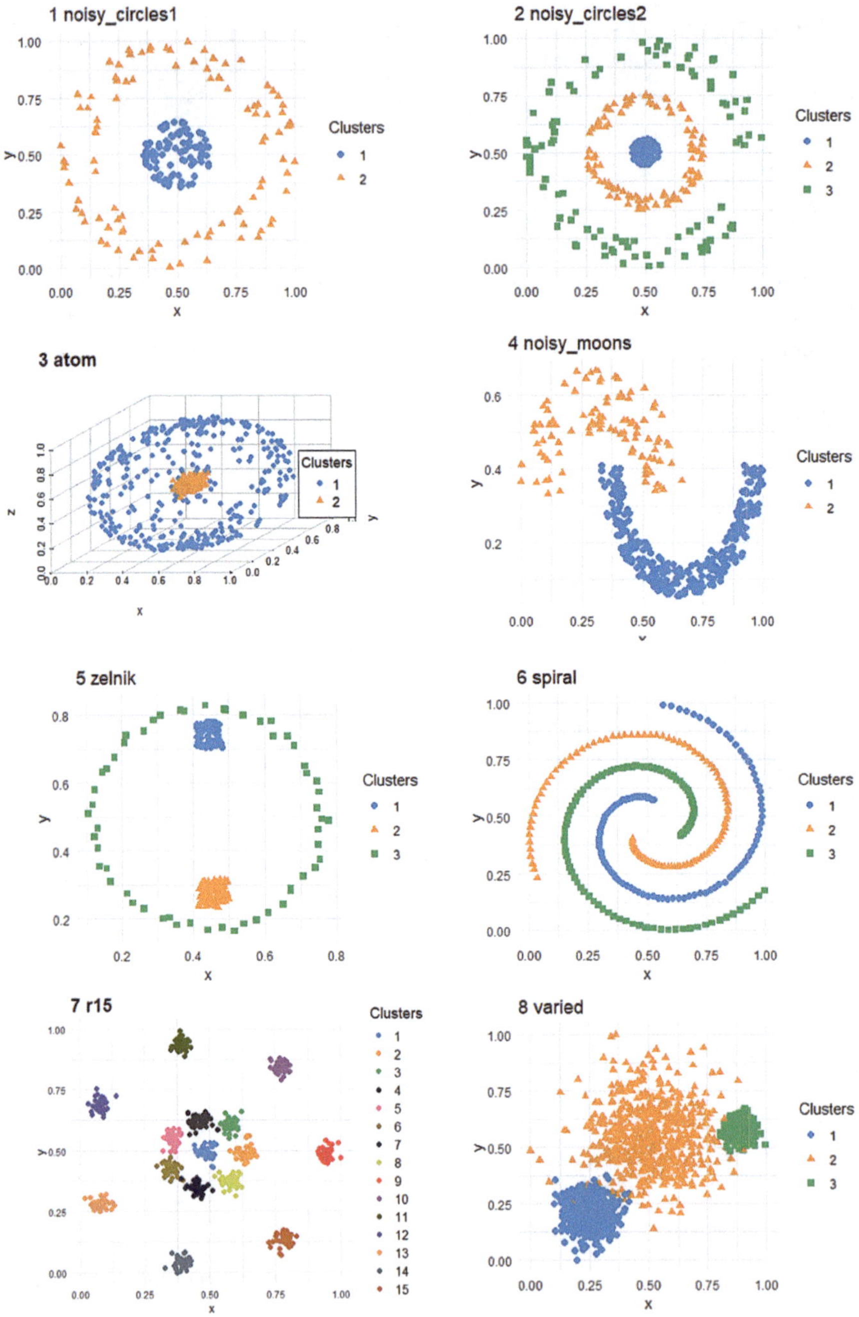

Fig. 8 Datasets employed (1–8)

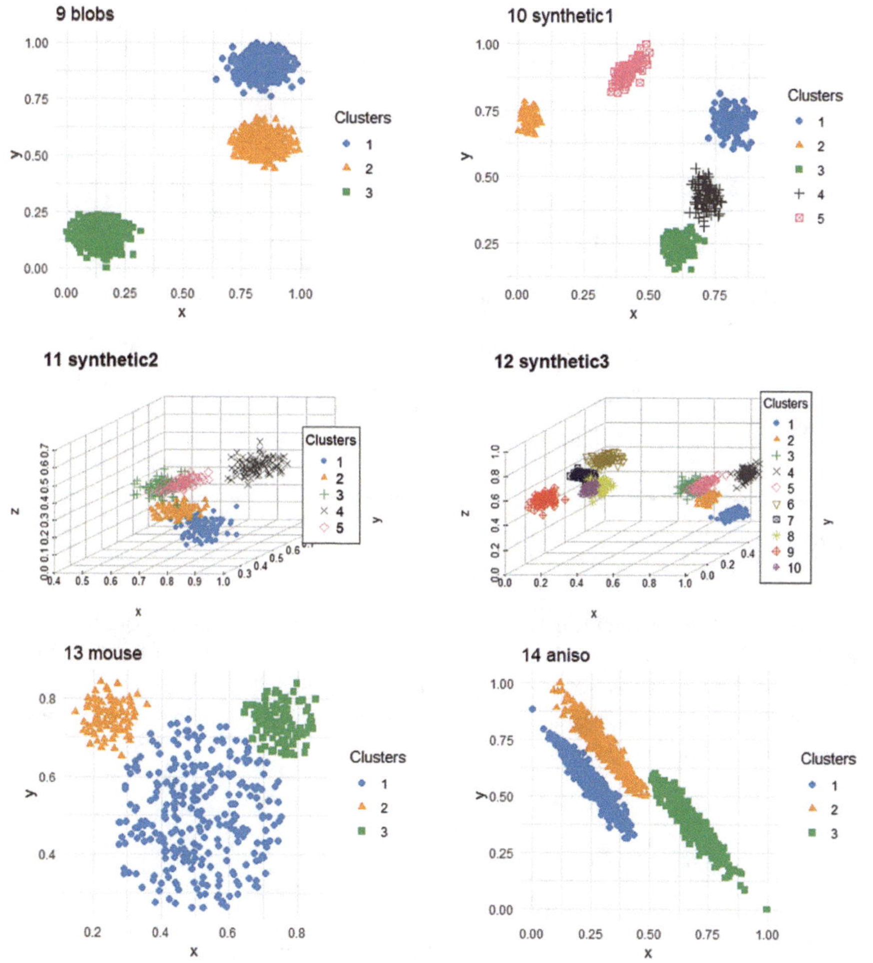

Fig. 9 Datasets employed (9–14)

4. Steps 1–3 are repeated until $i = M$ for each dataset, in order to provide M tests that validate the evaluation.
5. A mean absolute error is determined by Eq. 14.

$$MAE = \frac{1}{M} \sum_{i=1}^{MAE} (Q_i) \tag{14}$$

The above procedure was executed with the proposed method and with the DBSCAN [16], Kernel K-means [25], and Spectral Clustering [65] algorithms to have a comparison against matched methods. DBSCAN since it is the best known density-based clustering method, Kernel K-means, because it is a version of K-means

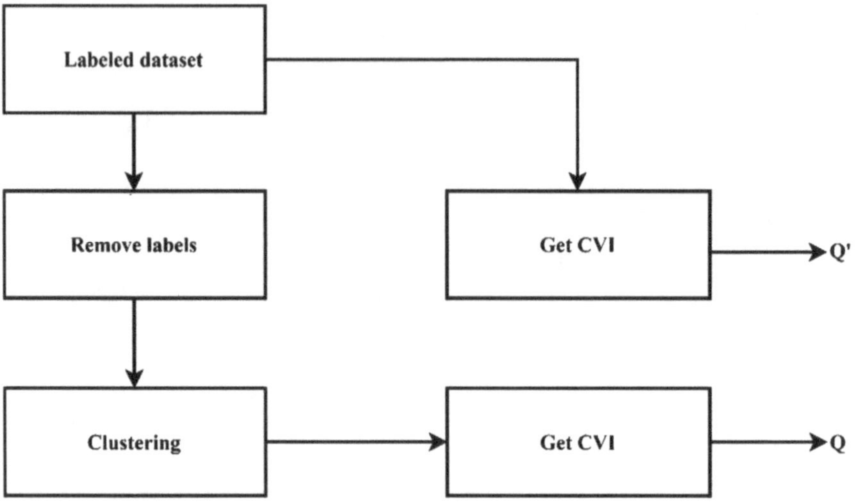

Fig. 10 Getting CVI Values

that implements the kernel functions, K-means is the baseline method for clustering, and Spectral Clustering, which also produces a non-linear separation between clusters. In this evaluation, we set $M = 100$, which indicates that with the 14 datasets and the 4 clustering methods, a total of 5600 executions were performed (100 executions per each dataset per each clustering method).

4.3 Parameters Setting

As mentioned, EGA implies the establishment of several parameters. In this work, EGA was executed with the following parameter values: $P_c = 0.89$, $P_m = 0.05$, $\Theta = 100$, $G = 500$. The values are based on a preliminary study, which showed that from a statistical viewpoint, EGA converges to optimal solutions around such values when the problems are demanding (a large and complex solution space) [46]. An additional parameter corresponding to the k-nearest neighbors used by CVDD was set in 7. This value is suggested by the authors of CVDD [55].

Parameters setting on comparison methods

Besides the fact that our proposed clustering method includes parameters that must be established, the compared clustering methods also have their own parameters. The values used for each compared method are described and justified below.

DBSCAN

We rely on the heuristics for the DBSCAN parameters selection:

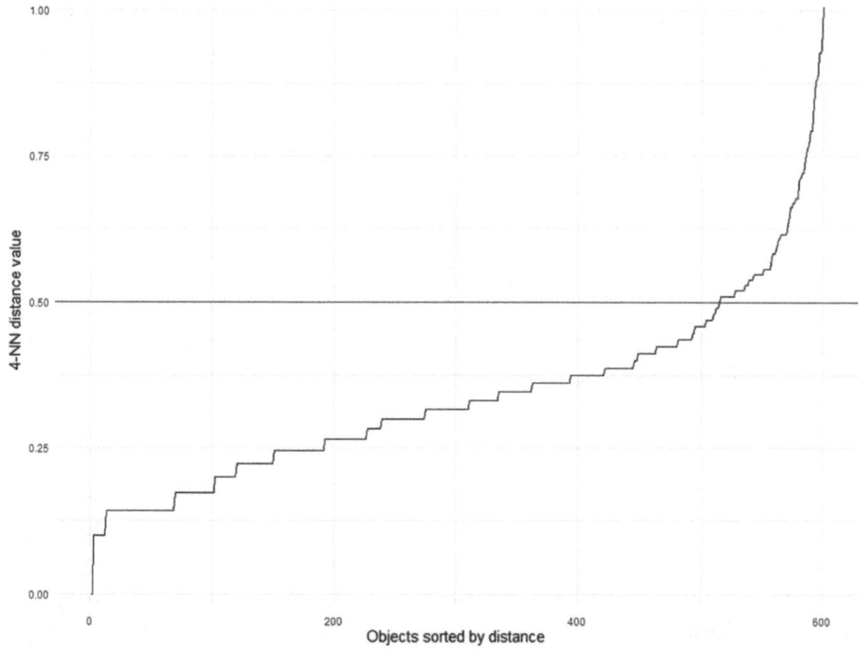

Fig. 11 Example of sorted 4-nearest-neighbor distances plot whose elbow is marked with a *red line*

- Keep the default value of *minPts* = 4 for two-dimensional data [16].
- For more than 2 dimensions use *minPts* = 2d [66].

Once the value of *minPts* is chosen, the *eps* value can be determined by plotting the sorted k-nearest-neighbor distances for each $\overrightarrow{x} \in X$ with $k =$ *minPts* and locating the *elbow* value on the plot, for example, in Fig. 11 the *elbow* is around a distance value of 0.5. Although the *minPts* values mentioned above work for most datasets, if the dataset has a high level of noise, is large, or is high-dimensional, the results can be improved by increasing the *minPts* value [67]. Therefore, we perform a search with *minPts* = *{4, 5, 10, 15, 20, 25, 30, 35}* and *eps* at its *elbow* value for each dataset to consider the most appropriate selection of these parameters. The *R* library used to perform the clustering with DBSCAN is *Density Based Clustering of Applications with Noise* (DBSCAN) and *Related Algorithms* [68].

Kernel K-means and Spectral Clustering

To perform the clustering with the Kernel K-means and Spectral Clustering methods, we used the R library *Kernel-Based Machine Learning Lab* (kernlab [69]). Among other methods, *kernlab* includes Kernel K-means, Spectral Clustering, Support Vector Machines, Kernel PCA, and more.

Like the standard K-means method, Kernel K-means receives as input parameters the dataset X and the value of the initial *k*-means. In addition, it requires to indicate the kernel function to be applied, as well as the values of the chosen kernel parameters. *Kernlab* provides a wide of kernel functions that can be used by setting the kernel parameter. In this evaluation, we chose the same radial basis kernel function that is used in our method, and in order to select the values of its parameters, we rely on a heuristic provided by *kernlab* to determine an appropriate value. The k-values were taken from Table 1 for each dataset X.

We use the same *kernlab* library to perform the Spectral Clustering. For this method, the input parameters are the dataset X, and the number of desired clusters k, the values of k were also taken from Table 1. Spectral Clustering also uses a kernel function to calculate a similarity matrix and perform the clustering process. We choose the radial basis kernel function and the heuristic provided by *kernlab* for the adjustment of the kernel parameters.

4.4 Results

Table 2 shows the mean absolute error (*MAE*) results given by the CVDD index for each dataset. The μ column represents the MAE of the 100 tests performed with each dataset, and the σ column denotes the standard deviation obtained. The best values of μ and σ are in bold for each dataset. Note that the standard deviation of DBSCAN is 0 since its definition states that its results only differ when an $\vec{x} \in X$ is density-reachable for more than one cluster [16]; thus, the results are strongly linked to the input parameters. Given the values in Table 2, it is possible to calculate the statistical significance of the results using the statistical technique known as the *confidence interval*. This measures the probability that a value lies within a confidence range. A significance level of $\alpha = 0.05$ was used for this analysis $\mu \pm z_{\alpha/2} \cdot \frac{\alpha}{\sqrt{M}}$ with $z_{\alpha/2} = 1.96, M = 100$). Table 3 shows the performance results in terms of the overall error μ and overall standard deviation σ of the 100 tests in the 14 datasets using the proposed method, DBSCAN, Kernel K-means (KK-means), and Spectral Clustering.

4.5 Discussion

Based on the results in Table 2, we can see that both our proposal and those based on density exhibit a low performance in contrast to Kernel K-means when they face problems with arrangements similar to the dataset *8 varied*. This represents a case in which there are dense data regions mixed with the sparse region. The boundaries between regions with very different density will affect any algorithm guided by density, including our proposal. Indeed, regions with deceptive density could be tricky unless the clustering was guided only by a proximity measure, as it is the case of Kernel K-means. That is why Kernel K-means got the best result for dataset *8 varied*. However, for all the rest datasets of Table 2, our proposed method

Table 2 Mean absolute errors and standard deviations

	Proposal		DBSCAN		KK-means		Spectral	
Dataset	μ	Σ	μ	σ	μ	σ	μ	σ
(1) Noisy_circles1	*0.0000*	*0.0000*	*0.0000*	*0.0000*	0.3423	0.4665	*0.0000*	*0.0000*
(2) Noisy_circles2	*0.0000*	*0.0000*	*0.0000*	*0.0000*	0.7883	0.1726	0.3323	0.3956
(3) Atom	*0.0000*	*0.0000*	*0.0000*	*0.0000*	0.9762	0.0136	*0.0000*	*0.0000*
(4) Noisy_moons	*0.0000*	*0.0000*	*0.0000*	*0.0000*	0.8442	0.2129	*0.0000*	*0.0000*
(5) Zelnik	0.0633	0.1034	*0.0000*	*0.0000*	0.6625	0.2749	0.3677	0.1763
(6) Spiral	*0.0000*	*0.0000*	*0.0000*	*0.0000*	0.9948	0.0021	0.1652	0.3535
(7) r15	*0.0142*	*0.0066*	0.2383	0.0000	0.9950	0.0281	0.6998	0.2448
(8) Varied	3.7422	0.0962	5.7632	0.0000	*1.1035*	*0.8195*	3.2927	1.0164
(9) Blobs	*0.0000*	*0.0000*	*0.0000*	0.0000	0.6294	0.4151	0.2033	0.3948
(10) Synthetic1	*0.0000*	*0.0000*	0.1601	0.0000	0.6919	0.4094	0.4782	0.4781
(11) Synthetic2	0.1448	0.0000	*0.0004*	*0.0000*	0.5470	0.3582	0.5306	0.3952
(12) Synthetic3	*0.0000*	*0.0000*	0.0692	0.0000	0.9498	0.1660	0.7036	0.3228
(13) Mouse	*0.0569*	*0.0350*	0.6506	0.0000	0.6405	0.2863	0.2662	0.3519
(14) Aniso	*0.0016*	*0.0009*	0.1841	0.0000	0.9586	0.0484	0.1574	0.3507

Table 3 Global error and confidence interval by algorithm

Algorithm	μ	σ	Lower limit	Upper limit
Proposal	*0.2873*	*0.0173*	*0.2783*	*0.2964*
DBSCAN	0.5047	0.0000	0.5047	0.5047
KK-means	0.7946	0.2624	0.6571	0.9321
Spectral	0.5140	0.3200	0.3464	0.6817

outperforms Kernel K-means and the rest of the algorithms. Moreover, as presented in Table 3, our algorithm provides the lowest overall error and standard deviation.

4.6 Complementary Experiments

We also carried out experiments considering real-world datasets to evaluate the effectiveness of our method. To this end, we used the datasets shown in Table 4, which are related to classification problems, whose class labels can be used as a reference to evaluate the effectiveness of a partition found by the clustering method. As part of the pre-processing stage of these datasets, we complete missing information (usually encoded as blanks, *NaNs,* or other placeholders) using multivariate imputation. For categorical features, we apply an ordinal encoding. At this point, the datasets are numerically valued but representing both categorical and numerical features. Then, we encode categorical features by means of one hot encoding.

In order to make a comparison between our proposal and different clustering approaches, we use again DBSCAN, Kernel K-means, and Spectral. The evaluation is performed in terms of the so-called *Adjusted Rand Index* (ARI), which is a performance measure of the agreement between two partitions: the partition found by each approach and the partition induced by the class labels of the used dataset. In Table 5, we show the results of our proposed approach and the three benchmark clustering

Table 4 Summary of the real-world datasets used in experiments

Dataset	Numerical	Categorical	Instances	Classes
German	6	14	1000	2
Australian	6	8	690	2
Hepatitis	6	13	155	2
Cleveland	5	8	303	5
Heart Statlog	6	7	270	2
Credit	6	9	690	2
Horse	7	15	300	2
Breast cancer	0	9	286	2
Audiology	0	69	200	24

Table 5 Average ARI using different encoding approaches

Dataset	Proposal	DBSCAN	KK-means	Spectral
German	0.0144	0.0574	0.0015	−0.0007
Australian	0.4320	0.0168	0.2142	0.3644
Hepatitis	0.3199	0.3161	0.0110	0.1978
Cleveland	0.2186	0.2856	0.1103	0.0841
Heart Statlog	0.3143	0.0080	0.2313	0.3020
Credit	0.4711	0.0026	0.2541	0.3595
Horse	0.0279	0.0257	−0.0074	−0.0002
Breast Cancer	0.1467	0.0105	0.0089	0.1073
Audiology	0.2928	0.0000	0.0881	0.2600
Average	*0.2486*	0.0803	0.1013	0.1860

methods for the performed experiments with 9 different real-world datasets. For each real-world dataset, we compute the average value of the partitions obtained by each clustering method (including our proposal) after 100 executions. It is worth noting that a negative value implies that the expected agreement between partitions is lower than a random result. As shown in Table 5, our approach outperforms the other clustering approaches in the experiments performed with most of the real-world datasets and achieving the highest average performance.

5 Conclusions and Future Work

In this chapter, we presented a clustering method from an optimization perspective, where the aim is to find better values for kernel parameters to optimize a density index, which measures the quality of candidate partitions. We apply a genetic algorithm called *Eclectic Genetic Algorithm* to perform the search for such parameters that induce a partition that maximizes the density index. The analysis of the results shows that the proposed clustering method has similar behavior to DBSCAN on the evaluated datasets; however, based on the probabilistic analysis, it was found that in most cases, the proposed method exceeds the performance of DBSCAN, Kernel K-means, and Spectral clustering.

The main limitation of the proposed method is its computational complexity. Important elements of this complexity are: (1) the calculation of the validity index and (2) the parameters associated with the used heuristic, specifically the number of individuals and the number of iterations to find an appropriate partition. Another limitation is to properly choose the kernel function because the performance of a kernel depends on the problem being addressed. Using an incorrect kernel function for a problem can result in worse partitions than with traditional clustering algorithms.

However, our method, even using solely the radial basis kernel, is still able to adapt its parameters to find an acceptable solution in a range of datasets.

As future work, we propose the study of alternative kernel functions to the radial basis one in order to include them in the method as a set of possible eligible kernel functions. In addition, the calculation of the validity index and the adaptation process of the individuals of the heuristic is intended to be carried out concurrently without affecting the clustering process. Another research line we will develop in the near future is the addition of distance criteria that could help to enhance the accuracy for cases when there are dense data regions mixed with sparse regions. As we have discussed, those kinds of instances could be difficult to solve for the proposed version.

Appendix A

Eclectic Genetic Algorithm

For those familiar with the methodology of genetic algorithms, it should come as no surprise that a number of questions relative to the best operation of the algorithm immediately arose. The Simple Genetic Algorithm [70] frequently mentioned in the literature leaves open the optimal values of, at least, the following parameters:

1. Probability of crossover (P_c).
2. Probability of mutation (P_m).
3. Population size.

Additionally, premature and/or slow convergence is also of prime importance. For this, the EGA includes the following characteristics:

1. The best (overall) n individuals are considered. The best and worst individuals ($1 - n$) are selected; then, the second best and next-to-the-worst individuals (2 – [n − 1]) are selected, etc.
2. Crossover is performed with a probability P_c. Annular crossover makes this operation position independent. Annular crossover allows for unbiased building block search, a central feature to GA's strength. Two randomly selected individuals are represented as two rings (the parent individuals). Semi-rings of equal size are selected and interchanged to yield a set of offspring. Each parent contributes the same amount of information to their descendants.
3. Mutation is performed with probability P_m. Mutation is uniform and, thus, is kept at very low levels. For efficiency purposes, we do not work with mutation probabilities for every independent bit. Rather, we work with the expected number of mutations, which, statistically is equivalent to calculating mutation probabilities for every bit. Hence, the expected number of mutations is calculated from $l * n * p_m$, where l is the length of the genome in bits, and n is the number of individuals in the population.

In what follows, we present the pseudocode of EGA.

Algorithm 1: Eclectic Genetic Algorithm

Data:
n = Number of individuals
p_c = Crossover probability
p_m = Mutation probability
ℓ = Length of the Individual
$b2m = \ell * n * p_m$ number of bits to mutate
Result: The top n individuals
Generate a population P of size n whose $n\ell$ bits are randomly set:
$initialize(P)$
$evaluate(P)$
Sort individuals from best to worst based on their fitness:
$sort(P)$
while *convergence criteria are not met* **do**
$duplicate(P)$
for i = **1 to** n **do**
Generate a random number R
if $R > p_c$ **then**
Generate a random integer $locus \in [1, \ell]$
Interchange the semi-ring starting at $locus$ for individuals i and $n - i + 1$:
$crossover(P(i), P(n - i + 1))$
end
end
Mutate the population in $b2m$ randomly selected bits:
$mutate(P)$
$sort(P)$
Eliminate the worst n individuals from P
Return P
end

References

1. Schwenker, F., Trentin, E.: Pattern classification and clustering: a review of partially supervised learning approaches. Pattern Recogn. Lett. **37**, 4–14 (2014). https://doi.org/10.1016/j.patrec.2013.10.017
2. Zelevinsky, V.V., Tunkelang, D., Knabe, F.C., Saji, M.Y., Tzanov, V.K.: Method and system for information retrieval with clustering. US Patent 8, 676, 802 (2014)
3. Kim, H., Kim, H.K., Cho, S.: Improving spherical k-means for document clustering: fast initialization, sparse centroid projection, and efficient cluster labeling. Expert Syst. Appl. **150**, 113288 (2020). https://doi.org/10.1016/j.eswa.2020.113288

4. Yang, Y., Xu, D., Nie, F., Yan, S., Zhuang, Y.: Image clustering using local discriminant models and global integration. IEEE Trans. Image Process. **19**(10), 2761–2773 (2010). https://doi.org/10.1109/TIP.2010.2049235
5. Yang, J., Parikh, D., Batra, D.: Joint unsupervised learning of deep representations and image clusters. In: 2016 IEEE Conference on Computer Vision and Pattern Recognition, pp. 5147–5156 (2016)
6. Wazarkar, S., Keshavamurthy, B.N.: A survey on image data analysis through clustering techniques for real world applications. J. Vis. Commun. Image Represent. **55**, 596–626 (2018). https://doi.org/10.1016/j.jvcir.2018.07.009
7. Abdallah, A., Maarof, M.A., Zainal, A.: Fraud detection system: a survey. J. Netw. Comput. Appl. **68**, 90–113 (2016). https://doi.org/10.1016/j.jnca.2016.04.007
8. Al-Yaseen, W.L., Othman, Z.A., Nazri, M.Z.A.: Multi-level hybrid support vector machine and extreme learning machine based on modified k-means for intrusion detection system. Expert Syst. Appl. **67**, 296–303 (2017). https://doi.org/10.1016/j.eswa.2016.09.041
9. MacQueen, J.: Some methods for classification and analysis of multivariate observations. In: Proceedings of the Fifth Berkeley Symposium on Mathematical Statistics and Probability, vol. 1, pp. 281–297. Oakland (1967)
10. Jain, A.K.: Data clustering: 50 years beyond k-means. Pattern Recogn. Lett. **31**(8), 651–666 (2010). https://doi.org/10.1016/j.patrec.2009.09.011
11. Dunn, J.C.: A fuzzy relative of the isodata process and its use in detecting compact well-separated clusters. J. Cybern. **3**(3), 32–57 (1973). https://doi.org/10.1080/01969727308546046
12. Nayak, J., Naik, B., Behera, H.S.: Fuzzy c-means (FCM) clustering algorithm: a decade review from 2000 to 2014. In: Jain, L.C., Behera, H.S., Mandal, J.K., Mohapatra, D.P. (eds.) Computational Intelligence in Data Mining, vol. 2, pp. 133–149. Springer, New Delhi (2015)
13. Kaufman, L., Rousseeuw, P.J. (eds.): Finding Groups in Data: An Introduction to Cluster Analysis. Wiley, Inc. (1990). https://doi.org/10.1002/9780470316801
14. Murtagh, F., Contreras, P.: Algorithms for hierarchical clustering: an overview, II. WIREs Data Min. Knowl. Discov. **7**(6), e1219 (2017). https://doi.org/10.1002/widm.1219
15. Vijaya, Sharma, S., Batra, N.: Comparative study of single linkage, complete linkage, and ward method of agglomerative clustering. In: 2019 International Conference on Machine Learning, Big Data, Cloud and Parallel Computing (COMITCon), pp. 568–573 (2019)
16. Ester, M., Kriegel, H.P., Sander, J., Xu, X.: A density-based algorithm for discovering clusters in large spatial databases with noise. In: Proceedings of the Second International Conference on Knowledge Discovery and Data Mining, AAAI Press, pp 226–231 (1996)
17. Hinneburg, A., Gabriel, H.H.: Denclue 2.0: fast clustering based on kernel density estimation. In: Berthold, M.R., Shawe-Taylor, J., Lavrac, N. (eds.) Advances in Intelligent Data Analysis VII, pp. 70–80. Springer (2007). https://doi.org/10.1007/978-3-540-74825-0_7
18. Ankerst, M., Breunig, M.M., Kriegel, H.P., Sander, J.: Optics: ordering points to identify the clustering structure. SIGMOD Record **28**(2), 49–60 (1999). https://doi.org/10.1145/304181.304187
19. McLachlan, G.J., Rathnayake, S.: On the number of components in a gaussian mixture model. WIREs Data Min. Knowl. Discov. **4**(5), 341–355 (2014). https://doi.org/10.1002/widm.1135
20. Oboh, B.S., Bouguila, N.: Unsupervised learning of finite mixtures using scaled Dirichlet distribution and its application to software modules categorization. In: 2017 IEEE International Conference on Industrial Technology (ICIT), pp. 1085–1090 (2017)
21. Zhou, Y., Cheng, H., Yu, J.X.: Graph clustering based on structural/attribute similarities. Proc VLDB Endow **2**(1), 718–729 (2009). https://doi.org/10.14778/1687627.1687709
22. Peng, B., Zhang, L., Zhang, D.: A survey of graph theoretical approaches to image segmentation. Pattern Recogn. **46**(3), 1020–1038 (2013). https://doi.org/10.1016/j.patcog.2012.09.015
23. Ben-Hur, A., Horn, D., Siegelmann, H.T., Vapnik, V.: Support vector clustering. J. Mach. Learn. Res. **2**, 125–137 (2001)
24. Cortes, C., Vapnik, V.: Support-vector networks. Mach. Learn. **20**(3), 273–297 (1995). https://doi.org/10.1023/A:1022627411411

25. Schölkopf, B., Smola, A., Müller, K.R.: Nonlinear component analysis as a kernel eigenvalue problem. Neural Comput. **10**(5), 1299–1319 (1998). https://doi.org/10.1162/089976698300017467
26. Huang, H., Chuang, Y., Chen, C.: Multiple kernel fuzzy clustering. IEEE Trans. Fuzzy Syst. **20**(1), 120–134 (2012)
27. Son, L.H.: A novel kernel fuzzy clustering algorithm for geo-demographic analysis. Inf. Sci. **317**(C), 202–223 (2015). https://doi.org/10.1016/j.ins.2015.04.050
28. MacDonald, D., Fyfe, C.: The kernel self-organising map. In: KES 2000. Fourth International Conference on Knowledge-Based Intelligent Engineering Systems and Allied Technologies. Proceedings (Cat. No.00TH8516), vol. 1, pp. 317–320 (2000). https://doi.org/10.1109/KES.2000.885820
29. Filippone, M., Camastra, F., Masulli, F., Rovetta, S.: A survey of kernel and spectral methods for clustering. Pattern Recogn. **41**(1), 176–190 (2008). https://doi.org/10.1016/j.patcog.2007.05.018
30. Xu, D., Tian, Y.: A comprehensive survey of clustering algorithms. Annals Data Sci. **2**(2), 165–193 (2015). https://doi.org/10.1007/s40745-015-0040-1
31. Bergstra, J., Yamins, D., Cox, D.: Making a science of model search: hyperparameter optimization in hundreds of dimensions for vision architectures. In: Proceedings of Machine Learning Research, vol. 28, pp. 115–123. Atlanta (2013)
32. Tao, Z., Huiling, L., Wenwen, W., Xia, Y.: Gasvm based feature selection and parameter optimization in hospitalization expense modeling. Appl. Soft Comput. **75**, 323–332 (2019). https://doi.org/10.1016/j.asoc.2018.11.001
33. Wang, Y., Zhang, H., Zhang, G.: cpso-cnn: an efficient pso-based algorithm for fine-tuning hyper-parameters of convolutional neural networks. Swarm Evol. Comput. **49**, 114–123 (2019). https://doi.org/10.1016/j.swevo.2019.06.002
34. Aloise, D., Deshpande, A., Hansen, P., Popat, P.: NP-hardness of euclidean sum-of-squares clustering. Mach. Learn. **75**(2), 245–248 (2009). https://doi.org/10.1007/s10994-009-5103-0
35. Glover, F., Laguna, M.: Tabu Search. Springer, Boston (1998). https://doi.org/10.1007/978-1-4613-0303-9_33
36. Kirkpatrick, S., Gelatt, C.D., Vecchi, M.P.: Optimization by simulated annealing. Science **220**(4598), 671–680 (1983). https://doi.org/10.1126/science.220.4598.671
37. Dorigo, M., Maniezzo, V., Colorni, A.: Ant system: optimization by a colony of cooperating agents. IEEE Trans. Syst. Man Cybern. Part B (Cy-bernetics) **26**(1), 29–41 (1996). https://doi.org/10.1109/3477.484436
38. Kennedy, J., Eberhart, R.: Particle swarm optimization. In: Proceedings of ICNN'95—International Conference on Neural Networks, vol. 4, pp. 1942–1948 (1995). https://doi.org/10.1109/ICNN.1995.488968
39. Xue, B., Zhang, M., Browne, W.N., Yao, X.: A survey on evolutionary computation approaches to feature selection. IEEE Trans. Evol. Comput. **20**(4), 606–626 (2016). https://doi.org/10.1109/TEVC.2015.2504420
40. Maheswaranathan, N., Metz, L., Tucker, G., Choi, D., Sohl-Dickstein, J.: Guided evolutionary strategies: augmenting random search with surrogate gradients. In: Proceedings of Machine Learning Research, vol. 97, pp. 4264–4273. PMLR, Long Beach (2019)
41. Contreras-Cruz, M.A., Ayala-Ramirez, V., Hernandez-Belmonte, U.H.: Mobile robot path planning using artificial bee colony and evolutionary programming. Appl. Soft Comput. **30**, 319–328 (2015). https://doi.org/10.1016/j.asoc.2015.01.067
42. Karakatic, S., Podgorelec, V.: A survey of genetic algorithms for solving multi depot vehicle routing problem. Appl. Soft Comput. **27**, 519–532 (2015). https://doi.org/10.1016/j.asoc.2014.11.005
43. Kar, A.K.: Bio inspired computing—a review of algorithms and scope of applications. Expert Syst. Appl. **59**, 20–32 (2016). https://doi.org/10.1016/j.eswa.2016.04.018
44. Rudolph, G.: Convergence analysis of canonical genetic algorithms. IEEE Trans. Neural Netw. **5**(1), 96–101 (1994). https://doi.org/10.1109/72.265964

45. Hruschka, E.R., Campello, R.J.G.B., Freitas, A.A., Carvalho, A.C.: A survey of evolutionary algorithms for clustering. IEEE Trans. Syst. Man Cybern. Part C (Applications and Reviews) **39**(2), 133–155 (2009)
46. Kuri-Morales, A., Aldana-Bobadilla, E.: The best genetic algorithm I. In: Castro, F., Gelbukh, A., González, M. (eds.) Advances in Soft Computing and Its Applications, pp 1–15. Springer (2013)
47. Kuri-Morales, A.F., Aldana-Bobadilla, E., Lopez-Pena, I.: The best genetic algorithm II. In: Castro, F., Gelbukh, A., González, M. (eds.) Advances in Soft Computing and Its Applications, pp. 16–29. Springer, Heidelberg (2013)
48. Kuri, A., Villegas-Quezada, C.: A universal eclectic genetic algorithm for constrained optimization. In: Proceedings 6th European Congress on Intelligent Techniques and Soft Computing, EUFIT 98 (1998)
49. Arbelaitz, O., Gurrutxaga, I., Muguerza, J., Pérez, J.M., Perona, I.: An extensive comparative study of cluster validity indices. Pattern Recogn. **46**(1), 243–256 (2013). https://doi.org/10.1016/j.patcog.2012.07.021
50. Calinski, T., Harabasz, J.: A dendrite method for cluster analysis. Commun. Stat. **3**(1), 1–27 (1974). https://doi.org/10.1080/03610927408827101
51. Davies, D.L., Bouldin, D.W.: A cluster separation measure. IEEE Trans. Pattern Anal. Mach. Intell. **PAMI-1**(2), 224–227 (1979). https://doi.org/10.1109/TPAMI.1979.4766909
52. Dunn, J.C.: Well-separated clusters and optimal fuzzy partitions. J. Cybern. **4**(1), 95–104 (1974). https://doi.org/10.1080/01969727408546059
53. Rousseeuw, P.J.: Silhouettes: a graphical aid to the interpretation and validation of cluster analysis. J. Comput. Appl. Math. **20**, 53–65 (1987). https://doi.org/10.1016/0377-0427(87)90125-7
54. Halkidi, M., Batistakis, Y., Vazirgiannis, M.: On clustering validation techniques. J. Intell. Inform. Syst. **17**(2–3), 107–145 (2001). https://doi.org/10.1023/A:1012801612483
55. Hu, L., Zhong, C.: An internal validity index based on density-involved distance. IEEE Access **7**, 40038–40051 (2019). https://doi.org/10.1109/ACCESS.2019.2906949
56. Quaintance, J., Gould, H.W.: Combinatorial Identities for Stirling Numbers. World Scientific (2015). https://doi.org/10.1142/9821
57. Ultsch, A.: Clustering with som: U*c. ProcWorkshop on Self-Organizing Maps, pp. 75–85 (2005)
58. Jain, A.K., Law, M.H.C.: Data clustering: a user's dilemma. In: Pal, S.K., Bandyo-padhyay, S., Biswas, S. (eds.) Pattern Recognition and Machine Intelligence, pp. 1–10. Springer, Heidelberg (2005)
59. Zelnik-Manor, L., Perona, P.: Self-tuning spectral clustering. In: Saul, L.K., Weiss, Y., Bottou, L. (eds.) Advances in Neural Information Processing Systems 17, pp. 1601–1608. MIT Press (2005)
60. Chang, H., Yeung, D.Y.: Robust path-based spectral clustering. Pattern Recogn. **41**(1), 191–203 (2008). https://doi.org/10.1016/j.patcog.2007.04.010
61. Veenman, C.J., Reinders, M.J.T., Backer, E.: A maximum variance cluster algorithm. IEEE Trans. Pattern Anal. Mach. Intell. **24**(9), 1273–1280 (2002)
62. Pedregosa, F., Varoquaux, G., Gramfort, A., Michel, V., Thirion, B., Grisel, O., Blondel, M., Prettenhofer, P., Weiss, R., Dubourg, V., et al.: Scikit-learn: machine learning in python. J. Mach. Learn. Res. **12**, 2825–2830 (2011)
63. Fasy, B.T., Kim, J., Lecci, F., Maria, C.: Introduction to the R package TDA. CoRR abs/1411.1830 (2014)
64. Venables, W.N., Ripley, B.D.: Modern Applied Statistics with S, 4th edn. Springer, New York (2002)
65. Ng, A.Y., Jordan, M.I., Weiss, Y.: On spectral clustering: analysis and an algorithm. In: Proceedings of the 14th International Conference on Neural Information Processing Systems: Natural and Synthetic, pp. 849–856. MIT Press, Cambridge (2001)
66. Sander, J., Ester, M., Kriegel, H.P., Xu, X.: Density-based clustering in spatial databases: the algorithm gdb scan and its applications. Data Min. Knowl. Discov. **2**(2), 169–194 (1998). https://doi.org/10.1023/A:1009745219419

67. Schubert, E., Sander, J., Ester, M., Kriegel, H.P., Xu, X.: Dbscan revisited, revisited: why and how you should (still) use dbscan. ACM Trans. Database Syst. **42**(3), 19:1–19:21 (2017). https://doi.org/10.1145/3068335
68. Hahsler, M., Piekenbrock, M., Doran, D.: DBSCAN: fast density based clustering with R. J. Stat. Softw. **91**(1), 1–30 (2019). https://doi.org/10.18637/jss.v091.i01
69. Karatzoglou, A., Smola, A., Hornik, K., Zeileis, A.: Kernlab—an S4 package for kernel methods in R. J. Stat. Softw. **11**(9), 1–20 (2004)
70. Holland, J.H.: Adaptation in Natural and Artificial Systems: An Introductory Analysis with Applications to Biology, Control and Artificial Intelligence. MIT Press, Cambridge (1992)

Students Satisfaction Description Based on Classical and Multivalent Discovery Techniques

Susana Beatriz Ruiz, Rafael Alejandro Espin-Andrade, and Myriam Beatriz Herrera

Abstract Like most of the ongoing organizations, educational ones need to evaluate the quality of their provided services. "Quality" is understood as to how this organization satisfies its student's expectations to contribute to the image those students have about it. Although the perceived quality in the received service is strongly related to students' needs, it is possible to extract objective indicators like their satisfaction as an essential factor in addressing the higher education quality. This paper presents the results of an exploratory study about the satisfaction of 112 students of the Facultad de Ciencias Exactas, Físicas y Naturales de la Universidad Nacional de San Juan during the 2017 year. Through the use of data mining techniques and bivalent and multivalent logic-based procedures, we discovered relationships between linguistic states of a data set (fuzzy predicates) that met the form $\boldsymbol{P \rightarrow Satisfaction}$. We consider that the information found is valuable for making institutional decisions to improve educational quality.

Keywords Students · Satisfaction · Fuzzy relationships · Linguistic variables · Association rules · Eureka-universe

1 Introduction

Nowadays, most of the institutions, organizations, or companies need to analyze the quality of the services or products they provide. Just as educational institutions certainly do. Quality is understood as an abstract concept so comprehensive in definition and an application that every organization can understand it from their interests [1]. The concept includes how the company satisfies every specific need since it contributes to the image the clients keep in their minds about the company [2]. A structure and organizational management must support the perception of the quality

S. B. Ruiz (✉) · M. B. Herrera
Universidad Nacional de San Juan, J5402 San Juan, Argentina

R. A. Espin-Andrade
Universidad Autónoma de Coahuila, 25280 Saltillo, Coahuila, Mexico

W. Pedrycz et al. (eds.), *Computational Intelligence for Business Analytics*,
Studies in Computational Intelligence 953,
https://doi.org/10.1007/978-3-030-73819-8_20

of the service offered in such a way that philosophy or service may be set to overcome the clients' expectations [3]. It should be noted that the quality perceived is closely related to the adequacy of the characteristics of the objects to the individuals' needs. Despite this, the contact of the individual with the service received and with the agents that offer it can bring objective data [4].

Universities, as educational institutions, also should have this perspective. Studies on student satisfaction in universities as indicators to evaluate the educational quality are particularly important because the students' satisfaction improves academic performance, reduces the academic dropouts and career shifting, and it is a requirement to succeed in the process of learning. Likewise, a well-known quality of education reinforces the institutional prestige [5]. To value the students' satisfaction and to determine which are the variables associated with this aspect contributes to the assertive decision making in the management of the university education quality.

2 Background

The availability of high volumes of data and the generalized use of information and communication technology have transformed the analysis of data guiding it towards global specialized techniques called Data Mining (DM). The DM techniques follow the automatic discovery of the knowledge contained in the information stored in a well-ordered model in a vast database. What we aim for is to detect consistent behavioral patterns or relations between the different variables to apply them to new sets of data [6]. Not only the statistics techniques but also the techniques of artificial intelligence are quite compelling. In some cases, these are only two different approaches to give a solution to the same problem. However, in other cases, these are complementing techniques because they solve problems of different nature [7].

Within DM, we find discovery strategies of rules of association based on classic logic, and they aim to locate sets of elements that co-occur together frequently, in a database [8, 9]. The application of this technology, also known as Association mining rules, can generate a considerable quantity of rules. For this reason, the need to select those rules of association relevant from the specific perspective of studying arises.

On the other side, Fuzzy Logics is a non-probabilistic approximate reasoning method that could be defined as an extension of Multivalued Logics which significantly facilitates the modeling of qualitative information approximately. Its success is mainly due to the possibility of solving very complex problems that are hard to resolve within traditional methods.

The Fuzzy Logics is the concept of the fuzzy set under which we set the idea that the element on which the human thought is built are not numbers but linguistic labels. The fuzzy logic allows us to represent the collective knowledge (that it is mostly qualitative linguistics and not necessarily quantitative) in a mathematics language utilizing the fuzzy set theory and its characteristic functions or associated with them.

We can see a clear example of its application in expressions like "the student is very pleased with his career" or "the student has a good relationship with peers".

Nevertheless, it is not easy to define the concept of "pleased" and "good relationship" here because it is challenging to specify how a student is satisfied in a place or how a student has good relationships with classmates.

The Fuzzy Logics works with data sets where there are no marked limits. It uses expressions that are not entirely true or completely false. It may be applied to concepts that can take any precisions value into a set of values that extends between two extremes: Absolute truth and complete falsehood. The general idea is that things are not black or white, but there are infinite shades of grey color.

A way of implementing the "principle of gradualism", which is an essential property of Fuzzy Logics, is to define logics where the predicates are functions of the universe X in the interval [0, 1]. The operations of conjunction, disjunction, negation, and implication are defined in a way that when the domain [10] is restricted, the Boolean logic is obtained. The various ways of defining operations and their properties determine different multivalent logics that are a part of the Fuzzy Logics Paradigm [11].

The concept of the linguistics variable plays an essential role in the imprecise representation of knowledge.

The structure of multiple linguistic variables often constitutes a system and a complete description of some knowledge. For instance, a person can be considered a linguistic variable that can be composed with another linguistic variable "age", "height", "weight", "physical appearance", and others. The values of a linguistic variable may consist of initial terms such as "young", "old" into the category of "age" and those terms from the use of modifiers or linguistic terms like "very", "slightly", "more or less", "extremely", etc. and the logic connectors "no", "or", "and" [12].

The formal definition of linguistic variables is a quintet $(X, T(X), U, G, M)$. $T(X)$ is the set of labels that can take X, U is the subjacent domain, G is the grammar to generate the linguistic labels, and M is the semantic rule that links each label of $T(X)$ to the set of values in U.

The transformation of value to a linguistic estate is called Fuzzification, while the reverse process is called Defuzzifier.

A function of membership or belonging in a space characterizes a fuzzy set. Each value x is associated with it, to a real number into the interval [0, 1], which represents the membership of x to the set A. The functions of belonging may be continuous or discrete functions, depending on the universe of the discourse.

A predicate affirms or negates an object. The classic predicates lead us to a clear division of the universe A according to the function of the predicate $P(x)$. Despite this, people daily use predicates that cannot be reduced to this division: fuzzy or vague predicates. Predicates are flexible structures that facilitate learning, and they are formed by the operators and the estates or linguistic variables. They could be represented by tree diagrams employing the type of logic. We can calculate the real value of the fuzzy operators: conjunction, disjunction, negation, and order.

Compensatory Fuzzy Logic (CL) constitutes a branch of the Fuzzy Logic (FL). It is a new multivalent system that breaks into the traditional axiomatic of this type of system, to achieve a better system than the classic ones, by the semantic point of view.

In general, all the models based on this logic combine experience and knowledge with numeric data; for this reason, it could be seen as a "grey box". It was made by a group of multidisciplinary Scientifics Company: Company Management in the Uncertain Investigation and Services.

3 Proposal

We can establish that, in general, direct indicators of the educational qualitative is the satisfaction of the career [13–16]. The satisfaction of university students constitutes a key factor in the approach of quality in higher education. We show a study made with 112 students of Exact and Natural Sciences Faculty of the National University of San Juan, Argentine, in 2017. Associated variables that may characterize in a precise way the satisfaction of the students were chosen applying techniques of data mining, procedure based on the bivalent logic. Sufficient logical conditions for Satisfaction were obtained by processing a survey.

4 Methodology

We want to characterize the satisfaction of students organized in the careers they follow at the Facultad de Ciencias Exactas, Físicas y Naturales at UNSJ (Argentina) using results provided from a survey.

The survey is about risk elements and quality of life and was made with the web tool of surveys at online EncuestaFacil.com, and it is divided into various sections. The variables considered can be arranged into Variables characterizing the School, university, and the career that the students attend, variables representing personal information, and of the family. Other variables related to performance, effort, and motivation of the students (e.g., amount of hours they study, attendance to university, the level of satisfaction in the current career, etc.) are included in the study. The survey can be checked at https://www.encuestafacil.com/RespWeb/Qn.aspx?EID=2197195. The questions considered are not mutually exclusive between them. For that reason, each question is a variable itself; the alternative answers of each section are mutually exclusive, and each of these answers is a modality (items, category, or linguistic label) of the qualitative variables they belong to. The resulting information is conveniently pre-processed, using the Excel software for the application of the various processes that are specified later. In this study, starting from the questionnaire, we considered a total of 16 linguistic variables and 56 linguistic estates, items, or categories for the investigation.

Given the importance of the theme to develop and the nature of the variables involved in the survey to discover relevant relations between linguistic variables in the shape of the logic form P, this paper is organized in two parts. In the first part, we work with the Boolean Logics. We take into account the results of a preliminary analysis, including the data of the survey in which we observe the existence of variables and linguistics states whose implications are right in a percentage that may vary between 50 and 90% of the cases. So, we decided to apply a standard procedure of data mining, the algorithm of search Apriori, to discover relations between variables, and to define the rules of the association. This task is done with the help of free software R, version 3.5.2.

In the second part, considering the weak nature of the linguistic variables involved in the survey, we applied algorithms of the discovery of the knowledge-based on Fuzzy Logics, such that they allow estimating a set of parameters to maximize the truthfulness of the implications in a universal way. At this final stage of research, given the importance of the searching, we use a recently designed system called Eureka-Universe 2.4.6, for the "discovery" of fuzzy predicates, as well as the estimate of optimal parameters and the assessment of the truthfulness of the predicates.

Between the procedures based on the classic logics, in the DM, the problem of discovering associations from de data consists of identifying groups of variables that are strongly correlated themselves. We count on a set of items and a big set of transactions that are a subset of those items. The Association Rules Mining aims to find relations between the items, the so-called rules of the association, from the frequent presence of various items into de transactions.

Formally, we consider the following concepts to approach the problem of the searching of rules of association: $D = d_1, d_2, \ldots d_m$ is a set of items; $T = t_1, t_2, \ldots t_n$ his a set of transactions, where each transaction t_i is a set of items such that $t_i \subseteq D, 1 \leq i \leq n$. The implication $X \rightarrow Y$ is a Rule of Association where $X \subset D$, $Y \subset D$, $X \cup Y = \emptyset$ y $X \cup Y \subseteq t_i$. That is to say, the sets X and Y are mutually exclusive, t_i is a set of items formed by those corresponding to the antecedent or to consequent of the rule of association. The set $X \cup Y$ must be included or must be equal to some of the transactions belonging to T.

A basic procedure to find frequent itemset to obtain boolean rules of association is the algorithm Apriori. The algorithm applies an iterative approximation known as the smart search of level, where the k-itemset is used to explore the $k + 1$ level. To improve the efficiency applied to the prior property that indicates that all the subsets of and frequent itemset must be frequent. It indicates that if a set cannot pass a test, the supra-sets derived from it could not do it either.

In this paper, the set D is formed by the set "yes answers" (items) to be responded in the survey, whereas each transaction is identified as an "answer yes set" (set of items) of a particular student. There are as many items as "yes answer" of the questionnaire and as many transactions as survey respondents.

Once the frequent itemsets of the transactions in the database are found, we proceed to select the rules of a strong association. The most popular means are support, confidence, and lift [17].

In the context of support-trust, formerly [18–20], the search of the rules of association adopts the factors support and trust to assess the rules discovered.

The support of an item is the frequency it is found in the transactions, divided for the number of transactions. That is, if X is an item, then:

$$\text{Support(X)} = \frac{\text{number of transactions containing the items X}}{\text{total number of transactions of the database}} \quad (1)$$

To obtain the support of a rule of association X → Y, we do the following:

$$\text{Support(X} \to \text{Y)} = \frac{\text{number of transactions containing items X and items Y}}{\text{total number of transactions of the database}} \quad (2)$$

The confidence measure of an association rule $X \to Y$ is:

$$\text{Conf(X} \to \text{Y)} = \frac{\text{Support(X} \to \text{Y)}}{\text{Support(X)}} \quad (3)$$

The confidence can be interpreted as an estimator of the probability of finding the right part of the conditioned rule that can be found in the left part as well [21].

The confidence is aimed to and calculate different values for the rules "X → Y" and "Y → X".

According to Romero Morales [22], confidence is not capable of detecting statics independence. The same author expresses the following: It is usually thought that the higher the trust is, the better the set of elements are, but this might not always be true, because a set of elements with high support could be a source of deception, due to they appear in almost all transactions. The restrictions possibly are shown by the support factors and confidence, plus the need to recover the new rules of a set probably numerous of generated rules, make the activity of the experts quite hard in any field DM is applied. Following this, many authors have developed other ways of measuring to assess the importance of the generated rules. Among these measurements, we recover the so-called measure of independence or lift. The lift factor represents a test to measure the statistics dependence, and it is defined as:

$$\text{lift(X} \to \text{Y)} = \text{lift(Y} \to \text{X)} = \frac{\text{P(X} \cup \text{Y)}}{\text{P(X) P(Y)}}$$

This factor sets a relationship between simultaneous occurrences of X and Y, whenever the sets of items conforming to the antecedent and de consequent of the rule are statistically independent. As the lifted item is symmetric, this value only measures the level e l dependence and not the implication in both directions.

According to Hassler [23], the rules recovered utilizing support and confidence should be filtered using their values of lift since the values of lift larger than 1 indicates the association between items. In contrast, the values minor than one should not be bared in mind for making decisions.

To apply the procedure, using the software R, we need the previous implementation of the bundles "arules" and "arulesViz" [24].

The program designed follows, in general terms, these steps:

1. Reading of survey data and installation of the bundles "arules" and "arulesViz".

2. Construction of transactions.
3. Application of the function "apriori" for the definition of the frequent itemsets.
4. Construction of the graphics to explore itemset, with the help of the command 'plot' and the methods "graph" and "scatterplot".
5. Filtering of frequent itemsets of the manner P Satisfaction, y means of the application of the function "apriori" considering a minimum value of support (70% of the data) and minimal trust of 0.7.
6. Selection of the rules of strong association utilizing assessment of support Eq. (2), confidence Eq. (3) and lift Eq. (4).

For the second part of the work, we apply techniques of discovery based on Compensatory Fuzzy Logic (CL) and the logic of Zadeh [25].

An interpretable theory is a feasible approach to reach a transdisciplinary theory. CL is a paradigm of an interpretable logical theory [26].

The CL [27] takes into account the following axioms.

Let $x = (x_1, x_2, \ldots, x_i, \ldots, x_j, \ldots, x_n)$ and $y = (y_1, y_2, \ldots, y_i, \ldots, y_j, \ldots, y_n)$ successively be any elements of the Cartesian product in the domain $[0, 1]^n$. A quartet of the continuous operators: conjunction, disjunction, negation and order (c, d, n, o) constitute a CL if the following axioms are fulfilled:

- Compensation: $\min(x_1, x_2, \ldots, x_i, \ldots, x_j, \ldots, x_n) \leq c(x_1, x_2, \ldots, x_i, \ldots, x_j, \ldots, x_n) \leq \max(x_1, x_2, \ldots, x_i, \ldots, x_j, \ldots, x_n)$
- Commutability: $c(x_1, x_2, \ldots, x_i, \ldots, x_j, \ldots, x_n) = c(x_1, x_2, \ldots, x_i, \ldots, x_j, \ldots, x_n)$
- Strict increasing: if $x_1 = y_1$, $x_2 = y_2$, ..., $x_{i-1} = y_{i-1}$, $x_{i+1} = y_{i+1}$, ..., $x_n = y_n$ except for $x_i > y_i$ then $c(x_1, x_2, \ldots, x_i, \ldots, x_j, \ldots, x_n) > c(y_1, y_2, \ldots, y_i, \ldots, y_j, \ldots, y_n)$.
- Axiom of veto: if $x_i = 0$ for some i, then $c(x) = 0$.
- The Compensatory Fuzzy Logic based on the Geometric Mean Based Compensatory Logic (GMBCL) [27, 28] is such that: $c_1(x_1, x_2, \ldots, x_n) = (x_1 \cdot x_2 \ldots x_n)^{\frac{1}{n}}$; $d_1(x_1, x_2, \ldots, x_i, \ldots, x_j, \ldots, x_n) = 1 - [(1-x_1)\cdot(1-x_2)\ldots(1-x_n)]^{\frac{1}{n}}$; $o_1[x, y] = 0.5[c_1(x) - c_1(y)] + 0.5$ and $n(x_i) = 1 - x_i$.

The universal and existential operators defined for the GMBCL are stated in the form (in case of discrete variables):

$$\left| \underset{x \in U}{\forall} p(x) = \underset{x \in U}{\wedge} p(x) = \sqrt[n]{\prod_{x \in U} p(x)} = \begin{cases} exp\left(\frac{1}{n} \sum\limits_{x \in U} ln(p(x))\right), \\ 0 \end{cases} \right. \tag{4}$$

$for x : p(x) \neq 0, in any other case$

$$\left| \underset{x \in U}{\exists} p(x) = \underset{x \in U}{\vee} p(x) = 1 - \sqrt[n]{\prod_{x \in U}(1 - p(x))} \right.$$

$$= \begin{cases} 1 - \exp\left(\frac{1}{n} \sum_{x \in U} \ln(1 - p(x))\right) \\ 1 \end{cases}, \tag{5}$$

for x: p(x) $\neq$ 1, in any other case

The continuous versions of these formulas read as follows:

$$\forall x\ p(x) = \begin{cases} e^{\frac{\int_X \ln(p(x))dx}{\int_X dx}} & if\, p(x) > 0\ \ for\ any\ x \int X \\ 0 & in\ any\ other\ case \end{cases} \tag{6}$$

$$\forall x\ p(x) = \begin{cases} 1 - e^{\frac{\int_X \ln(p(x))dx}{\int_X dx}} & if\, p(x) > 0\ \ for\ any\ x \int X \\ 0 & in\ any\ other\ case \end{cases} \tag{7}$$

While Diffuse Compensatory Logic Based on the Arithmetic Mean (AMBCL) [29] (in case of discrete variables): $c_2(x_1,x_2,\ldots,x_n)=\left[\min(x_1,x_2,\ldots,x_n)\frac{\sum_{i=1}^{n} x_i}{n}\right]^{\frac{1}{2}}$ and $d_2(x_1,x_2,\ldots,x_n)= 1-\left[min(1-x_1,\, 1-x_2,\ldots,1 - x_n)\frac{\sum_{i=1}^{n}(1-x_i)}{n}\right]^{\frac{1}{2}}$.

The continuous versions of these formulas read as follows: $\forall x p(x)=\left[\underset{X}{min}(p(x))\frac{\int_X p(x)dx}{\int_X dx}\right]^{\frac{1}{2}}$ and $\exists x p(x)=1 - \left[\underset{X}{min}(1-p(x))\frac{\int_X (1-p(x))dx}{\int_X dx}\right]^{\frac{1}{2}}$.

A model of a CL system includes data input, a fuzzification module, a core with the base of knowledge, and a motor of interference that generates knowledge from the rules, a defuzzifier module, and data output.

Using a CL system, we can assess predicates, discover knowledge, and make inferences. The Eureka-Universe software is a unique system that allows us to solve tasks and combine them in solutions for decision making in a simple way. It can resolve tasks of assessment, discovery, and inferences. In assessments, it calculates the values of the truth of a fuzzy predicate for a set of data. It builds a predicate y selecting the linguistic variables and the operators. It assesses every register in the set of data, and it obtains a truth value, here, we also calculate the existential operator and the universal one. During the discovery, we look for relations between the linguistic states of a set of data (fuzzy predicates) that fulfills the users' requirements. In search of fuzzy predicates, we use genetic algorithms and the parameter adjustments of the functions of membership defined in the linguistic states. Genetic algorithms are popular metaheuristic approaches because of their efficiency in addressing complex real-world problems [30–32].

Eureka-Universe is programmed in Java. It allows working with different types of logic; among them, we find the Geometric Mean Based Compensatory Logic,

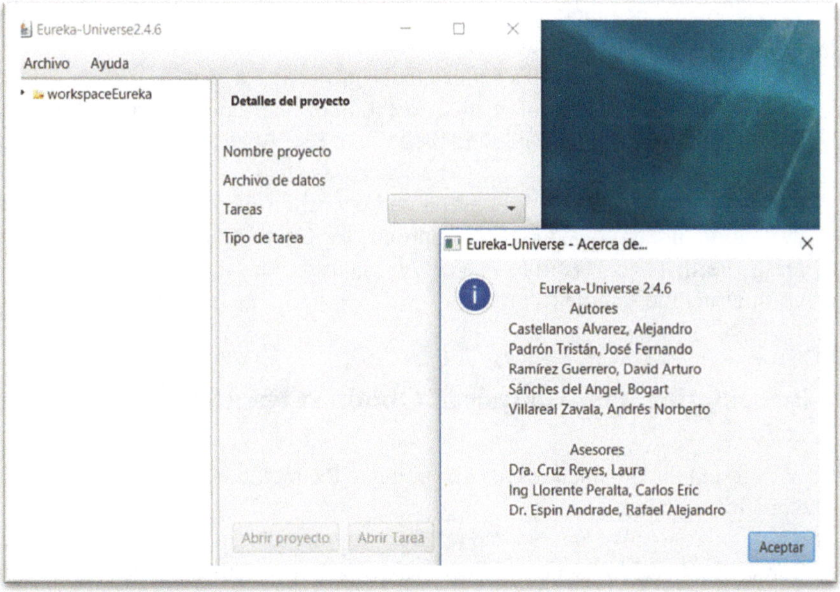

Fig. 1 Initial image Eureka-Universe system

The Arithmetic Mean based Fuzzy Logic, and the Zadeh Logic. Figure 1 offers an initial image of the system for its application where, at the moment of activating the helping window, it shows the group of scientists and researchers who took part in the development of the version 2.4.6.

The Linguistic states in Eureka-Universe are an association between the name of a column of the data set and a set of membership functions and the corresponding labels or states. During the tasks of discovery, the linguistic states are defined without specifying the function of belonging of the mentioned states for the system to use Functions of Generalised Membership (FPG) with variable parameters.

At this final stage of work, as the initial step, the database of the survey is divided, at random, in a sample of the discovery of relationships between variables formed by the answers of 78 students (70%) of the surveyed students) and a sample test (30% of the students) to assess results. Among the selected variables, we find ones considered in work at the initial stage of the research. Here we applied the Eureka 2.4.6 system for searching relevant relations between linguistic states if the shape $P \rightarrow Satisfaction$. In this paper, we consider a relation P-Satisfaction relevant if the truthfulness of the universal predicate $(\forall x)(P(x) \rightarrow Satisfaction(x))$ results greater than or equal to 0.9. After that, we do the following steps:

1. Initial step: Reading of database for discovery. Creation of Discovery Projects to look for relevant relations between linguistic states, for simple P predicates (includes only a linguistic state). The logic applied is GMBCL, AMCL, and

Zadeh Logic. The functions of belonging of the linguistic states are not specified in any project.
2. Execution of discovery tasks defined in the project created at the previous stage. Analysis and assessment of results. Selection of relevant relations.
3. Assessment of results considering the sample test. Analysis and comparison of the results.
4. Developing a hypothesis. Definition and execution of discovery projects and evaluation, under-considered assumptions, for Complex P predicates, considering the discovery sample—selection, analysis, and comparison of results considering the sample test too.

5 Presentation and Analysis of Obtained Results

Now we present the obtained results according to the methodology of work detailed in the previous section.

First stage results: We want to find relations among items of the form X [Satisfaction -career- yes], inside a set of transactions defined by the students' answers. Then, the reading of the survey data and installation of the bundles "arules" and "arulesViz", in R, define the transaction considering all the information coming from the database to apply the function a priori later. In Fig. 2 we can observe 21 items between the ones considered in this work, in which the first 19 are part of transaction 1.

As we can see in Fig. 3, 307 general rules of association were found for the database considered, with a minimum value of support of 0.71. Figures 4 and 6 show the ranking of more frequent items, among the first 20. Graph of Fig. 5 offers a graphical characterization of the values of confidence, among which the item of "satisfaction of student = yes" is included.

Figures 7 and 8 show results after the filter of the frequent itemsets, considering as a minimum value of support 0.7 and the minimum value of confidence 0.7. a total of 10 rules of association of the form X [satisfaction-career-yes] is obtained. Among the 10 of the obtained rules, those stronger rules (with values of lift larger than 1) are selected.

- {2} [stress-studio-yes] →[Satisfaction_career_yes]
- {3} [career-better-future-yes] [satisfaction-career-yes]
- {4} [relationships-classmates-good] [satisfaction-career-yes]
- {5} [average-good]-[satisfaction-career-yes]
- {7} [career-better-future-yes-average-good]-satisfaction-career-yes]
- {8} [Average-good, relationships-classmates-good]-[satisfaction-career-yes].

For the obtained rule {2}, the value of support expresses that in 71.4% of the transactions, students that reveal they suffer from stress because of studying and others are satisfied with their career 81.6% of the stress of the transaction due to studying, also expressed satisfaction with the career.

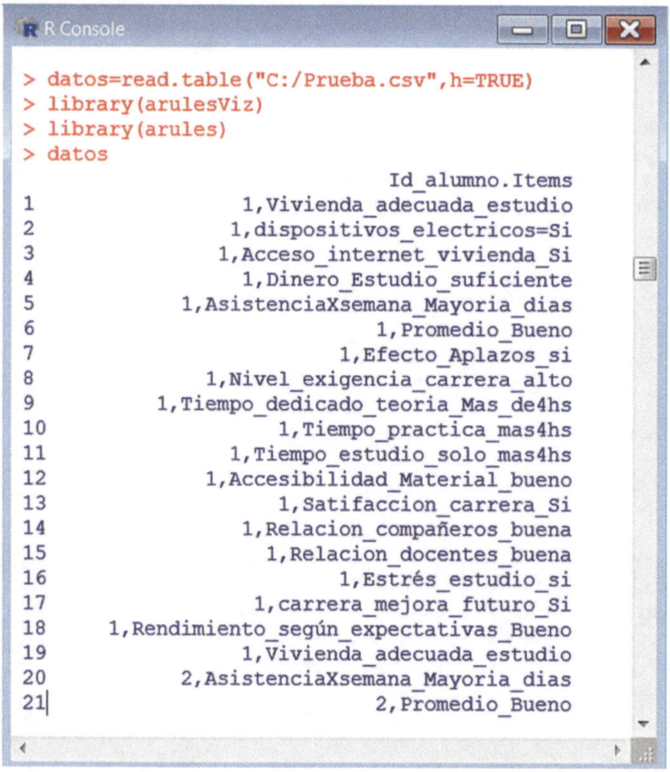

Fig. 2 Display of items by installing of the bundles "arules" and "arulesViz", in R

The lift value larger than one indicates association (statistics dependence) between the items "stress-study-yes" and [satisfaction-career-yes]. Similarly, we can interpret the rules {3}, {4}, {5}, {7}, {8}. We concluded "stress for studying", "student's perception that the career will change his future lifestyle", "good relationship with classmates" and "good average during the career" are associated with the item of "students' satisfaction for the career".

Second stage results: After the reading of the database in Excel, included in Fig. 9, we define a different project to discover relations among fuzzy linguistic variables, interested in studying, using the Eureka-Universe system. The first line of the data matrix of the survey contains a list of linguistic variables; among them, the ones worked in the first part of the work are included. Meanwhile, the remaining lines (112 lines) represent the answers of the students in the survey. The marks of each answer are converted to minimum values (see Fig. 9), such that it can facilitate the processing of the data system (Fig. 10).

Table 1 shows results that allow us to characterize the discoveries obtained when executing the system ten times, for projects with many logics and using the sample of discovery. On each execution, by default, the system makes some discoveries

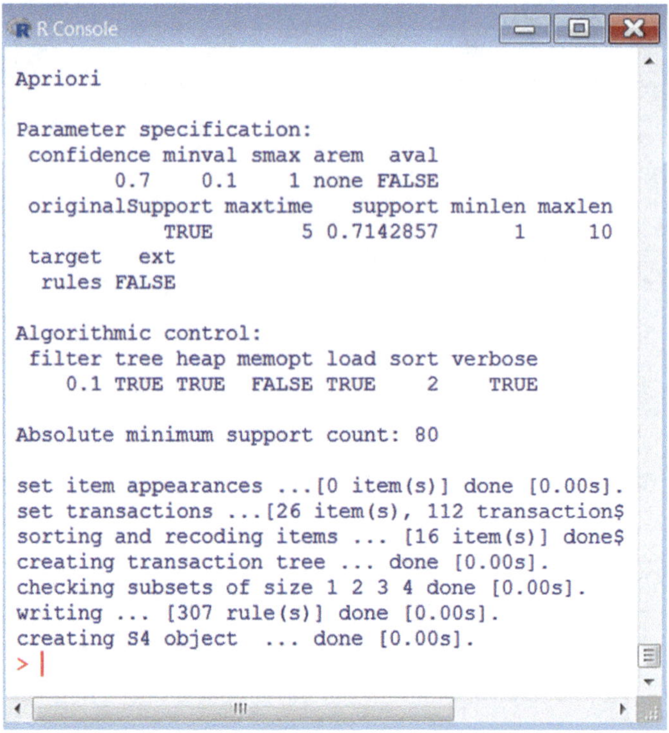

Fig. 3 Application of the function ‘apriori’ for the definition of the frequent itemsets

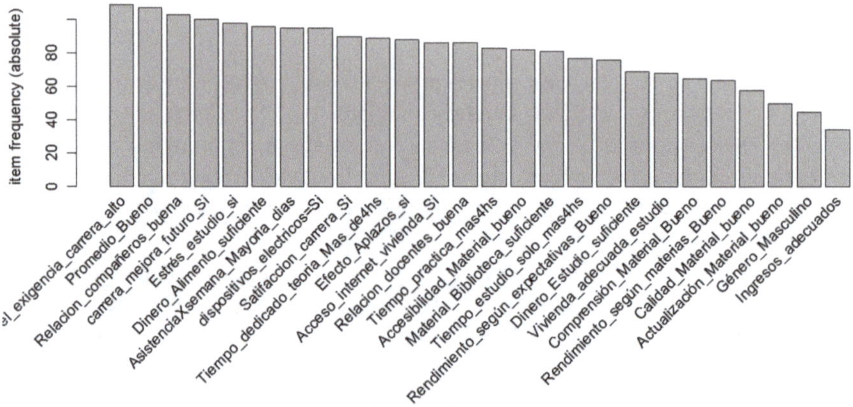

Fig. 4 More frequent items, when applying apriori

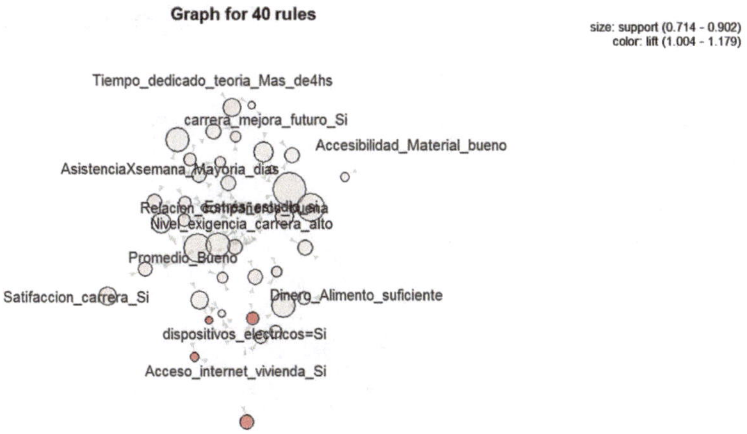

Fig. 5 Graphic characterization of the confidence values

Fig. 6 Relationship between confidence, support and lift values

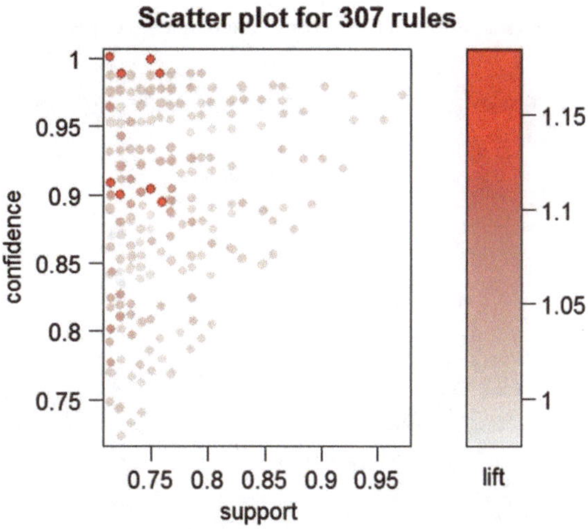

applying a genetic algorithm using 50 iterations. We can observe that, among the set of a discovered set, the maximum value of truthfulness found is y applying GMBCL. The optimal value is reached by setting a relation among variables "Money for food" and "satisfaction for a career" using estimating the parameters of the generalized functions of belonging, as shown in Fig. 11 On average, with the Zadeh logic, we obtain values of optimization superior to the ones obtained with the LDC. Whereas the GMBLC, in general terms, a higher number of predicates relevant by execution. With the logic of Zadeh, 11 linguistic variables are discovered related to the student's satisfaction by the career. These are "Money to study" "Weekly attendance", "Relationships with classmates", "The house conditions to study", "Students age",

```
Apriori

Parameter specification:
 confidence minval smax arem  aval originalSupport maxtime   support
        0.7    0.1    1 none FALSE            TRUE       5 0.7142857
 minlen maxlen target   ext
      1     10  rules FALSE

Algorithmic control:
 filter tree heap memopt load sort verbose
    0.1 TRUE TRUE  FALSE TRUE    2    TRUE

Absolute minimum support count: 80

set item appearances ...[1 item(s)] done [0.00s].
set transactions ...[26 item(s), 112 transaction(s)] done [0.00s].
sorting and recoding items ... [16 item(s)] done [0.00s].
creating transaction tree ... done [0.00s].
checking subsets of size 1 2 3 4 done [0.00s].
writing ... [10 rule(s)] done [0.00s].
creating S4 object  ... done [0.00s].
```

Fig. 7 The result after the filter of the frequent itemsets

```
    support          confidence          lift            count
 Min.   :0.7143   Min.   :0.7982   Min.   :0.9933   Min.   :80.0
 1st Qu.:0.7232   1st Qu.:0.8041   1st Qu.:1.0007   1st Qu.:81.0
 Median :0.7366   Median :0.8168   Median :1.0165   Median :82.5
 Mean   :0.7473   Mean   :0.8146   Mean   :1.0137   Mean   :83.7
 3rd Qu.:0.7723   3rd Qu.:0.8195   3rd Qu.:1.0199   3rd Qu.:86.5
 Max.   :0.8036   Max.   :0.8421   Max.   :1.0480   Max.   :90.0

mining info:
          data ntransactions   support confidence
 transacciones           112 0.7142857        0.7
> inspect(reglas_satisfaccion_carrera_bueno)
     lhs                            rhs                        support confidence      lift count
[1]  {}                          => {Satifaccion_carrera_Si} 0.8035714  0.8035714 1.0000000    90
[2]  {Estrés_estudio_si}         => {Satifaccion_carrera_Si} 0.7142857  0.8163265 1.0158730    80
[3]  {carrera_mejora_futuro_Si}  => {Satifaccion_carrera_Si} 0.7321429  0.8200000 1.0204444    82
[4]  {Relacion_compañeros_buena} => {Satifaccion_carrera_Si} 0.7410714  0.8058252 1.0028047    83
[5]  {Promedio_Bueno}            => {Satifaccion_carrera_Si} 0.7857143  0.8224299 1.0234683    88
[6]  {Nivel_exigencia_carrera_alto} => {Satifaccion_carrera_Si} 0.7767857  0.7981651 0.9932722    87
[7]  {carrera_mejora_futuro_Si,
      Promedio_Bueno}            => {Satifaccion_carrera_Si} 0.7142857  0.8421053 1.0479532    80
[8]  {Promedio_Bueno,
      Relacion_compañeros_buena} => {Satifaccion_carrera_Si} 0.7232143  0.8181818 1.0181818    81
[9]  {Nivel_exigencia_carrera_alto,
      Relacion_compañeros_buena} => {Satifaccion_carrera_Si} 0.7232143  0.8019802 0.9980198    81
[10] {Nivel_exigencia_carrera_alto,
```

Fig. 8 Association rules obtained by applying the apriori algorithms, of the form [satisfaction-race-yes]

"Time devoted only to practice", "Low grades effects", "Gender and academic performance", "Relations with teachers". Whereas with the LDCMA, we discovered the two linguistic variables related to satisfaction "Time devoted to practice" and "Time devoted to the theory".

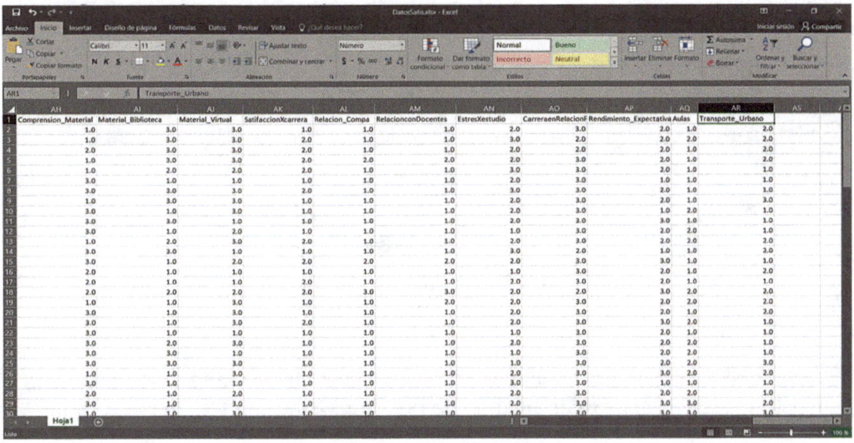

Fig. 9 Database in excel, to apply the Eureka-Universe system

Fig. 10 Memberships function estimated for the variable "money food" with Eureka-Universe

Table 1 The table with results that characterize the discoveries obtained using the discovery sample, for different logics

Universal predicates discovered	Logic of Zadeh	GMBCL	AMBCL
The maximum value of truth	0.954919	0.979428	0.908960
Average of maximum values of predicate veracity	0.9308126	0.908542	0.873402
A maximum number of relevant predicates discovered per execution	3	5	1
An average number of relevant predicates discovered per execution	1.5	1.8	0.2
The median number of relevant predicates discovered by execution	2	1.5	0
Total of relevant variables discovered in 10 executions	11	13	2

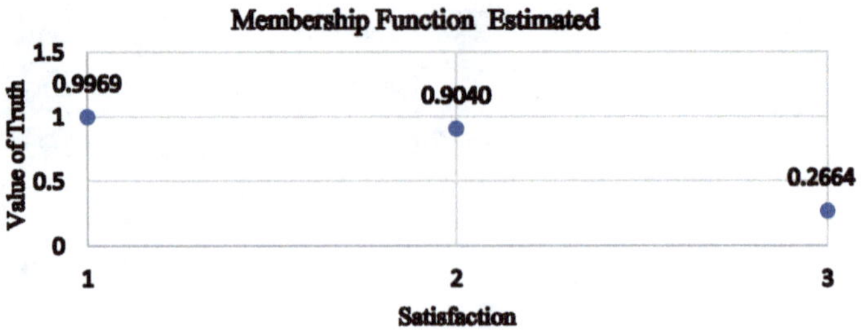

Fig. 11 Memberships function estimated for the variable "satisfaction" with Eureka-Universe

From the analysis of the results in Table 2 we take into account the truthfulness of universal predicates $GP(x)$ defined in the first row of the chart, to the truthfulness of

Table 2 Table shows relevant rules for the discovery sample as well as the sample test, discovered with the Eureka-Universe system using GMBCL

$G_P(x)$:'$(\forall x)(P(x) \rightarrow Satisfaction(x))$'				
Ri: Estimated parameters of the generalized membership function for the hypothesis $P \rightarrow Satisfaction$	"P:FPG(gamma, beta, m)" Estimated parameters of the generalized membership function for the thesis	"Satisfaction: FPG(gamma, beta, m)"	Discovery sample	Sample test
			Truth value of $GP(x)$	Truth value of $GP(x)$
R1	P: money-food g = 1.349403 b = 1.043866 m = 0.922501	g = 2.60282 b = 1.028884 m = 0.017727	0.979151	0.983254
R2	P: weekly attendance g = 3.432042 b = 1.084790 m = 0.903678	g = 2.701103 b = 1.039786 m = 0.019924	0.963075	0.961204
R3	P: relations with classmates g = 3.912455 b = 1.012025 m = 0.644238	g = 2.965499 b = 1.024826 m = 0.011069	0.97095	0.934230
R4	P: career to succeed in the future g = 2.504814 b = 1.451060 m = 0.000329	g = 2.77201 b = 1.005739 m = 0.013742	0.940794	0.931273

the universal predicates defined in the first row in the table, for the sample of discovery as well the sample test. For both samples, the values of truthfulness found y means of tasks of assessment of the predicates, for the sample and the universal predicates corresponding to the relations R1, R2, R3, and R4 are all larger than 0.9. For each relation Ri discovered, we obtained that the truthfulness found for the sample of discovery and the sample test softly differs from in each case, an indicator of good quality. The implications found to express that the linguistic variables "Money for food", "Weekly attendance", "Relations with classmates," and "career to succeed in the future" are related to the linguistic variable "Satisfaction for the career".

The first column of Table 2 shows new relevant rules for the discovery sample as well as the sample test, discovered with the Eureka-Universe system using GMBCL.

To interpret relations Ri of Table 2, we make graphics representations of the respective FPG estimated in every case, with the help of the Eureka-Universe system, where we keep in mind the states of the linguistic variables involved. In this way, for example, to interpret the relation R1, we analyze graphics of the function of belongings estimated in Figs. 11 and 12. Here we consider states of the linguistic variables "Money for food" and "Satisfaction with career" defined in the questionnaire. States for the variable "Money for Food" are: "1. enough", "2 more or less", "3. insufficient", whereas the states of "satisfaction with career" are "1. satisfied", "2. doubt" and "3. dissatisfied".

Relation R1 can be approximately interpreted as if a student says that the money is more or less enough, then the student is satisfied with the career or has doubts about it. Similarly proceeding for the rest of the relevant relations of Table 2, they can be interpreted as R2: if a student does not attend classes during the week, then he/she must be happy or must have doubts about it". R3: If a student has bad relations

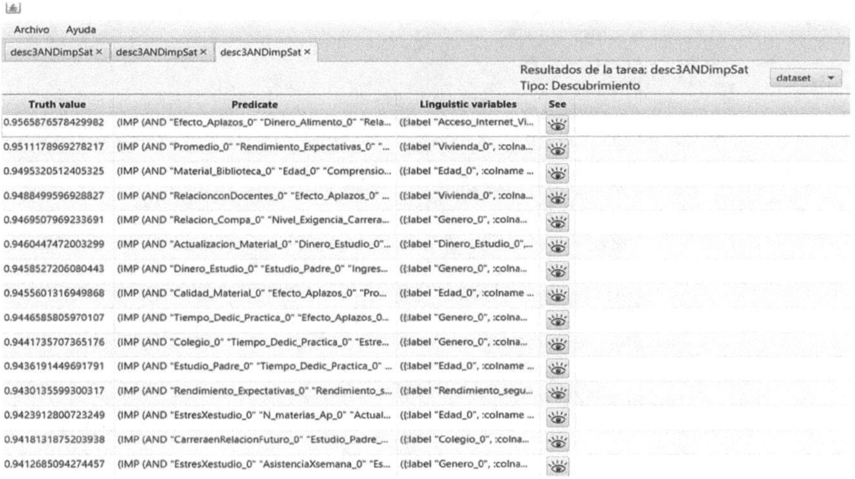

Fig. 12 Visualization of relevant relationships discovered between linguistic variables, assuming values of the FPG parameters of the thesis as true, through Eureka-Universe

or no relations at all with classmates then, he/she should be satisfied, not satisfied, or has doubt about the career. R4: if a student claims that the career about the future does not make it, then he/she can have doubts about it.

Assuming the parameters of the FPG of the thesis, for the R1, as real, we proceed to resume analysis for discoveries of relations applying Eureka-Universe. In the new instances, we consider the possibility of conjunctions in the antecedents of the implications. As it is seen in Fig. 12 we find more than ten new relevant relations among linguistic states. Consequently, we select one of them; the relation called R5, defined in Table 3. R5 may be interpreted approximately in the following way: "If all the student considers that the library material is more or less sufficient to sufficient, she is 25 years old or even older, she is a woman, she claims that the academic material is more or less updated to updated, the level of study reached by the father is tertiary or incomplete university or a superior level, then the student is satisfied with the career".

Table 3 Summary table of new discoveries, assuming values of the FPG parameters of the thesis as true

$G_P(x){:}'(\forall x)(P(x) \rightarrow Satisfaction(x))'$

Ri $P \rightarrow Satisfaction$	Estimated parameters FPG-hypothesis "P"	Estimated parameters FPG-thesis "Satisfaction"	Discovery Sample Truth value $GP(x)$	Sample Test Truth value $G_P(x)$
R5	P: *library material* g = 2.191349 b = 1.0113401 m = 0.014228 P: *years* g = 2.487906 b = 1.084615 m = 0.543219 P: *woman* g = 2.756108 b = 1.239369 m = 0.45679 P: *academic material* g = 2.487383 b = 1.058244 m = 0.601059 P: study reached by the father g = 3.615967 b = 1.178201 m = 0.811115	g = 2.602826 b = 1.028883 m = 0.017727	0.924598	0.903798

6 Concluding Remarks and Directions for Future Research

This report results from an exploring analysis of data coming from the survey, which results from applying different techniques of Data Mining. To characterize students, in terms of satisfaction with the careers they follow, we apply techniques of the discovery of relations among linguistic variables, of the form P, using procedures based on bivalent and multivalent logic.

From the bivalent logic, using the algorithm Apriori implemented in R, we could find the rules of strong association that allow relating variables statistically: stress caused by the study, better future, good relationships, and good marks with the variable Satisfaction of students with careers. Whereas from the Fuzzy Logic, using in the GMBCL, AMBCL, and the Zadeh logics, we made a study of relevant relations with the help of the Eureka-Universe system. We could understand other fuzzy variables, besides the ones previously found, related to the satisfaction of the student, for example: "Weekly attendance to classes", "House Conditions to study ", "Student age", "Time spent on practice", "Money for food", "Money to study", "quality of material", "level of demand", etc. In particular, the variable "Weekly attendance at the classes" shows that it is related to "Student Satisfaction for Careers", from the discovery of relevant relationships through the use of the Eureka-Universe system and using any of the three logics considered in this work (GMBCL and AMBCL and Zadeh).

From the analysis and interpretations of the relationships discovered through the Eureka-Universe system, taking into account the survey database mentioned in this study, it is observed that the results are predominantly linked to the linguistic state "is satisfied or doubtful of the career". In addition, in the case of the validity of the estimated parameters for the thesis of one of the relevant relationships discovered, new conclusions were drawn. These allowed to characterize the group of "satisfied or dubious students of the career", taking into account the states or attributes involved in the compound predicates discovered. Thus, as an example, it is discovered with a degree of veracity greater than 0.9, that 'adult students (over 25 years old) whose father reached a medium-higher level of study, and who express that the material of the Library is sufficient and the educational material is more or less updated, belong to the group of students who are satisfied or doubtful of the career.

In this study of the Eureka-Universe system, it resulted in a new tool, easy to use, and very valuable because of its high potential in the discovery of useful knowledge from the fuzzy-nature information using various paradigms of Fuzzy Logic. A high number of relevant relations are obtained by applying, followed by the Zadeh Logic. When these logics are interpreted, they show that the appraisal of educational quality in the offered careers from the students' perspective, they are related to personal aspects, attitudinal, institution, economic social, and cultural.

We believe that the information here utilizing the application of techniques based on the bivalent and multivalent logic results valuable at the time of making decisions at institutions and making the educational quality much better.

We consider as future lines of work to apply these techniques to identify profiles of the population of Greater San Juan related to the Nation, based on information on needs and interests at the social, educational, economic and cultural levels.

References

1. Traba, L., Barletta, M., Velázquez, J.: Teoría y práctica de las organizaciones: herramienta para la gestión de la calidad. Universidad Nacional del Litoral, Santa Fe (2010)
2. Fernández Ziegler, R.: Planificación y Control de Gestión. Universidad de Quilmes, Buenos Aires (2014)
3. Botero, M.M., Peña, P.: Calidad en el servicio: el cliente incógnito. Suma Psicológica **13**(2), 217–228 (2006)
4. Salvador, C., Pozo C., Alonso, E.: Percepción del cliente de los predictores de calidad en el sector servicios. Boletín de Psicología. https://www.uv.es/seoane/boletin/previos/N94-5.pdf (2008). Accessed 20 Nov 2018
5. Surdez, E.G., Sandoval, M. del C., Lamoyi, C.L.: Satisfacción estudiantil en la valoración de la calidad educativa universitaria. Educación y Educadores, **21**(1), 9–26. http://www.scielo.org.co/pdf/eded/v21n1/0123-1294-eded-21-01-00009.pdf (2018). Accessed 10 Feb 2019. https://doi.org/10.5294/edu.2018.21.1.1
6. Marín Llanes, L.A., Carro Cartaya, J.C.: La Minería de Datos como Herramienta en el Proceso de Inteligencia Competitiva. Consultoría Biomundi, Dirección de Inteligencia Corporativa, Instituto de Información Científica y Tecnológica, del Ministerio de Ciencia, Tecnología y Medio Ambiente. https://docplayer.es/7929027-La-mineria-de-datos-como-herramienta-en-el-proceso-de-inteligencia-competitiva.html (2000). Accessed 11 Aug 2020
7. Rojas, J., Chavarro, J., Moreno, R.: Técnicas de lógica difusa aplicadas a la minería de datos. Scientia et technica **3**(40), 1–6 (2008). https://doi.org/10.22517/23447214.3095
8. Agrawal, R., Imielinki, T., Swami, A.: Mining association rules between sets of items in large databases. In: ACM SIGMOD International Conference on Management of Data. ACM Press, pp. 207–216. Washington D.C. (1993). https://doi.org/10.1145/170036.170072
9. Hipp, J., Güntzer, U., Nakhaeizadeh, G.: Algorithms for association rule mining—a general survey and comparison. ACM SIGKDD Explorations Newsl **2**(1), 58–64 (2000). https://doi.org/10.1145/360402.360421
10. Marin Ortega, P.G.Á.L.: Contribución a la modelación de una arquitectura empresarial, para soluciones de inteligencia de negocios. Universidad Central "Marta Abreu" de Las Villas (2009)
11. Dubois, D., Prade, H.: Fuzzy sets and systems: theory and applications. In: Mathematics in Science and Engineering, vol. 144. Academic Press, New York (1980). https://doi.org/10.2307/2273604
12. Wang X., Ruan D., Kerre E.E.: Fuzzy inference and fuzzy control. In: Mathematics of Fuzziness—Basic Issues. Studies in Fuzziness and Soft Computing, vol. 245. Springer, Berlin, Heidelberg (2009). https://doi.org/10.1007/978-3-540-78311-4_6
13. Alves, H., Raposo, M.: La medición de la satisfacción en la enseóanza universitaria: El ejemplo de la Universidade da beirainterior. Int. Rev. Publ. Nonprofit Market. **1**(73) (2004). https://doi.org/10.1007/BF02896618
14. Salinas, A., Martínez, P.: Principales factores de satisfacción entre los estudiantes universitarios. Revista Internacional de Ciencias Sociales y Humanidades, SOCIOTAM **17**(1), 163–192 (2007)
15. Salinas, A., Morales, J., Camblor, P.: Satisfacción del estudiante y calidad Universitaria: Un análisis explicatorio en la Unidad Académica Multidisciplinaria Agronomía y Ciencias de la Universidad Autónoma de Tamaulipas. Universidad de Sevilla, Tesis Doctoral (2007)

16. Fernández, J., Fernández, S., Álvarez, A., Martínez, P.: Éxito Académico y Satisfacción de los Estudiantes con la Enseñanza Universitaria. Revista Electrónica de Investigación y evaluación Educativa **13**(2), 203–214 (2007). https://doi.org/10.7203/relieve.13.2.4207
17. Brown, M.: Data mining for dummies. Wiley. Hoboken, New Jersey (2014)
18. Lucas, J.P.: Métodos de clasificación basados en asociación aplicados a sistemas de recomendación. Universidad de Salamanca, Tesis Doctoral (2010)
19. Brin, S., Motwani, R., Ullman, J.D., Tsur, S.: Dynamic itemset counting and implication rules for market basket data. In: ACM SIGMOD International Conference on Management of Data, pp. 255–264. Tucson, Arizona (1997). https://doi.org/10.1145/253262.253325
20. Silverstein, C., Brin, S., Motwani, R.: Beyond market baskets: generalizing association rules to dependence rules. Data Mining Knowl. Discov. **2**, 39–68 (1998). https://doi.org/10.1023/A:1009713703947
21. Hipp, J., Ulrich Güntzer, U., Nakhaeizadeh, G.: Algorithms for association rule mining—a general survey and comparison. SIGKDD Explorations Newslett. **2**(1), 58–64 (2000). https://doi.org/10.1145/360402.360421
22. Romero, C.: Aplicación de técnicas de adquisición de conocimiento para la mejora de cursos hipermedia adaptativos basados en Web. Tesis Doctoral. Universidad de Granada. E.T.S, Ingeniería Informática (2003)
23. Hahsler, M., Hornik, K., Reutterer, T.: Implications of probabilistic data modeling for mining association rules. In: Spiliopoulou, M., Kruse, R., Borgelt, C., Nürnberger, A., Gaul, W. (eds.), From Data and Information Analysis to Knowledge Engineering. Studies in Classification, Data Analysis, and Knowledge Organization. Springer, Berlin, Heidelberg (2006). https://doi.org/10.1007/3-540-31314-1_73
24. Amat, R.: Reglas de asociación y algoritmo Apriori con R. https://rpubs.com/Joaquin_AR/397172 (2018). Accessed 2 Feb 2019
25. Zadeh, L.: A. Fuzzy sets. Inf. Control **8**(3), 338–353 (1965). https://doi.org/10.1016/S0019-9958(65)90241-X
26. Espin Andrade, R., Gonzalez, E., Pedrycz, W., Fernandez, E.: An interpretable logical theory: the case of compensatory fuzzy logic. Int. J. Comput. Intell. Syst. **9**(4), 612–626 (2016). https://doi.org/10.1080/18756891.2016.1204111
27. Espin Andrade, R., Fernández González, E.: La Lógica Difusa Compensatoria: Una Plataforma para el Razonamiento y la Representación del Conocimiento en un Ambiente de Decisión Multicriterio. In: Análisis Multicriterio para la Toma de Decisiones: Métodos y Aplicaciones. Coedición: editorial Plaza y Valdes / editorial Universidad de Occidente (2009)
28. Espín Andrade, R., Marx Gómez, J., Mazcorro Téllez, G., Fernández González, E.: Compensatory logic: a fuzzy approach to decision making. In: 4th International Symposium on Engineering of Intelligent Systems (EIS' 2004). Island of Madeira, Portugal (2004)
29. Bouchet, A; Pastore, J.; Brun, M.; Ballarin, V.: Logica Difusa Compensatoria basada en la media aritmética y su aplicación en la Morfología Matemática Difusa. IEEE TRIC IV Cuarto Torneo Regional de Inteligencia Computacional (2010)

30. Rivera, G., Cisneros, L., Sánchez-Solís, P., Rangel-Valdez, N., Rodas-Osollo, J.: Genetic algorithm for scheduling optimization considering heterogeneous containers: a real-world case study. Axioms **9**(1), 27 (2020). https://doi.org/10.3390/axioms9010027
31. Alvarado, O., Castro, B., González, L., Rivera, G., Rodas-Osollo, J., Sánchez-Solís, J.: Metaheuristic-based optimization of treated water distribution in a Mexican City. Aplicaciones Recientes en la Investigación de Operaciones, pp. 19–30. Universidad Autónoma de Coahuila, Coahuila (2018)
32. Rivera, G., Rodas-Osollo, J., Bañuelos, P., Quiroz, M., Lopez, M.: A genetic algorithm for surgery scheduling optimization in a Mexican public hospital. Recent advances in artificial intelligence research and development. In: Aguiló, I., Alquézar, R., Angulo, C., Ortiz, A. (eds.) Frontiers in Artificial Intelligence and Applications, vol. 300, pp. 269–274. IOS Press, Amsterdam (2017). https://doi.org/10.3233/978-1-61499-806-8-269

Is Economic Performance Affected by Social Conditions and Rights? The Case of the Central Region of San Luis Potosí, Mexico

Juan Carlos Yáñez-Luna and Leonardo David Tenorio-Martínez

Abstract This work aims to prove that the relationships of some variables related to social backgrounds have a positive influence on the economic development of a certain region. This study focuses on the central region of the State of San Luis Potosí, Mexico. We built a data set based on the last economic census in Mexico (2010). The data set contains information about the demographic and social status of several towns belonging to the municipalities of the central area of San Luis Potosí. We designed a PLS-SEM (Partial Least Squares—Structural Equations Model) based model to measure the main variables in this study. The model was tested in SmartPLS 2, and some measures were tested in Microsoft Excel. Conclusions are considered in this paper; principal results indicate that this region has improved in the economic and social conditions in terms of education; consequently, it could be attractive for local or foreign investors due to the privileged localization.

Keywords Regional economic growth · Regional economic development · Socioeconomic situation · Subregional study

1 Introduction

One of the first projects that studied the effects of human conditions (such as education implications on economic performance) is the works of [1, 2]. Those works pointed out that the economic returns are obtained by the result of higher qualification and specialization in education. Based on this theory, international organizations such as UNESCO (United Nations Educational, Scientific and Cultural Organization) or OECD (Organization for Economic Co-operation and Development) aims to prove that certain socio-economic conditions influence educational performance. In the same order, this variable is influenced by the educational system normative of each region. This relation could explain the differences in terms of economic and regional performance and development related to human capital.

J. C. Yáñez-Luna (✉) · L. D. Tenorio-Martínez
Faculty of Economics, Autonomous University of San Luis Potosi, 78213 San Luis Potosí, Mexico
e-mail: jcyl@uaslp.mx

W. Pedrycz et al. (eds.), *Computational Intelligence for Business Analytics*, Studies in Computational Intelligence 953,
https://doi.org/10.1007/978-3-030-73819-8_21

This paper aims to prove if there is any relationship between the theory of human capital, through socio-economic conditions, and the inverse situation in the central region of the state of San Luis Potosí (SLP) in Mexico. To perform this hypothesis, we consider the National Dataset from municipalities and all their nearby towns.[1] In order to evaluate the hypothesis, we will propose a statistical model based on structural equations using the technique of partial least squares (PLS).[2] This technique will allow us to make multiple associations of socioeconomic variables to explain the relationship raised above.

This document is designed in four sections. In the first section, we will present a theoretical review of how economic development is explained by social and educational conditions. Also, we will show a brief review of the actual conditions in the center region of SLP. The following two sections outline the model and methodology used to address the problem. The analysis of the model was based on the statistical technique of route analysis [3], which allows us to know the causal relationships that exist between several basic variables for the application of the Partial Least-Squares (PLS). We can highlight that PLS "is the specification of causal and/or predictive relationships in terms of predictors (conditional expectations), followed by the estimation by least-squares of the variables" [4]. In the final section, we will expose our conclusions about the results of the proposed model, in this order; we will show the implications on the economic development of the selected region.

2 Background and Objective. Educational Status, Economic Performance, and Regions

As we described in the previous section, the main objective of this research is to prove if there is any relationship between the theory of human capital and the inverse situation in the central region of the state of San Luis Potosí (SLP) in Mexico. For the case, we built an overview of the economic status of the evaluated region; in the next paragraphs, we will show a detailed analysis.

The economics of education is the result of the big question of economic growth and the theory of production issues. The concept of Human Capital arises from the influential works of [5]. This idea is understood as "the future product of the present investment in education" and depends on two conditions: efficiency and equity.

The degree of returns that education generates and its effects in terms of economic and/or social benefits were drawn from the condition of efficiency. Whereas, in terms of equity is important to point out some topics: First, we must identify who obtains the greatest number on levels of learning. In this stage, the homogeneity and heterogeneity are shown and describes the conditions and the immersion of the actors in the educational environment. Second, the participation of the State is important and mandatory. The government must promote an integral investment for allowing the

[1] Data obtained from the 2010 population and housing census, INEGI.

[2] This analysis tool is known as the Partial Least Squares (PLS).

process of educational homogenization [6]. Third, the government must restrict the actual inequalities by region. Most of them are the product of the disparate historical-social conditions, starting from programs focused in the less favored regions, which leads to a process of an increasingly homogeneous society.

The foregoing shows that education plays a main role in expanding opportunities for individuals, regions, and countries. It will depend directly on the structure of the education system of the region and on the levels of education and better exploitation of the abilities that children and young people could attain. The abilities will derive from the social conditions, i.e., housing area (rural or urban), family structure and values, the economic family structure, labor accessibility, etc.

Based on the Third Regional Comparative and Explanatory Study (TERCE)[3] carried out by UNESCO, some conclusions can be drawn about the strong relationship that exists between education and economic performance. The study describes how are the conditions in which children and young people learn and the strong and direct effects on educational and economic results. First, the study shows that socio-economic level is the variable with the greatest impact on learning and that there is a direct relationship between the inequality level of any region and the school systems of the region [7]. Second, a negative relationship in those variables has negative results, such as low social mobility and, in consequence, intergenerational transmission of social status and poverty [8].

Poverty and inequality are two variables associated with each other; this may cause that the objectives of the educational system of a region should not have high levels. This is because poverty causes restrictions in terms of food and healthy nutrition, adequate housing and services, and welfare. This situation also distinguishes the opportunity cost of sending children to school [7],[4] the topic of gender should be mentioned, because, in some regions, girls are disadvantaged due to cultural issues.

In addition, to explain the learning level of a region, we must include the socioeconomic variable and the dependent variables such as education level of the parents, the form of employment, income level, housing details (floor, services, etc.), access to libraries, etc. In the case of Mexico, there is a correlation between the level of learning and the social condition; if the first one has a high level by consequence, the second one should have it. This asseveration is confirmed in [7], which establishes a direct relationship between the economic situation and the degree of academic achievement.

The Rural area is a variable considered explaining school achievement. It is evident there is a close relationship between it and poverty, inequality, and inequity. In this study, we have to mark that in Mexico, the average rural population has the lowest levels of socio-economic development, while urban areas have the highest [7].

In relation, the social and economic needs in the population of developing countries show high levels. Therefore, young people are compelled to study and work; in the worst-case scenario, most of them leave their studies early. A recommendation

[3] TERCE, in its Spanish acronym of Tercer Estudio Regional Comparativo y Explicativo.

[4] According to this study, Mexico has been increasing the percentage of its population below the poverty line, from 31.7 to 41.2 from 2006 to 2014.

for UNESCO is to suppress the processes that affect the educational performance (for children, especially girls), such as compulsory quotas, selection processes, among others [7].

The New Economic Geography (NEG) and spatial economics, establish a relationship between education, work, and remuneration in regions. That means, a region with a considerable distance between consumer markets and the inputs supply, is more propenseity to induce low levels of income for those with more favorable conditions. These have critical implications for the economic heterogeneity between regions because it is common for the less favored (economically) regions that are characterized as the regions with the lowest level of education [9]. Meanwhile, the concept of local economic development defines if the economic growth of a region is the result of a productive system and how it allows the creation of economies of scale and productivity for improving the competitiveness of an economic and social system. This relation serves as a basis for development and a local administrative political system for supporting production and drives sustainable development [10].

The case of migration is another phenomenon associated with poverty and the lack of regional opportunities. This concept had been identified indirectly by Kaldor [11] and directly by Romer [12] in which skilled and unskilled workers tend to migrate from low-income to high-income countries. According to Garduño-Rivera [13], people who are seeking well-being will migrate to regions with greater economic and social opportunities, determining differentiated levels of growth and productivity [14].

Likewise, the NEG sustains that market size, transportation costs, and economies of scale will be the main determinants of the concentration of activities [15], which refers to the issue of infrastructure and the availability of productive factors to explain the level of regional economic development.

2.1 The Central Area of San Luis Potosi

According to the State Competitiveness Index 2016 of the Mexican Institute of Competitiveness (IMCO),[5] there are six factors for increasing the competitiveness of the federal entities of the country: Economy, which refers to the capacity of exporting and the index of Direct Foreign Investment (FDI); A high activity in the manufacturing industry; Connection to energy networks (infrastructure); High percentages of the workforce (employment) for formal companies and competitive salaries; High number of large companies that generate formal and quality jobs and better education systems that generate human capital related with the productive structure of the state [16].

San Luis Potosí has remained almost in the same place in the latter two evaluations—place 19—. The state had shown a growth in the FDI index, particularly in the automotive and auto parts sectors. This result improves the perspective of

[5] IMCO, Spanish acronym of Instituto Mexicano de la Competitividad.

growth in terms of foreign trade, investment of large companies, formality, and jobs with well-paid wages. Furthermore, the participation in the manufacturing sector will keep the entity with economic growth close to the national average, where the central sub-region of the state is the principal concentrator of FDI.

Based on the IMCO study, we identify some areas of opportunity in the entity: the supply of natural gas as an energy source. This area is a determining factor for the increase of investment and will solve some limitations for the installation of large companies. Other opportunities are related to investments in terms of quality, i.e., infrastructure, transport, roads, and highways, that have been stagnated or not grown with the necessary speed.[6] This negatively affects competitiveness in terms of transport costs and the movement of people and materials.

The educational aspect is another opportunity area in the entity. Since 2010, no municipality stood out among those with the highest education averages, and in fact, the state school average is 8.58 years, very close to the national average of 8.63 years.

The relationship between education and innovation is established from the concentration and training of human capital in a region. In this regard, companies are allowed to adopt and develop new technologies [18]. This is a fact in the studied entity, nowadays the investment in training of professionals and specialists that has reached a high percentage. The results have been reached through governmental programs that include scholarships for postgraduate studies. Those programs place the State of San Luis Potosí in the first position at the national level, as well as the fact that more and more undergraduate and postgraduate programs are certified by external evaluating bodies.

According to FCCyT [19], the growth of the competitiveness of the State is reached by the companies' investment in R and D in their production process based on the science, technology, and innovation laws. The state of San Luis Potosí is ranked in the 13th place with—0.067 points,[7] which means that the state is very close to the national average, but at the bottom. The variables evaluated in this ranking are the infrastructure for research, scientific productivity, population with professional and postgraduate studies, economic and social environment, ICTs, and human resources trainers, where the entity obtains results close to the national average, which establishes the areas of opportunity to reach the first places in growth and economic development at the national level.

This study also uses other variables to issue a ranking such as the number of patents granted, percentage of the population with postgraduate, the average number of years studied, illiteracy, internet connectivity, the postgraduate and undergraduate teaching staff, among others. In this respect, San Luis Potosí was ranked in the eighth

[6] Perrotti and Sánchez [17] explain that public investment declined in the 1990s as the role of the state in Latin American and Caribbean economies was limited by seeking an increase in private sector participation, which did not happen and resulted in a process of deterioration of public infrastructure with effects to date.

[7] With information from the National Ranking of Science, Technology, and Innovation (CTI) of the Scientific and Technological Advisory Forum (FCCyT).

position according to the number of academic researchers members of the National System of Researchers (SNI)[8] [19], which represents a fortress.

At the sub regional level, San Luis Potosí is divided into four geographical zones: Central, Middle Area, Huasteca and Plateau Area. We will concentrate in this research on the Central Region which is composed of 11 municipalities namely: Ahualulco, San Luis Potosí, Armadillo de los Infantes, Cerro de San Pedro, Mexquitic de Carmona, Soledad de Graciano Sánchez, Villa de Arriaga, Villa de Reyes, Santa María del Río, Villa de Zaragoza and Tierra Nueva.

The Central Zone of the State includes the capital that has the same name as the State. As is common in capitals, San Luis Potosí concentrates the largest and most important infrastructure in the region. In this regard, the Report San Luis Potosí 2017 indicates that eight municipalities in the Central Zone of the State represent 84.1% of GDP and concentrate 42% of the population [20].

This can be identified as the advantages of economies of scale, scope, and concentration. Numerous sectors and industries have been taken advantage of it, such as the automotive industry and auto-parts sector, manufacturing industry, the food sector, and the specialized medical services or logistics, among others.

In addition, the expected investments for 2019 are around 90 billion dollars by railway and logistics companies. In the same context, the Inter-American Development Bank (IDB) considers the city of San Luis Potosí a strategic city (given its geographical position) for logistics platform, dry port, and logistics corridor, as an area of influence of the main ocean ports of the Pacific (Manzanillo and Lazaro Cardenas) and the Gulf of Mexico (Altamira, Tuxpan and Veracruz).

In terms of employment, the generation of job opportunities is directly related to the investment of companies in the automotive industry. This sector showed an increase of 47.2% in their production in 2018 in comparison with 2017 which means 30% of the GDP of the state. In addition, this industry generates the development of other industries and with positive externalities such as the specialization of the local human factor, by achieving the formal installation, through a decree of law, and consolidation of dual education in the entity. These activities allow the joint participation between companies and education centers of higher and middle levels and promote the specialization and expansion of qualified workers in the region.

3 The Method. Building the Proposed Model

The proposed model allowed us to explain the economic performance of the central region of the State of San Luis Potosí (SLP) with the following variables: Economy, Basic Education, Secondary and Higher Education, Marginalization, Migration, and Health. These variables were selected from two secondary data: the 2010 National

[8] SNI, Spanish acronym of Sistema Nacional de Investigadores.

Census of Population and Housing[9] and the Human Development Index of the State of San Luis Potosí. At the same time, the study was limited only to those localities in the different municipalities that make up the central zone of the State. Due to circumstances out of our control, some localities do not present data, that might be because of the social and geographical characteristics of the region. We concentrate on those localities that had enough data to prepare our model. We aim to construct a scenario in which we could observe the behavior of the variables mentioned above and their relationship with the economic development of the locality.

With this in mind, we built a model based on route analysis to calculate the relationship between each variable and its contribution to the region's economic development. The model focused on two aspects: (a) economic development and (b) social marginalization. As external variables, we consider the population's access to health services and access to the education sector (basic, intermediate, and higher levels).

As was pointed in the previous section; the central region is the subregion with the greatest economic development due to its high economic concentration. This region is the main recipient of public expenditure (investments in infrastructure, transport, health, education, etc.). However, this does not contribute to the fact that all the population has access to the services. Presumably that in this region the marginalization index will be quite low concerning the other regions of the State, i.e., the population that has not been able to access health services will have a direct implication for the education variable; as a result, the percentage of marginalization will increase; that is, if there is a certain number of populations in the region studied that does not have access to basic health services, it is very probable that their performance and development in education will probably be null. In this sense, Marginalization can be considered in the concept of Camberos [21]:

> Marginalization is defined as a situation in which a group of individuals and families living within a locality or municipality, urban or rural, do not meet the needs considered basic, according to criteria determined by institutions recognized as the United Nations Development Programme (UNDP) and the World Bank.

The authors also point out that: "Marginalized population is also understood as the sector of society that, for reasons of socio-economic and political organization, excludes it from access to consumption and enjoyment of goods and services, and participation in political affairs". In our case study, we can point out that if basic health services are reduced in the central zone of the State, it will mean an increase in the marginalization of the region. The study on the Absolute Marginalization Index of CONAPO[10] [22] mentions the main indicators that influence marginalization among them Education. Based on the foregoing, the following assumptions may be made:

- BasicEducation = (Health).
- HigherEducation = (BasicEducation, Health).

[9] The National Population and Housing Census is conducted by the National Institute of Statistics, Geography, and Informatics (INEGI), among other studies of economic and social connotation.

[10] CONAPO acronym of Consejo Nacional de Poblaciôn. A Mexican government bodies.

- Marginalization = (HigherEducation, Health).

The above expressions demonstrate the relationship between main variables; however, we can define them through a linear function expressing our hypotheses in the following way:

$$BasicEducation = \beta_1 Health + \varepsilon \tag{1}$$

$$HigherEducation = \beta_2 BasicEducation + \beta_3 Health + \varepsilon \tag{2}$$

$$Marginalization = \beta_4 HigherEducation + \beta_5 Health + \varepsilon \tag{3}$$

The linear functions can be represented graphically by a path diagram, as shown in Fig. 1. This diagram shows a direct causal relationship between the Health variable and the Basic and High Education variables, as well as a direct relationship between the Health variable and Marginalization.

Once the model has been designed, the next step consists of involving other variables that influence on the economic development of the entity. In the first instance, the variables Marginalization and Education influence the economic development variable, as mentioned in previous paragraphs. In the development of our model, we decided to establish a causal relationship between Higher Education and Economic Development, with the understanding that the level of education also has important

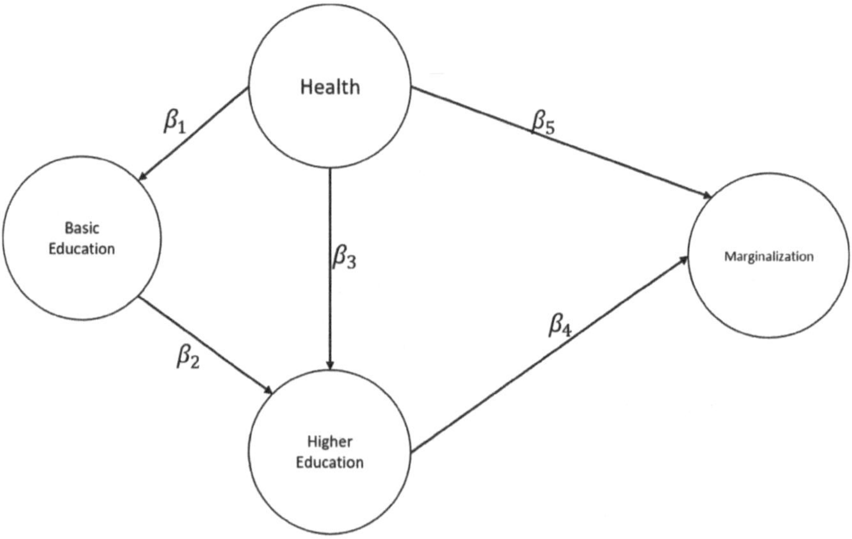

Fig. 1 Relationship of social marginalization with the education and health sector. *Source* Own elaboration

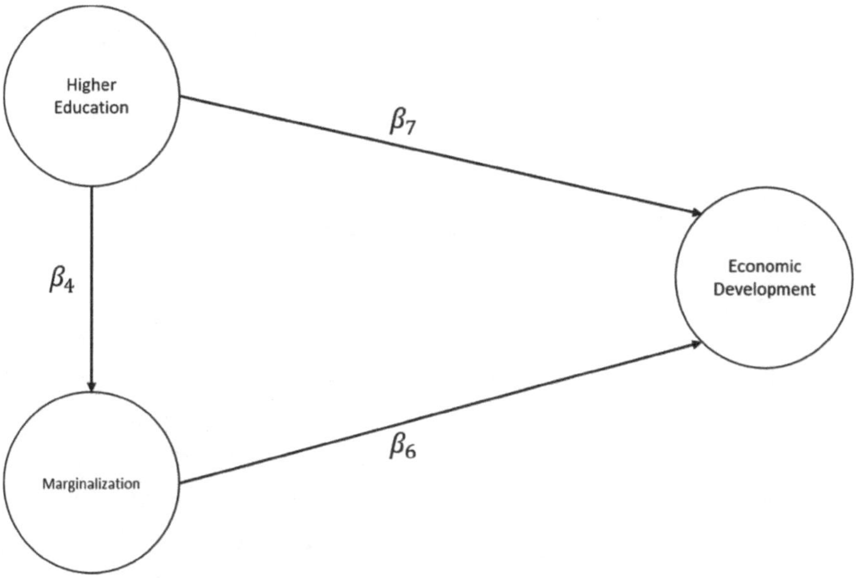

Fig. 2 Relationship between the study variables social marginalization and higher education with economic development. *Source* Own elaboration

influences; that is, if the level of education in a region is higher, its economic development will gradually increase, and the level of marginalization will be reduced in the same sense. So, we can suggest that there is a causal relationship between these two variables, such as:

$$EconomyDevelopment = \beta_6 Marginalization + \beta_7 HigherEducation + \varepsilon \quad (4)$$

Equation 4 is represented graphically, as shown in Fig. 2.

In another aspect, migration, as stated in [23], is considered as the "geographical displacement of individuals or groups, usually for economic or social reasons",[11] this definition ties in with our objectives, we can identify the migration variable as an economic and social indicator with implications from the education sector and at the same time from marginalization. As already mentioned, if the population settled in a region with few educational opportunities and therefore better-paid workplaces will generate marginalization, however, not only would this negative social gap be established, but it would also mean potential population mobility; that is, a migrant population in search of better opportunities. We can, therefore, suggest that there is a causal relationship between these variables, as indicated:

$$Migration = \beta_8 Basic\ Education + \beta_9 Higher\ Education$$

[11] Authors pointed out that is a concept of the Real Academia Española Dictionary.

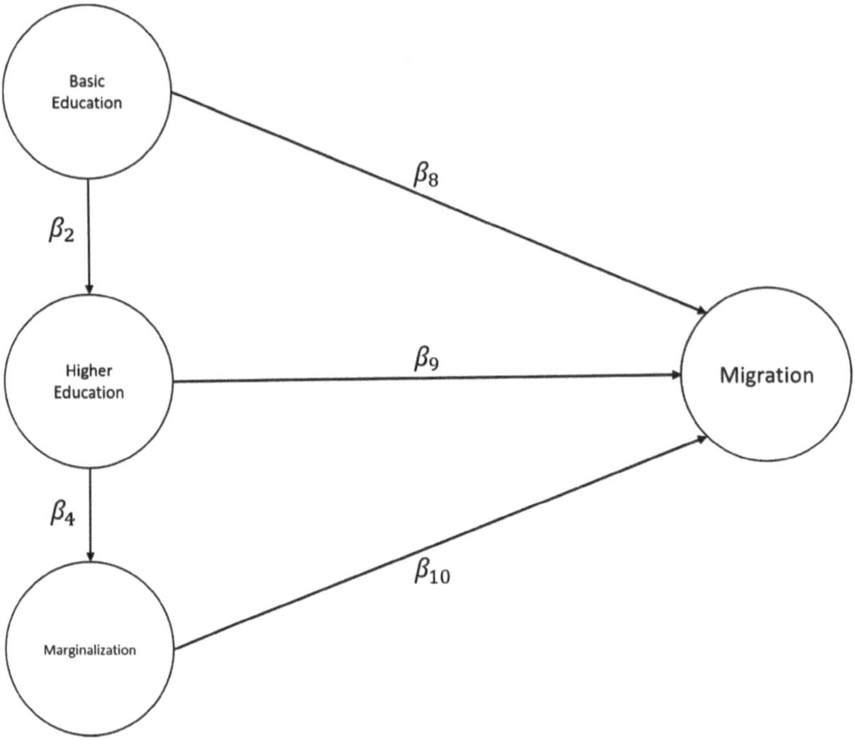

Fig. 3 Relationship between the study variables social marginalization and basic and higher education with migration. *Source* Own elaboration

$$+ \beta_{10} Marginalization + \varepsilon \quad (5)$$

Equation 5 is represented graphically, as shown in Fig. 3.

4 Analysis of the Results

As it was observed in the previous section, our research model is based on quantitative analysis. This model will allow us to make a representation of the reality of the research object, that is, the relationship that exists between the latent variables and the manifest variables of the proposed model. To evaluate the related variables, we must perform a statistical analysis to obtain the value of the regression and its coefficient of determination.

In quantitative research, the main input is the dataset. Furthermore, a provision should be made for statistical evaluation. Two classic types of assessment can be referenced: the reliability of variables, the validity of each of the elements, and the measurement of the model. For example, in [24] is pointed out that:

> Reliability is related to the elimination of contingent distortions in the application of the instrument (from the 'presence' of the interviewer to the context of the interview, the sample quality is the most important issue) and validity, as a correspondence relation between the measurement and the measured value.

In this context, Yong Varela [25] pointed out that there are two statistical techniques for data analysis in quantitative research. The first one relates to the description and organization of the data. This technique allows to 'describe' the relationship of the research variables. The second technique evaluates the inference of the research variables in correlation with the population sample. This method will allow the researcher to evaluate the consistency of the proposed model and the hypothesis.

In the case of our proposed model, the causal relationships between various variables can be observed in Figs. 2 and 3. A statistical technique for evaluating this type of relationship between variables is the 'multiple regression' based on the 'path analysis'. In this respect, Streiner [26] points out that this type of instrument allows complex models to be analyzed and compared. However, these kinds of statistical techniques cannot be used to establish causality or to determine whether a model is correct; by contrast, they can only consolidate the consistency of the data with the proposed model. In the same context, in [27] is pointed out that structured equation techniques are multivariate techniques based on path analysis. These techniques will allow determining the validity of the empirical research, analyzing groups of variables and their respective cause-effect relationship.

These instruments of a statistical analysis based on structural equations have been used in areas of economic and social research. For example, two of the techniques most commonly used in this context are Models based on covariances and partial least squares models [28]. Methodological differences can be distinguished between these two techniques: Partial Least Squares maximizes the explained variance of endogenous latent variables by estimating the relationships of the partial models in an iterative sequence of ordinary least squares; on the other hand, models based on CB-SEM tend to estimate the model parameters in such a way as to minimize the discrepancy between the covariance and sample matrices [29].

Rodríguez [30] points out that the PLS technique is characterized by its ability to analyze multivariable models based on linear regression with high degrees of dimensionality, multicollinearity, and few observations. As we describe above, the selection of the analysis instrument will depend on the objectives and context of the research [31]. Due to the complexity of the proposed model, in this research, we adopted the partial least squares (PLS-SEM) technique to analyze the consistency.

The analysis consists mainly of two phases: In the first phase, the reliability of the constructs is analyzed, whereas, in the second phase, the hypothesis will be validated using PLS-SEM. The PLS-SEM algorithm, T-tests, and predictive relevancies were performed using a computer modeling software: SmartPLS v2.0 [32]. For descriptive analysis and graphics in general, the Microsoft Excel 2016 office application was used.

Table 1 Sampling for statistical power at 95%

f tests—linear multiple regression: fixed model, R^2 deviation from zero		
Analysis	A priori: compute required sample size	
Input	Effect size f^2	=0.15
	α err prob	=0.05
	Power ($1-\beta$ err prob)	=0.95
	Number of predictors	=3
Output	Noncentrality parameter λ	=17.8500000
	Critical F	=2.6834991
	Numerator df	=3
	Denominator df	=115
	Total sample size	=119
	Actual power	=0.9509602

Source Own elaboration

One of the most common discussions regarding the use of the PLS methodology is the determination of the sample size. Commonly, the "finger" rule is suggested [28]. This rule indicates that the total of manifest variables that have the same path should be counted, as well as the number of exogenous variables of the causal relationship with the greatest number of them and multiplied by 10. The use of this rule has been questioned in the academic literature because it presents inconsistencies when making the statistical decision to test hypotheses committing error type II. To avoid this type of error, a methodology for determining the sample size suggested in [33] is statistical power analysis.

To perform the calculation suggested above, we use the statistical software G-Power with an average size $f^2 = 0.15$ and assigning as the construct of the proposed model "Migration" with 3 causal relationships (Basic Education, Higher Education, Marginalization) for an optimal value of 95% in the level of confidence, the acceptable value in the area of social sciences [34]; resulting in a minimum sample size of 119 elements as can be seen in Table 1.

5 Model Measurement

In the previous sections, the sampling methodology was explained, and the model proposed for this research was proposed. The models based on path analysis should be evaluated according to their causal modality, that is, how the relationships between the various variables in the model will be considered. For this research, the relationships between all variables were considered as a reflexive mode. Roldán [35] recommended evaluating this type of relationship measurement procedure with the convergent validation, discriminant, and structural evaluation of the model.

In the analysis of convergent validity, some rules must be considered; for example, in [36] is pointed out that this type of analysis indicates the degree of correlation that exists between the elements of a scale, which should show a strong correlation. One of the most widely used social science methods for this analysis (reliability of a scale and its internal consistency) is Cronbach's alpha. The acceptable value in academic literature for this unit of measure is $\alpha \geq 0.7$ and $\alpha \leq 0.9$ [37].

The values obtained in the model for Cronbach's Alfa are higher than 0.7 in each of the variables studied, so it can be considered to have an acceptable internal consistency. To ensure that the model is consistent, Fornell and Larcker [38] propose to evaluate the composite reliability of the constructs (in the social science field). In this case, the values proposed for this type of analysis must be greater than 0.7. In addition to this evaluation measure, the convergent validity should be measured through the Average Variance Extracted (AVE) for each construct of the model. The recommendation for this measure is AVE > 0.5. For our case study, all the constructs proposed in the model exceed the values mentioned; this assumes that there is an internal consistency in the model.

The discriminant validity of constructs suggests that the degree to which the measured variable should not reflect some other variable, i.e., there should be no correlation between them [36]. To perform this test, it is suggested to use the square root of the Average Variance Extracted from each construct on the correlation matrix. Fornell and Larcker [38] suggest that the result of this operation should be greater than the correlations of the group's constructs. Table 2 shows the results of this operation for the proposed model. The square root of the AVE (values in bold) is

Table 2 Convergent and discriminant validity analysis

	AVE		Rel. Comp		R^2	α Cr
Economic development	0.9782		0.9926		0.9296	0.9303
Basic education	0.8554		0.922		0.6253	0.9303
Higher education	0.9154		0.9558		0.8061	0.9303
Marginalization	0.8404		0.9133		0.1189	0.9303
Migration	0.8989		0.9468		0.6151	0.9303
Health	0.8276		0.9505		0	0.9303
Correlations						
	Eco	Bas_Ed	High_Ed	Marg	Migr	Health
Economic development	0.989					
Basic education	0.8548	0.9249				
Higher education	0.9522	0.8705	0.9568			
Marginalization	0.4492	0.2826	0.3211	0.9167		
Migration	0.7479	0.76	0.7271	0.3682	0.9481	
Health	0.8123	0.7907	0.823	0.3356	0.7457	0.9097

Source Own elaboration

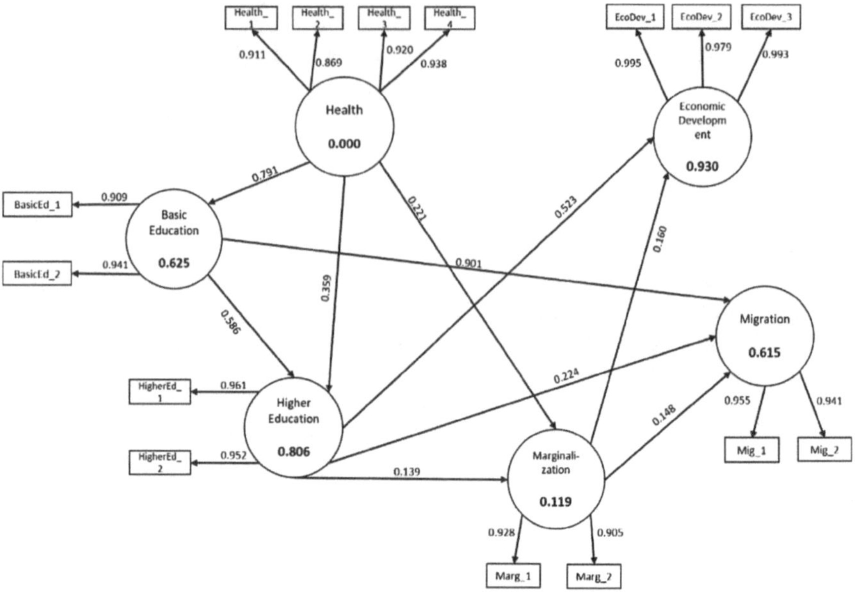

Fig. 4 Results of the structural model in SmartPLS. *Source* Own elaboration

greater than the loads of each construct in the group. Figure 4 shows the loads and weights of the relationships in the structural model in the SmartPLS 2 software.

6 Evaluation of the Structural Model

As mentioned above, it is important to perform the analysis of the structural model by statistical significance through Bootstrapping. This technique relies on random sampling with replacement, which means, replaces the original sample to generate a new sample providing standard errors and T statistics for corroboration of the hypotheses raised.

Regularly the minimum value accepted for resampling is 500, and another aspect of validity indicates that the number of cases should be the same as the number of observations of the original sample. For this study, we considered the observations of Hair et al. [39]. The authors suggest increasing the number of samples from 500 to 5000. The analysis of the confidence intervals we considered the values proposed in [35], authors pointed out values of $\rho < 0.05$; $\rho < 0.01$; $\rho < 0.001$. In relation to the above, the T statistics will be obtained for one-tailed test defined as: $t(N - 1) \rightarrow t(5000 - 1) \therefore t(0.05; 4999) = 1.6451$, $t(0.01; 4999) = 2.3270$, $t(0.001; 4999) = 3.0902$. Table 3 shows the results of the structural model. We observed that the relations existing between the variables Middle and Higher Education with the variables Marginalization and Migration, in the same way the relation Health and

Table 3 Results of the structural model (total effects). Based on one-tailed test t(N − 1)

Structural model results (total effects)					
Hyp		Sug eff	Path cohef	t-value bootstrap	Supp
H1	BAS_ED → HIGH _ED	+	0.5863	8.6888***	Y
H2	BAS_ED → Mig	+	0.5228	2.8958***	Y
H3	High_Ed. → Eco	+	0.9008	39.5713***	Y
H4	HIGH _ED. → Mrg	+	0.1391	0.9533	NS
H5	HIGH _ED. → Mig	+	0.2244	1.4861	NS
H6	Mrg. → Eco	+	0.16	4.2445***	Y
H7	Mrg. → High_Ed	+	0.1484	2.2171**	Y
H8	Health → Bas_Ed	+	0.7907	11.2239***	Y
H9	Health → High_Ed	+	0.3593	5.134***	Y
H10	Health → Mrg	+	0.2212	1.5271	NS

NS not significant
$^{*}p < 0.05 = 1.64$, $^{**}p < 0.01 = 2.32$, $^{***}p < 0.001 = 3.092$

Marginalization are not significant since they did not reach the minimum criterion of the distribution for $p < 0.05$.

The measurement of the structural model requires an analysis of confidence intervals as a measure of uncertainty index to avoid standard errors arising from resampling. To calculate resampling confidence intervals, we adopted the method suggested in [35]. This analysis suggests the use of percentile analysis with minimum values of 2.5% and maximums of 97.5%. Table 4 shows that for this study, the hypotheses

Table 4 Structural model results (Percentile bootstrap method at 97.5% confidence interval, n = 5000 samples)

Hypo		Sug. eff	Path cohef	Lower (2.5%)	Upper (97.5%)	Supp
H1	Bas_Ed → High_Ed	+	0.5863	0.4473	0.7109	Y
H2	Bas_Ed → Mig	+	0.5228	0.1205	0.8191	Y
H3	High_Ed. → Eco	+	0.9008	0.8477	0.9368	Y
H4	High_Ed. → Mrg	+	0.1391	−0.1583	0.4186	NS
H5	High_Ed. → Mig	+	0.2244	−0.0382	0.5549	NS
H6	Mrg. → Eco	+	0.1600	0.0586	0.2380	Y
H7	Mrg. → Mig	+	0.1484	0.0304	0.2917	Y
H8	Health → Bas_Ed	+	0.7907	0.6326	0.8985	Y
H9	Health → High_Ed	+	0.3593	0.3175	0.5027	Y
H10	Health → Mrg	+	0.2212	-0.0427	0.5271	NS

Source Own elaboration

Table 5 Effects on endogenous variables

Variable	Hypo	R^2	Q^2	Effect	Correl	Expl variance
Economy		0.9296	0.9018			
H3	High_Ed → Eco			0.9008	0.9522	0.8577
H6	Mrg. → Eco			0.1600	0.4492	0.0719
Basic education		0.6253	0.5174			
H8	Health → Bas_Ed			0.7907	0.7907	0.6252
Higher education		0.8061	0.7114			
H1	Bas_Ed → High_Ed			0.5863	0.8705	0.5104
H9	Health → High_Ed			0.3593	0.8230	0.2957
Migration		0.6151	0.5274			
H2	Bas_Ed → Mig			0.5228	0.7600	0.3973
H5	High_Ed → Mig			0.2244	0.7271	0.1632
H7	Mrg. → Mig			0.1484	0.3682	0.0546
Marginal		0.1189	0.0967			
H4	High_Ed. → Mrg			0.1391	0.3211	0.0447
H10	Health → Mrg			0.2212	0.3356	0.07423

Source Own elaboration

H4, H5, and H10 are not statistically significant coinciding with the previous test manifested in Table 3.

The most accepted consideration for explaining variables in a model is the R^2. However, another statistical criterion can be considered: The Stone-Geisrer test Q^2 on dependent constructs [34]. The test of predictive relevance suggests the cross-validation calculation of the components for redundancy and communality, and these must be greater than 0. In Table 5 the model has predictive relevance $Q^2 > 0$.

As a base value for exposing the model statistically, we considered the values of R^2. In our model, we adopted the recommendations of Hair et al. [39]. Authors suggest an R^2 values of 75% for substantial, 50% as a moderate, and 25% as a weak explanatory value. Table 5 shows that our proposed model explains that the factors studied (Education, Health, and Marginalization) have an important implication in Migration and Economic Development in the Central Zone of the State. The study suggests that if the population has less access to basic services, such as Health and Education, then it is probable that economic development will decrease, and migration could be increased. In our model, Economic Development has a direct effect on Middle and Higher Education and Marginalization, which together explain that 92.9% of these factors influence an acceleration or deceleration in the development of the entity. This data could also explain the current situation in the country where, according to the Education Panorama [40], 16% of adults (25–64 years) have higher education studies. In this way, the next point of the study is the Migration of the population, in our analysis, the variable Migration showed a 61.5% of explanation on the variables Middle and Higher Education 16.3 and Basic Education 39.7%, it is logical to think that too few opportunities of Education there will be less developed and therefore the Migration will increase, the reason why for this zone of the State the results suggest that people with a higher level of education tend to live in their work zone.

In the case of the relationship of the Health and Education variables, we can observe that both have a direct relationship. These ratios represent 62.5% for Basic Education and 29.5% for Higher Education; these data show us a clear panorama in which we can synthesize that if the population does not have access to Basic Health services, it is very probable that they will not complete their Basic and Middle-Upper studies.

This may be since the population is not necessarily influenced by higher educational status or health services to be considered marginalized, as mentioned above, the population's access to health services, education, and added sources of employment may justify low marginalization in the central zone of the State.

7 Conclusions

The results of the proposed model show that there is a direct and positive relationship between explanatory variables (health, education, and marginalization) with the explained variables (migration and economic performance). In this regard, we can assume that as economic and social conditions have improved (i.e., health, access to basic services, such as water, electricity, drainage, and housing), it will improve in terms of quality and quantity in education condition. This improvement could be attractive for more investment in the region and, for that matter, better jobs, and business opportunities, decreasing marginalization.

The central region is the most developed area and has the largest infrastructure of San Luis Potosí. Likewise, this region has financial investments and human capital formation, which allows the existence of economies of concentration and the formation of the virtuous circles of social development linked to growing economic development. However, this area has been delayed in terms of infrastructure; also, the participation of government and business sectors presented some troubles in coordination and communications. This asseveration could be studied in future research lines to allow the virtuous circle to continue expanding.

The lack of planning is evident in the limitations in terms of human capital and workforce. This situation explains the behavior of firms to hire labor force from neighboring regions; in addition, most of the firms must invest in training of the hired unskilled labor, increasing expenditures. Likewise, the challenges in infrastructure and coordination capacities between the economic and sociocultural political systems should be considered to increase the comparative and competitive advantages of the entity, to strengthen local economic development.

References

1. Schultz, T.: Capital formation by education. J. Polit. Econ. **68**(6), 571–583 (1960). https://doi.org/10.1086/258393
2. Becker, G.: Human capital: a theoretical and empirical analysis, with special reference to education. The University of Chicago Press, New York (1964)
3. Wright, S.: The method of path coefficients. Ann. Math. Stat. **5**(3), 161–215 (1934). https://doi.org/10.1214/aoms/1177732676
4. Wold, H.: 11—Path models with latent variables: the NIPALS approach. Quant. Sociol. 307–357 (1975). https://doi.org/10.1016/B978-0-12-103950-9.50017-4
5. Schultz, T.: Investment in human capital. Am. Econ. Rev. **51**(1), 1–17 (1961)
6. Leyva, S., Cárdenas, A.: Economía de la educación: capital humano y rendimiento educativo. Análisis Económico. **XVII**, 79–106 (2002)
7. Treviño, E., Villalobos, C., Baeza, A.: Recomendaciones de Políticas Educativas en América Latina en base al TERCE. Organización de las Naciones Unidas para la Educación, la Ciencia y la Cultura y UNESTO, Santiago (2016)
8. Clark, H., Grynspan, R.: Informe Regional sobre Desarrollo Humano para América Latina y el Caribe 2010. Actuar sobre el futuro: romper la transmisión intergeneracional de la de-sigualdad. Programa de las Naciones Unidas para el Desarrollo, Nueva York (2010)
9. López-Rodríguez, J., Faíña, J., López-Rodríguez, J.: Human capital accumulation and geography: empirical evidence from the European Union. Reg. Stud. **41**(2), 217–234 (2007). https://doi.org/10.1080/00343400601108440
10. Vázquez, A.: Desarrollo económico local y descentralización: aproximación a un marco conceptual. Comisión Económica para América Latina y El Caribe (CEPAL), Santiago (2000)
11. Kaldor, N.: Capital accumulation and economic growth. In: Lutz, F., Hague, D. (eds.) The Theory of Capital, pp. 177–222. Martin's Press, New York (1961)
12. Romer, P.: Capital accumulation in the theory of long run growth. In: Barro, J. (ed.) Modern Business Cycle Theory, pp. 51–127. Harvard University Press, New York (1989)
13. Garduño-Rivera, R., Baylis, K.: Effect of tariff liberalization on Mexico's income distribution in the presence of Migration. Agric. Appl. Econ. Assoc. Seattle (2012). https://doi.org/10.22004/ag.econ.124740
14. Kaldor, N.: Causas del lento ritmo de crecimiento del reino unido. Investig Económica **43**(167), 9–27 (1984)
15. Krugman, P., Elizondo, R.: Trade policy and the third world metropolis. J. Dev. Econ. **49**(1), 137–150 (1996). https://doi.org/10.1016/0304-3878(95)00055-0
16. Diez, V., González, T., Newell, R., Barranza, J., Carrillo, E., et. al.: Un puente entre dos Méxicos. Índice De Competitividad Estatal 2016
17. Perrotti, D.E., Sánchez, R.J.: La brecha de infraestructura en América Latina y el Caribe. In CEPAL - Serie Recursos naturales e infraestructura. Santiago de Chile (2011)
18. Simón, B., Aixala, J., Giménez, G., Fabro, G.: Determinantes del crecimiento económico. La interrelación entre el capital humano y tecnológico en Aragón (2004)
19. Diagnóstico en Ciencia, Tecnología e Innovación. Foro Consultivo Científico y Tecnológico. Distrito Federal (2012)
20. Plan Estatal de Desarrollo 2015–2021. Eje 1. San Luis Potosí. COPLADE (2016)
21. Camberos, M., Bracamontes, J.: Marginación y políticas de desarrollo social: un análisis regional para Sonora. Probl Desarro **38**(149), 113–135 (2007). https://doi.org/10.22201/iiec.20078951e.2007.149.7666
22. Almejo, R., Téllez, Y., López, J.: Índice absoluto de marginación 2000–2010. CONAPO, D.F. (2013)
23. Rojas, G., Fritsch, R., Castro, A., Guajardo, V., Torres, P., Díaz, B.: Trastornos menta-les comunes y uso de servicios de salud en población inmigrante. Rev. Med. Chil. **139**, 1298–1304 (2011). https://doi.org/10.4067/S0034-98872011001000008
24. Canales, M.: Metodologias de investigacion social Introducción a los oficios. Lom Ediciones, Santiago (2006)

25. Yong, L.: Modelo de aceptación tecnológica (tam) para determinar los efectos de las dimensiones de cultura nacional en la aceptación de las tic. Rev. Int. Ciencias Sociales y Humanidades, SOCIOTAM **XIV**(1), 131–171 (2004)
26. Streiner, D.: Finding our way: an introduction to path analysis. Can. J. Psychiatry **50**(2), 115–122 (2005). https://doi.org/10.1177/070674370505000207
27. Chen, H., Tseng, H.: Factors that influence acceptance of web-based e-learning systems for the in-service education of junior high school teachers in Taiwan. Eval. Program. Plann. **35**(3), 398–406 (2012). https://doi.org/10.1016/j.evalprogplan.2011.11.007
28. Caballero, A.: SEM vs. PLS : Un enfoque basado en la práctica. In: IV Congreso de Metodología de Encuestas, pp. 57–66. Pamplona (2006)
29. Hair, J., Sarstedt, M., Ringle, C., Mena, J.: An assessment of the use of partial least squares structural equation modeling in marketing research. J. Acad. Mark. Sci. **40**, 414–433 (2012). https://doi.org/10.1007/s11747-011-0261-6
30. Rodríguez, E.: Mínimos cuadrados parciales con el método de descenso de mayor pendiente. Rev. Tecnocientífica URU, 29–38 (2012)
31. Hair, Jr., Sarstedt, M., Hopkins, L., Kuppelwieser, V.: Partial least squares structural equation modeling (PLS-SEM): an emerging tool in business research. Eur. Bus. Rev. **26**(2), 106–121 (2014). https://doi.org/10.1108/EBR-10-2013-0128
32. Cheung, R., Vogel, D.: Predicting user acceptance of collaborative technologies: an extension of the technology acceptance model for e-learning. Comput. Educ. **63**, 160–175 (2013). https://doi.org/10.1016/j.compedu.2012.12.003
33. Ringle, C., Sarstedt, M., Straub, D.: A critical look at the use of PLS-SEM in MIS quarterly. MIS Quartely (MISQ) **36**(1), iii–xiv (2012). https://doi.org/10.2307/41410402
34. Hair, J., Sarsted, M., Pieper, T., Ringle, C.: The use of partial least squares structural equation modeling in strategic management research: a review of past practices and recommendations for future applications. Long Range Plan. **45**(5–6), 320–340 (2012). https://doi.org/10.1016/j.lrp.2012.09.008
35. Roldán, J., Sánchez-Franco, M.: Variance-based structural equation modeling: guidelines for using partial least squares in information systems research. In: Mora, M., Gelman, O., Steenkamp, A., Raisinghani, M. (eds.) Research Methodologies, Innovations and Philosophies in Software Systems Engineering and Information Systems, pp. 193–221. IGI Global, Hershey (2012). https://doi.org/10.4018/978-1-4666-0179-6.ch010
36. Martinez-Torres, M., Toral, S., Barrero, F., Gallardo, S., Arias, M., Torres, T.: A technological acceptance of e-learning tools used in practical and laboratory teaching, according to the European higher education area. Behav. Inf. Technol. **27**(6), 495–505 (2008). https://doi.org/10.1080/01449290600958965
37. Oviedo, H., Arias, A.: Aproximación al uso del coeficiente alfa de Cronbach. Rev. Colombiana de Psiquiatría. **XXXIV**(4), 572–580 (2005)
38. Fornell, C., Larcker, D.: Evaluating structural equation models with unobservable variables and measurement error. J. Mark. Res. **18**(1), 39–50 (1981). https://doi.org/10.2307/3151312
39. Hair, J., Ringle, C., Sarstedt, M.: PLS-SEM: indeed a silver bullet. J. Mark. Theory Pract. **19**(2), 139–152 (2014). https://doi.org/10.2753/MTP1069-6679190202
40. ODE: Panorama de la Educación 2016—México (2016). https://doi.org/10.1787/eag-2016-es

Social Well-Being Analysis Using Interval-Valued Fuzzy Predicates

Diego S. Comas, Eugenio Actis Di Pasquale, Juan I. Pastore, Agustina Bouchet, and Gustavo J. Meschino

Abstract Social well-being is one of the main goals of public policy. In this work, it is proposed a new methodology based on fuzzy predicates and interval-valued fuzzy logic for its computation and analysis, applied to urban areas of Argentina. The social well-being level of a territory is described through a fuzzy predicate, considering properties of different social indicators, and exploiting the advantages of the interval-valued fuzzy logic to deal with vague concepts. Fuzzy predicates allow to include knowledge about the meaning of social well-being and how it is traditionally measured, as well as linguistic descriptions involving social indicators. As a result, the social well-being level of each urban area is described by an interval. A method for interval comparing previously used in data clustering is applied here to rank the urban areas according to their social well-being levels. Results are consistent with those obtained using the known weighted average method. The approach proposed solves the problem of subjectivity in the evaluation of instances for comparative purposes, since once the fuzzy predicates are determined, the mathematical computation is the same for all the urban areas. This approach may have other interesting applications in the context of the Social Sciences, where new case studies are expected to be explored in the future.

Keywords Interval-valued fuzzy logic · Social well-being · Decision-Making · Fuzzy predicates · Ranking

D. S. Comas (✉) · J. I. Pastore · A. Bouchet · G. J. Meschino
Institute of Scientific and Technological Research in Electronics (ICyTE), National University of Mar del Plata-CONICET, Mar del Plata, 2290 Buenos Aires, Argentina
e-mail: diego.comas@fi.mdp.edu.ar

E. Actis Di Pasquale
Studies of Work Group, School of Economic and Social Sciences, National University of Mar del Plata, Mar del Plata, 2290 Buenos Aires, Argentina

W. Pedrycz et al. (eds.), *Computational Intelligence for Business Analytics*, Studies in Computational Intelligence 953,
https://doi.org/10.1007/978-3-030-73819-8_22

1 Introduction

In Social Sciences, social well-being, which is one of the main goals of public policy, has not a unique meaning, making its analysis difficult. Consequently, in order to assess the social well-being level of a society, it is necessary to assume a definition of it and then to propose a method for its quantification, which should preserve a significant correspondence between what is seen in the real world and the conceptual world.

Traditional assessment methods are based on statistics, as the methods based on statistical regression techniques, principal component analysis, distance measures [1–3], and data enveloping analysis [4], among others. Some of them have rigid computing processes not allowing during the definition of the equations include either personal appreciations or the participation of qualified informants. From a purely statistical point of view, this feature can be beneficial, but it is not from methodological and conceptual approaches. These methods do not allow consulting qualified informants, and experts cannot use the benefits of considering the data quality or incorporate knowledge about society in assessing social well-being. Despite this, such methods are not subjectively neutral because they include the ideological baggage of how the data and indicators are related, which implies a way of perceiving how the society works.

Social well-being level assessment can be addressed as a decision-making problem in which different cases (territories) are studied and sorted from the obtained result. In such an approach, different variables (indicators), selected according to the adopted definition, are observed, and experts define which variables should be considered and how these are related.

Fuzzy Logic (FL) [5] is a suitable tool for dealing with linguistic expressions, modeling knowledge and working with vague concepts, which has been widely used in decision-making problems, including analysis of supply chains [6], risk assessment [7], churn prediction [8], evaluation of employee's performance [9], company's competitiveness [10], support for medical diagnosis based on image processing [11], among others. In FL, knowledge is expressed by means of fuzzy predicates and membership functions. A final decision is obtained by evaluating the predicates (i.e., obtaining a truth value) for each case and comparing their results [12].

Interval valued FL [13, 14] has the proven ability to deal with vague expressions and uncertainly in the truth values of predicates, using intervals of truth values. Specifically, if there is imprecision or vagueness, for instance, when the combination of different opinions of experts is needed, intervals allow modeling the minimum and maximum assessment between the opinions [14].

Based on the previous observations, in this paper, social well-being assessment is addressed using interval-valued fuzzy predicates. A ranking of the analyzed territories according to the social well-being level is obtained. The method proposed exploits the advantages of the intervals to deal with vague concepts and is applied to the analysis of the social well-being level of the twenty-nine urban areas of the Republic of Argentina. The fuzzy predicates allow evaluating the social well-being level in the territories, resulting in intervals that indicate the truth value in which each territory has a high level of social well-being. In order to obtain a ranking of the urban areas, it is applied the so-called measure of intervals of truth values proposed in previous works for data clustering [13, 14].

The rest of this paper is structured as follows. In Sect. 2, the main concepts of FL are presented. In Sect. 3, the proposed methodology is described in detail. Section 4 introduces the results, and the paper concludes in Sect. 5 by discussing the results.

2 Fuzzy Logic Basics

FL was introduced by Lofti Zadeh in 1965 [5] extending Boolean (classic) logic in order to deal with linguistic expressions and work with concepts containing vague or imprecise expressions. It considers truth values between 0 (false) and 1 (true) instead of using only 0 and 1 as in classic logic, which is known as the "principle of gradualism" [5]. Specifically, FL provides an effective conceptual framework for dealing with knowledge representation in environments of uncertainty and imprecision, as is the case of human reasoning [15]. For this reason, FL is appropriate for assessing social well-being from expert knowledge.

As FL concepts are well-known, only more relevant concepts for this paper are presented in this section, trying to keep it as short as possible.

The main limitation of the traditional FL, called type-1 FL, is that the truth values are single values in the [0, 1] interval which may not be suitable in solving problems defined by incomplete or imprecise information, data affected by noise or disagreement between opinions of experts [9, 14, 16, 17]. Interval-valued FL provides additional grades of freedom by defining degrees of truth using intervals called intervals of truth values [17]. The main concepts of this kind of FL will be presented in the next paragraphs.

Definition #1: An interval of truth values is an interval $A = [a_L, a_R]$, with $0 \leq a_L \leq a_R \leq 1 \wedge 0 \leq a_L \leq a_R \leq 1$ [13, 14], where a_L and a_R are respectively the left end and right end of the interval. An interval of truth values defines the degree of truth of a logic expression when interval-valued FL is used.

Definition #2: An interval-valued membership function $\overline{\mu_A}$ on a discourse universe U is a function $\overline{\mu_A} : U \rightarrow \chi$ [13, 14], where χ is the set of all the closed intervals contained in [0, 1], i.e. the set of all the possible intervals of truth values, and **A** is a property (an attribute). For a specific value $u \in U$, $\overline{\mu_A}(u)$ is an interval of truth values $A_{\overline{\mu_A}(u)} = \left[a_{\overline{\mu_A}(u),L}, a_{\overline{\mu_A}(u),R}\right]$, which defines the degree of truth in which the value u satisfies the property **A**.

Definition #3: A fuzzy predicate $p(x)$, where x indicates an object or a variable, is a declarative sentence that assigns one or more properties to the object x. Using interval-valued FL, the truth value taken by $p(x)$, noted by $\nu(p(x))$, is an interval of truth values $A_{p(x)} = \left[a_{p(x),L}, a_{p(x),R}\right]$, with $0 \leq a_{p(x),L} \leq a_{p(x),R} \leq 1$.

Definition #4: Let $p(x)$ be a fuzzy predicate with truth value $A_{p(x)} = \left[a_{p(x),L}, a_{p(x),R}\right]$. A linguistic modifier of dilation (also called dilation hedge) defines a new fuzzy predicate noted by $p_{dil}(x)$ with truth value:

$$\nu(p_{dil}(x)) = \left[\left(a_{p(x),L}\right)^{1/n},\ \left(a_{p(x),R}\right)^{1/n}\right], \quad (1)$$

where $n \in \mathbb{N}$ defined in each case. The hedge modifies the meaning of the properties assigned by p to the object x. For instance, if p assigns the property "*low*" to x, then a dilation hedge can be "*slightly*" and $p_{slightly}(x)$ assigns the property "*slightly low*".

The next distinction between fuzzy predicates is adopted in the present paper [18]:

Definition #5: Let x be a variable on universe U. A fuzzy predicate $p(x)$ is called a simple fuzzy predicate if its values for different values of x are obtained from a membership function. A simple fuzzy predicate $p(x)$ correspond to a sentence as "*the value of x satisfies the property* **A**" or simpler "*x is* **A**" and is equivalent to a membership function $\overline{\mu_A}$ defined on a universe U.

Definition #6: Two fuzzy predicates $p(x_1, x_2, \ldots, x_n)$ and $q(x_1, x_2, \ldots, x_n)$ are equivalent; this is $p(x_1, x_2, \ldots, x_n) \equiv q(x_1, x_2, \ldots, x_n)$, if and only if $\nu(p(x_1, x_2, \ldots, x_n)) = \nu(q(x_1, x_2, \ldots, x_n))$ for all the possible values of $x_1, x_2, \ldots, x_n$.

Definition #7: Let $x_1, x_2, \ldots, x_n$ be variables respectively defined on universes $U_1, U_2, \ldots, U_n$. A fuzzy predicate $p(x_1, x_2, \ldots, x_n)$ is a compound predicate if it is equivalent to a logic combination of simple predicates or others compound predicates defined on one or more of the variables $x_1, x_2, \ldots, x_n$ using logical operators such as "and" ($\wedge$), "or" ($\vee$), "not" ($\neg$), "implication" ($\Rightarrow$), and "double-implication" ($\Leftrightarrow$).

In order to know the truth value of a compound fuzzy predicate, it is necessary to operate with the truth values of the composing predicates using functions known as fuzzy aggregation operators or fuzzy operators. Exhaustive analysis of these operators is available in the classic papers [19] and [20]. Its selection should be made according to their properties and how predicates are interpreted and evaluated by the experts.

Considering interval-valued FL, compound interval-valued fuzzy predicates are evaluated, applying the fuzzy operators separately on the left ends and on the right ends of the intervals. As the mathematical formalism of conjunctions, disjunctions, and complements between interval-valued fuzzy predicates can be consulted in the literature [12–14], they are omitted here.

The methodology proposed, described in the next Section, enables to describe the social well-being level of an urban area using a compound fuzzy predicate considering properties on different social indicators. As a result, the social well-being level of each territory is described by an interval. Intervals must be compared in order to define a ranking of urban areas. As mentioned before, the measure of intervals of truth values proposed in [13, 14] is used, and its definition is now recalled and analyzed considering its application in the present paper.

Definition #8: Let χ be the set of all the closed intervals contained in [0, 1], i.e., the set of all the possible intervals of truth values. The function $f : \chi \to \mathbb{R}^+$ is the measure of the interval of truth values:

$$f(A) = f([a_L, a_R]) = \frac{a_L + a_R}{2} a_R, \tag{2}$$

where $A = [a_L, a_R]$ is an interval of truth values. The function f describes with a number the degree of truth represented by the interval of truth values, mapping from the interval space to $\mathbb{R}^+$. The higher the value of f, the higher the degree of truth. The reasons of combining the mean value and the maximum of the interval are: (a) if two intervals have the same mean value, then that with the higher maximum value (closer to 1) represents a higher degree of truth, and (b) in the case of two intervals with the same maximum, that with lower mean value represents a lower degree of truth. The reason (a) is very important for the application considered here because the closer the maximum of the interval to 1, the closer the variables described by the predicates to those which completely meet the attributes (those with truth value equal to 1).

Given two intervals of truth values $A = [a_L, a_R]$ and $B = [b_L, b_R]$, it is possible to rank the intervals through the values $f(A)$ and $f(B)$. Specifically, if $f(A) < f(B)$ then $A <_f B$ $A <_f B$, if $f(B) < f(A)$ then $A >_f B$, and if $f(A) = f(B)$ then $A =_f B$; where $<_f, >_f$ and $=_f$ is the ranking induced from f.

In addition, the measure f has the next properties, being A, B, and C intervals of truth values [13]:

- If $A = [0, 0] = 0$, i.e., the minimum interval of truth values, then $f(A) = 0$.
- If $A = [1, 1] = 1$, i.e., the maximum interval of truth values, then $f(A) = 1$.
- If $A = [a, a] = a$, $B = [b, b] = b$ then $f(A) = a^2$ and $f(B) = b^2$, following in these cases the ranking obtained using type-1 FL.

As f mapping to $\mathbb{R}^+$ and its value is used to ranking intervals, as $f(A) < f(B)$ and $f(B) < f(C)$ then $A <_f B$ and $B <_f C$, therefore $f(A) < f(C) \Rightarrow A <_f C$ being a transitive ranking.

3 Proposed Methodology

The methodology proposed is divided into two stages: (1) *Study of the social well-being concept*, (2) *Fuzzy model of social well-being, and ranking of urban areas*.

Hereinafter, the set $\{\boldsymbol{x}_i\}_{i=1,\dots,N}$ is the set of urban areas, with $N = 29$. A specific urban area $\boldsymbol{x}_i$ is associated with a d-uple $\boldsymbol{x}_i = (x_{i,1}, x_{i,2}, \dots, x_{i,d})$ relating the urban area with the values of social indicators. When a generic urban area is mentioned during the predicate definition, it is referred to as $\boldsymbol{x}$ with a generic d-uple associated $(x_1, x_2, \dots, x_d)$.

At the first stage, a study of the concept of social well-being was performed in order to establish a definition, including different existing approaches and relating them with social indicators. Then, fuzzy predicates and interval-valued membership functions were defined by experts. As a result, a compound fuzzy predicate $q(\boldsymbol{x})$: "*The urban area* $\boldsymbol{x}$ *has a high level of social well-being*" was obtained, describing

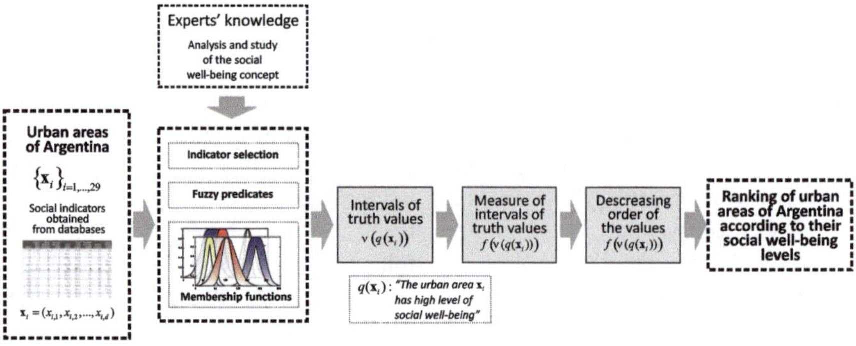

Fig. 1 Proposed methodology for social well-being level analysis in the urban areas of Argentina. From the definition of social well-being made at the first stage, descriptions explaining relations between social indicators and high social well-being level are made and fuzzy predicates and membership functions are defined. Then, the social well-being level is analyzed for each urban area, ranking the levels obtained by way of the measure of interval of truth values

necessary conditions for an urban area to achieve a high social well-being level. Finally, the compound predicate was evaluated for each urban area of Argentina, and the urban areas were ranked, enabling to compare and sort their levels of social well-being. A flowchart of the methodology is shown in Fig. 1. In the rest of this section, each stage is described in detail.

3.1 Study of the Social Well-Being Concept

Typical existing papers relate the social well-being to social objectives and, in consequence, enable to evaluate the achievement of the social well-being through indicators measured on a society. However, the indicator selection is not trivial and has been done according to the chosen approach.

In [21, 22], Actis Di Pasquale studied and defined the concept of social well-being, based on the approach of the theoretical basis of capabilities [23–25] and the approach of basic human needs [26]. According to this, the social well-being is a condition of social order in which all people successfully reach the achievements of good health, public safety, good education, decent work, suitable housing, and suitable income level, which promote both individual and social development. Individual development is considered because these achievements guarantee to live a dignified, healthy, and long life. Social development is also considered because, in a community, individual and group acts link with the social background. In this sense, there is a mutual dependence between the individual and society.

From the approach of Actis Di Pasquale, the level of social well-being in an urban area was assessed evaluating the grade in which the next achievements are satisfied: (a) *enjoy of good health*, (b) *enjoy public safety*, (c) *achieve an appropriate educational level*, (d) *have a decent work*, (e) *live in a suitable housing*, and (f) *have a suitable income level.*

Different social indicators available for the urban areas were analyzed in order to select those related to the achievements. As a result, it was possible to relate the grade of the fulfillment of each achievement with the indicators. This process was performed in four steps. First, indicators satisfying criteria of validity and reliability were pre-selected considering previous studies, recommendations of international organizations, and opinions of experts. As a result, 75 indicators were pre-selected. Next, databases of Argentina were conscientiously studied, both the free access databases and those requiring formal requests, trying to incorporate new relevant indicators to the pre-selected ones. Then, the informants of some public institutions were consulted in order to know the process used for defining each indicator and its limitations. Finally, once recognized the advantages and limitations of the different available databases and the processes involved in their construction, 9 indicators were selected, related to the scope of the achievements previously defined.

The indicators were obtained from databases of the *Instituto Nacional de Estadísticas y Censos* (INDEC) [National Institute of Statistics and Censuses], the *Dirección Nacional de Política Criminal—Ministerio de Justicia y Derechos Humanos* (DNPC/MJDDHH) [National Office of Criminal Policy—Ministry of Justice and Human Rights] [27], and the *Dirección de Estadística e Información de Salud/Ministerio de Salud* (DEIS/MS) [Department of Statistics and Health Information/Ministry of Health], which are governmental institutions of the Argentine Republic [28]. The indicators and their descriptions are summarized in Table 1. In all cases, scales of the indicators were analyzed jointly to experts and defined in accordance with recommendations of international organizations.

After the indicator selection, relations between them and the achievements were defined, obtaining the next relations:

- "*enjoy of good health*" is related with slightly low values of "*potential years of life lost*" (indicator #1);
- "*enjoy public safety*" requires slightly low values both of "*intentional-homicide rate*" and "*road traffic accident death rate*" (indicators #2 and #3);
- "*achieve an appropriate educational level*" implies high values both of "*mean percentage of school progress*", "*mean percentage of educational progress*", and -"*mean of schooling years*" (indicators #4 to #6);
- "*have a decent work*" is related to high values of "*decent work level*" (indicator #7);
- "*live in a suitable housing*" requires high values of the "*housing condition*" *index* (indicator #8);
- "*have a suitable income level*" is related to slightly high values of "*income level*" (indicator #9).

It is important to note that these relations associate the achievements adopted in this work for the social well-being assessment with attributes (properties) such as "*low*," or "*high*," "*suitable*" defined on the selected indicators. In some cases, the hedge "*slightly*" was used in the descriptions. As a result, it is possible to analyze the

Table 1 Selected indicators used for social well-being assessment. The range of the indicators and the year of the database are specified in each case

Index	Indicator name	Brief description
1	Potential years of life lost	Estimation of the average years that a person would have lived if not prematurely died, per 10,000 inhabitants (range: [0, 6000]; year: 2014).
2	Intentional-homicide rate	Unlawful death purposefully inflicted on a person by another person, per 100,000 inhabitants (range: [0, 80]; year: 2008).
3	Road traffic accident death rate	Number of deaths per 10,000 inhabitants in road traffic accident (range: [0, 80]; year: 2008).
4	Mean percentage of school progress	Mean percentage of school progress in relation with the theoretical age -from 6 to 17 years—(range: [0, 100]; year: 2014).
5	Mean percentage of educational progress	Mean percentage of educational progress in relation with the theoretical age -18–24 years—(range: [0, 100]; year: 2014).
6	Mean of schooling years	Years of schooling of the population of 25 years and over (range: [0, 20]; year: 2014).
7	Decent work level	A combination of remuneration, social security registration, holidays, stability, and working hours (range: [0, 12]; year: 2014).
8	Housing condition	An index made combining quality of the materials of the house, access to essential services, and housing regime (range: [0, 1], being 1 the ideal conditions; year: 2014).
9	Income level	The number of basic food baskets that can be accessed by person according to the level of household income (range: [0, 8]; year: 2014).

grade in which the achievements are jointly satisfied by analyzing the values of the indicators resulting in a way of assessing the social well-being level. In this paper, such a procedure was done by means of interval-valued fuzzy predicates.

3.2 *Fuzzy Model of Social Well-Being and Ranking of Urban Areas*

In this stage, a fuzzy model was obtained for modeling social well-being, defining a simple interval-valued fuzzy predicate for each of the indicators selected. Membership functions were obtained by consulting with experts.

Table 2 Meaning of the simple fuzzy predicates $p_j(\boldsymbol{x}),\ j = 1, \ldots, 9$

Predicate	Meaning
$p_1(\boldsymbol{x})$	"The potential years of life lost in the urban area $\boldsymbol{x}$ are low"
$p_2(\boldsymbol{x})$	"The intentional-homicide rate in the urban area $\boldsymbol{x}$ is low"
$p_3(\boldsymbol{x})$	"The road traffic accident death rate in the urban area $\boldsymbol{x}$ is low"
$p_4(\boldsymbol{x})$	"The mean percentage of school progress in the urban area $\boldsymbol{x}$ is high"
$p_5(\boldsymbol{x})$	"The mean percentage of educational progress in the urban area $\boldsymbol{x}$ is high"
$p_6(\boldsymbol{x})$	"The mean of schooling years in the urban area $\boldsymbol{x}$ is high"
$p_7(\boldsymbol{x})$	"The decent work level in the urban area $\boldsymbol{x}$ is high"
$p_8(\boldsymbol{x})$	"The housing condition index in the urban area $\boldsymbol{x}$ is high"
$p_9(\boldsymbol{x})$	"The income level in the urban area $\boldsymbol{x}$ is relatively high"

Unless it appears necessary to distinguish between urban areas, all the descriptions will be made considering a generic urban area $\boldsymbol{x}$, which is strictly represented by the d-uple $\boldsymbol{x} = (x_1, x_2, \ldots, x_d), d = 9$ and $x_1, x_2, \ldots, x_d$ are the 9 selected indicators in the order shown in Table 1. The 9 simple fuzzy predicates, each corresponding to one of the selected indicators and noted by $\{p_j(\boldsymbol{x})\}_{j=1,\ldots,9}$, are described in Table 2.

The membership functions $\{\overline{\mu_j}\}_{j=1,\ldots,9}$ each associated with one of the predicates $\{p_j(\boldsymbol{x})\}_{j=1,\ldots,9}$ are shown in Fig. 2. An analysis o f each predicate is given below:

1. $p_1(\boldsymbol{x})$ is related to low values of the indicator #1 "*potential years of life lost*". Until approximately 1000 years of life lost, the value of $p_1(\boldsymbol{x})$ stays high, considering that, in this case, are the best conditions for the urban area. Between 1000 and 4000, the truth value descends to subsequently keep low up to the maximum possible value of the variable (6000), the worst condition for this indicator.
2. The intentional-homicide rate takes values between 0 and 80. Values lower than 10 are considered low, resulting in high values of $\nu(p_2(\boldsymbol{x}))$. Between 10 and 50, the truth value descends until the worst condition for an urban area. For values higher than 50, $\nu(p_2(\boldsymbol{x})) \simeq 0$.
3. The road traffic accident death rate has similar behavior to the intentional-homicide rate. Therefore, the membership function of $p_3(\boldsymbol{x})$ was defined in a similar way of $p_2(\boldsymbol{x})$.
4. In the case of $p_4(\boldsymbol{x})$, values lower than 20 are considered low for the mean percentage of school progress. Values close to 100% correspond to the cases in which a high portion of the population finished high school studies. It should be noted that the high of the intervals in the membership function decreases when the percentage is closer to 100, meaning there is lower imprecision about the condition of the high mean percentage of school progress if it is close to its maximum.
5. A similar analysis to the previous one was done for the mean percentage of educational progress ($p_5(\boldsymbol{x})$). In this case, a percentage higher than 70 is considered high for this indicator. In addition, as this membership function has higher

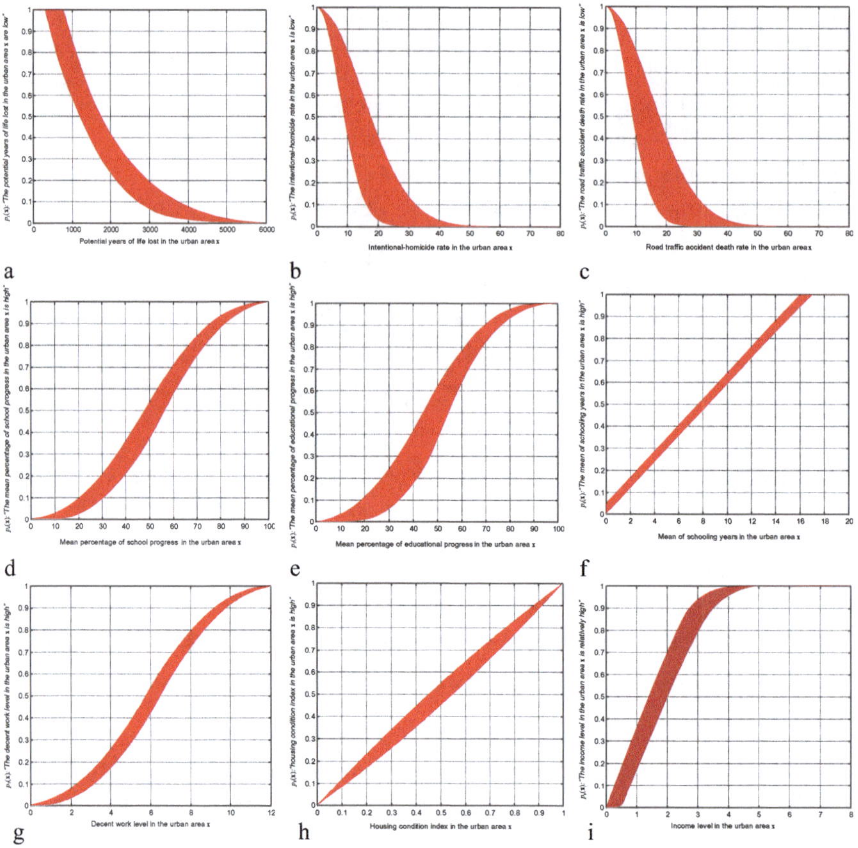

Fig. 2 Interval-valued membership functions associated with each of the simple fuzzy predicates. Both indicators and functions were defined by experts' knowledge. Y-axis: truth value of the predicate; x-axis: value of the indicator related to the predicate. **a** $p_1(\boldsymbol{x})$. **b** $p_2(\boldsymbol{x})$. **c** $p_3(\boldsymbol{x})$. **d** $p_4(\boldsymbol{x})$. **e** $p_5(\boldsymbol{x})$. **f** $p_6(\boldsymbol{x})$. **g** $p_7(\boldsymbol{x})$. **h** $p_8(\boldsymbol{x})$. **i** $p_9(\boldsymbol{x})$

intervals than the function of $p_4(\boldsymbol{x})$ there is higher imprecision between experts about when to consider high values of the percentage of educational progress comparing to the same analysis in the case of the percentage of school progress ($p_4(\boldsymbol{x})$).

6. In the case of a high mean of schooling years, a value higher than 16 is considered very high, therefore the value of $\nu(p_6(\boldsymbol{x}))$ are close to 1. In the other hand, values lower than 16 have assigned lower values of $\nu(p_6(\boldsymbol{x}))$, with 0 the worst condition with $\nu(p_6(\boldsymbol{x})) = 0$.
7. In the "*decent work level*" (indicator #7), the middle condition corresponds to a value of 6 of the indicators, which has assigned the highest imprecision with an interval centered in 0.5. Values closer to 0 (the minimum) or to 12 (the maximum) have lower imprecision in the membership function.

8. For the housing condition, it was considered a linear relationship between the value of the index and the value of the membership function, but with impression growing from the end of the range (0 and 1) to the middle (0.5).
9. The "*income level*" in the urban area is evaluated, considering the average number of basic food baskets that the household income level may secure. It was assumed that values higher than 4 are "*relatively high*," then in these cases $\nu(p_9(\boldsymbol{x}))$ is 1. From 4 to 0, the "*relative income level*" becomes smaller, corresponding to lower values of the membership function.

From the simple fuzzy predicates $\{p_j(\boldsymbol{x})\}_{j=1,\ldots,9}$ and the corresponding membership functions $\{\overline{\mu_j}\}_{j=1,\ldots,9}$, it was possible to define compound fuzzy predicates for the achievements defined in the first stage. As a result, 6 compound fuzzy predicates were defined: $p_A(\boldsymbol{x})$: "*In the urban area **x** people enjoy good health,*" $p_B(\boldsymbol{x})$: "*In the urban area **x** people enjoy public safety,*" $p_C(\boldsymbol{x})$: "*In the urban area **x** people achieve an appropriate educational level,*" $p_D(\boldsymbol{x})$: "*In the urban area **x** people have a decent work,*" $p_E(\boldsymbol{x})$: "*In the urban area **x** people live in a suitable housing*", and $p_F(\boldsymbol{x})$: "*In the urban area **x** people have a suitable income level*".

Considering the descriptions resulting of stage 1, the predicates $p_A(\boldsymbol{x})$ to $p_F(\boldsymbol{x})$ can be written in terms of $p_1(\boldsymbol{x})$ to $p_9(\boldsymbol{x})$ as follow:

$$p_A(\boldsymbol{x}) \equiv p_{1,\ slightly}(\boldsymbol{x}), \tag{3}$$

$$p_B(\boldsymbol{x}) \equiv p_{2,\ slightly}(\boldsymbol{x}) \ \wedge \ p_{3,\ slightly}(\boldsymbol{x}), \tag{4}$$

$$p_C(\boldsymbol{x}) \equiv p_4(\boldsymbol{x}) \ \wedge \ p_5(\boldsymbol{x}) \ \wedge \ p_6(\boldsymbol{x}), \tag{5}$$

$$p_B(\boldsymbol{x}) \equiv p_7(\boldsymbol{x}), \tag{6}$$

$$p_E(\boldsymbol{x}) \equiv p_8(\boldsymbol{x}), \tag{7}$$

$$p_F(\boldsymbol{x}) \equiv p_{9,\ slightly}(\boldsymbol{x}), \tag{8}$$

where the symbol $\wedge$ indicates the conjunction of the fuzzy predicates, implying the joint satisfaction of the simple predicates. As an example, the predicate $p_B(\boldsymbol{x})$: "*In the urban area **x** people enjoy public safety*" is linguistically interpreted as: $p_B(\boldsymbol{x})$ is "equivalent" to: "*The intentional-homicide rate in the urban area **x** is slightly low <u>and</u> the road traffic accident death rate in the urban area **x** is slightly low*". In the cases of $p_A(\boldsymbol{x})$, $p_B(\boldsymbol{x})$, and $p_F(\boldsymbol{x})$ the dilation hedge "*slightly*" was included.

As a result of the previous procedure, it was possible to evaluate the grade in which the different achievements related to the social well-being level are satisfied for an urban area, from the truth values of the predicates $p_1(\boldsymbol{x})$ to $p_9(\boldsymbol{x})$. Finally, the fuzzy predicate $q(\boldsymbol{x})$: "*The urban area **x** has a high level of social well-being*" was defined as follow:

$$q(\boldsymbol{x}) \equiv p_A(\boldsymbol{x}) \wedge p_B(\boldsymbol{x}) \wedge p_C(\boldsymbol{x}) \wedge p_D(\boldsymbol{x}) \wedge p_E(\boldsymbol{x}) \wedge p_F(\boldsymbol{x}). \quad (9)$$

In other words, the truth value of $q(\boldsymbol{x})$ defines the grade in which the social well-being level in an urban area $\boldsymbol{x}$ is "*high*", and it depends on the grade in which people enjoy good health, enjoy public safety, achieve an appropriate educational level, have a decent work, live in a suitable housing, and have a suitable income level. As higher the grade in which these achievements are jointly satisfied, as higher the grade in which the social well-being level in the urban area is high. Therefore, the social well-being level of an urban area is described (and can be measured) by the compound predicate $q(\boldsymbol{x})$, relating properties of the social indicators.

By computing the degree of truth of $q(\boldsymbol{x}_i)$ for the 29 urban areas of Argentina, it was possible to know the degree in which the social well-being level is high in each one. Then, the urban areas were ranked using the measure of intervals of truth values of Definition #8. The procedure followed is described below:

1. For each urban area $\boldsymbol{x} \in \{\boldsymbol{x}_i\}_{i=1,\ldots,29}$ the simple predicates $p_1(\boldsymbol{x})$ to $p_9(\boldsymbol{x})$ were evaluated using the membership functions $\{\overline{\mu_j}\}_{j=1,\ldots,9}$.
2. From the truth values of $p_1(\boldsymbol{x})$ to $p_9(\boldsymbol{x})$, the predicates $p_A(\boldsymbol{x})$ to $p_F(\boldsymbol{x})$ were computed for each urban area applying one fuzzy conjunction operator and obtaining the truth value in which each of the achievements is meet for the urban area. The dilation hedge, when necessary, was evaluated as explained in the Definition #4, considering for the present application $n = 2$.
3. The value of the predicate $q(\boldsymbol{x}_i)$, $i = 1, \ldots, 29$, was computed, getting for each case an interval of truth values $\nu(q(\boldsymbol{x}_i)) = A_{q(\boldsymbol{x}_i)} = \left[a_{q(\boldsymbol{x}_i),L}, a_{q(\boldsymbol{x}_i),R}\right]$.
4. The measure f was applied to the intervals $A_{q(\boldsymbol{x}_i)} = \nu(q(\boldsymbol{x}_i)), i = 1, \ldots, 29$; obtaining the set of values $\{f(A_{q(\boldsymbol{x}_i)})\}_{i=1,\ldots,29}$.
5. The set of values $\{f(A_{q(\boldsymbol{x}_i)})\}_{i=1,\ldots,29}$ is sorted in decreasing order inducing a ranking on the urban areas $\{\boldsymbol{x}_i\}_{i=1,\ldots,29}$ in accordance with the position of $f(A_{q(\boldsymbol{x}_i)})$ in the ordered set.

As a result, the urban areas of Argentina are ranked according to decreasing social well-being level, solving the problem addressed. Both compensatory as classic MIN-MAX fuzzy operators were tested [18] for the evaluation of the compound fuzzy predicates. Results are presented and analyzed in the next section.

4 Results

As a result of the proposed methodology, the social well-being levels of the 29 urban areas of Argentina were analyzed by means of fuzzy predicates and interval valued FL.

The fuzzy predicates were evaluated through both compensatory and MIN-MAX fuzzy operators [29, 30]. According to the opinions of experts, it was expected that compensatory operators based on the arithmetic mean reflect what they typically do to estimate indexes of social well-being, which typically consists of using weighted averages of the indicators [21, 22]. Consequently, compensatory reasoning appears to be suitable for social level assessment.

The analysis presented here is focused on the results obtained through compensatory operators based on arithmetic mean [29]. Raking differences respect to the other fuzzy operators are also included. In addition, as a way of external validation of the obtained raking, the results were compared with those corresponding to the weighted-average method proposed by Actis Di Pasquale [21, 22], which had already been applied to social well-being study of the urban areas of Argentina.

In Table 3, results obtained for the 29 urban areas are shown. For each urban area $\boldsymbol{x}_i$, the table indicates: ranking order, name, the interval of truth values associated with the predicate $q(\boldsymbol{x}_i)$: "*The urban area* $\boldsymbol{x}_i$ *has a high level of social well-being*", and the value of the measure of the interval of truth values $f(\nu(q(\boldsymbol{x}_i)))$ introduced in Sect. 2.

The results of the method proposed were similar to the ranking obtained by Actis Di Pasquale using the weighted-average method [21, 22], locating Ciudad de Buenos Aires and Ushuaia–Río Grande in the first two places and Concordia, Gran Resistencia and Santiago del Estero–La Banda in the last three positions. In particular, the urban areas that improved their positions in comparison with the weighted average method were Rio Cuarto, Mar del Plata–Batán, and the territories of Patagonia Argentina.

As the measure of intervals of truth values indicates the grade in which an urban area has a high level of social well-being, an additional analysis can be done by grouping the urban areas according to the value of this measure. Four clusters were defined:

- *Cluster #1* ($f(\nu(q(\boldsymbol{x}_i))) > 0.700$): It includes Ciudad de Buenos Aires and Ushuaia–Río Grande, the two best positions.
- *Cluster #2* ($0.500 < f(\nu(q(\boldsymbol{x}_i))) \leq 0.700$): It is formed by Comodoro Rivadavia–Rada Tilly, Bahía Blanca–Cerri, Neuquén–Plottier, Gran Paraná, Gran Rosario, Gran La Plata, Río Cuarto, Mar del Plata–Batán, San Luis–El Chorrillo, Gran Mendoza, Río Gallegos, La Rioja, Gran Córdoba, and Gran Catamarca.
- *Cluster #3* ($0.450 < f(\nu(q(\boldsymbol{x}_i))) \leq 0.500$): It includes Gran Santa Fe, Jujuy–Palpalá, Gran Tucumán–Tafí Viejo, Posadas, Partidos del GBA, and Salta.
- *Cluster #4* ($f(\nu(q(\boldsymbol{x}_i))) \leq 0.450$): It is formed by Gran San Juan, Formosa, Santa Rosa–Toay, Corrientes, Concordia, Gran Resistencia, Santiago del Estero–La Banda.

The clusters defined on the values of $f(\nu(q(\boldsymbol{x}_i)))$ were heuristically defined by experts. Defining thresholds and labeling the urban areas help experts to focus on specific features of the urban areas in each cluster (related to the indicators and the achievements analyzed in Sect. 3) which may assist in defining priority policies in order to improve the social well-being in the territories considering their main needs.

In addition, an analysis by qualitative methods was done by a group of experts in social well-being, which gave relevant knowledge about the characteristics and

Table 3 Ranking in decreasing order obtained for the social well-being level of the urban areas of Argentina using the proposed methodology

Ranking	Urban area name	$\nu(q(\boldsymbol{x}_i))$	$f(\nu(q(\boldsymbol{x}_i)))$
1	Ciudad de Buenos Aires	[0.809, 0.910]	0.782
2	Ushuaia–Río Grande	[0.816, 0.876]	0.742
3	Comodoro Rivadavia–Rada Tilly	[0.760, 0.835]	0.667
4	Bahía Blanca–Cerri	[0.676, 0.848]	0.646
5	Neuquén–Plottier	[0.668, 0.838]	0.632
6	Gran Paraná	[0.630, 0.840]	0.618
7	Gran Rosario	[0.651, 0.832]	0.617
8	Gran La Plata	[0.649, 0.828]	0.612
9	Río Cuarto	[0.622, 0.806]	0.576
10	Mar del Plata–Batán	[0.608, 0.788]	0.550
11	San Luis–El Chorrillo	[0.610, 0.786]	0.549
12	Gran Mendoza	[0.598, 0.790]	0.548
13	Río Gallegos	[0.496, 0.809]	0.528
14	La Rioja	[0.590, 0.763]	0.516
15	Gran Córdoba	[0.575, 0.766]	0.514
16	Gran Catamarca	[0.576, 0.754]	0.502
17	Gran Santa Fe	[0.509, 0.773]	0.495
18	Jujuy–Palpalá	[0.561, 0.743]	0.484
19	Gran Tucumán–Tafí Viejo	[0.562, 0.741]	0.483
20	Posadas	[0.554, 0.728]	0.466
21	Partidos del GBA	[0.546, 0.729]	0.464
22	Salta	[0.542, 0.720]	0.454
23	Gran San Juan	[0.519, 0.715]	0.441
24	Formosa	[0.508, 0.690]	0.413
25	Santa Rosa–Toay	[0.379, 0.738]	0.412
26	Corrientes	[0.492, 0.675]	0.394
27	Concordia	[0.335, 0.691]	0.355
28	Gran Resistencia	[0.453, 0.644]	0.354
29	Santiago del Estero–La Banda	[0.439, 0.633]	0.340

particularities of the different Argentinian urban areas. These experts agreed that results are consistent with those expected for the problem studied. This conceptual comparison could be made because of the limited number of areas. The expertise, experience, academic, and personal knowledge of each expert, together with the review of previous studies, allowed validating the results. It was applied the concept presented in [31–33], which refers to the grade in which a measuring instrument measures a variable, according to qualified opinions.

Regarding the results obtained with the others fuzzy operators, the ranking generated using compensatory operators based on the geometric mean swapped the order of Río Cuarto for Santa Rosa–Toay, Gran La Plata for Gran Paraná, and moved Gran Mendoza and Gran Santa Fe three positions to the bottom of the ranking. Main differences respect to the results shown in Table 3 were related to higher values of the intervals and significantly smaller ranges using geometric mean than using the arithmetic mean resulting in higher values of the measure of the interval of truth values. These effects can be associated with a higher compensatory effect of the geometric mean, resulting in an undesirable effect in the case analyzed in this paper.

On the other hand, MIN-MAX operators obtained a very different ranking, changing the relative position of several urban areas, which was reflected in the composition of possible clusters, producing not suitable results comparing with the weighted average result, which was assumed to be an external validation result. In addition, the values of the bounds of the interval and, in consequence, the values of the measure of the intervals were low. This, according to experts, does not correspond with the expected results of an analysis of the social well-being level.

5 Conclusions

It was proposed a new methodology for the analysis of the social well-being level of urban areas based on fuzzy predicates and interval valued FL, which was applied to the urban areas of Argentina.

The ranking results obtained for the urban areas were similar to those obtained by the weighted average method, a known method of estimation of social well-being level. In addition, consulted experts concluded that both the ranking results and the groups of urban areas obtained were consistent with the results they expected.

In this sense, the approach proposed presents an advantage against the method traditionally used for social well-being analysis: it enables including the available knowledge from previous studies, qualified informants, and experts and recommendations of international organizations into the model, including linguistic descriptions involving relations between social indicators, achievements and social well-being.

In this regard, the method proposed enabled to describe the social well-being level of an urban area using a fuzzy predicate, describing properties on different social indicators and exploiting the advantages of the interval-valued FL to deal with vague concepts as those related to social well-being. Both membership functions and fuzzy predicates were defined in accordance with the opinions of different experts, merging them into a unique fuzzy model. The model captured all the available knowledge about what social well-being means and how it is traditionally measured.

Considering that the consistency of the approach presented has been positively evaluated against other procedures previously proposed, the same methodology can be used in situations that require a similar analysis. After a detailed reliability analysis, an approach like the proposed method could be applied in new problems involving a large number of cases, which could give relevant results in different fields. For example, it could be used to rank thousands of candidates to get a job or to receive a scholarship.

The proposed approach solves the problem of subjectivity in the evaluation of instances for comparative purposes. Once defined the membership functions and the fuzzy predicates, the computation is the same for all instances, no matter how much they are.

This approach could have interesting applications in the context of the Social Sciences, where new case studies are expected to be discovered in future work.

Acknowledgements Authors acknowledge support from Consejo Nacional de Investigaciones Científicas y Técnicas (CONICET) from Argentina.

References

1. Espina, P.: Aproximación a la medición del bienestar social. Idoneidad del indicador sintético "Distancia-P(2)". Spanish J. Econ. Financ. **24**(68), 139–163 (1996)
2. Ivanovic, B.: Comment établir une liste des indicateurs de développement. Revue de statistique appliquée **22**, 37–50 (1974)
3. Trapero, J.: Problemas de la medición del bienestar y conceptos afines. Una aplicación al caso español. Presidencia del Gobierno. Instituto Nacional de Estadística de España, Madrid (1977)
4. Charnes, A., Cooper, W., Rhodes, E.: Measuring the efficiency of decision making units. Eur. J. Oper. Res. **2**(6), 429–444 (1978). https://doi.org/10.1016/0377-2217(78)90138-8
5. Zadeh, L.A.: Fuzzy sets. Inf. Control **8**, 338–353 (1965)
6. Pilevari, N., SeyedHosseini, S., Jassbi, J.: Fuzzy Logic Supply Chain Agility Assessment Methodology. IEEE International Conference on Industrial Engineering and Engineering Management, pp. 1113–1117, Singapore (2008). https://doi.org/10.1109/ieem.2008.4738043
7. Nunes, I., Simões-Marques, M.: Applications of Fuzzy Logic in Risk Assessment–The RA_X Case. In: Azeem, M.F. (ed.) Fuzzy Inference System–Theory and Applications, pp. 21–40. InTech, Rijeka (2012). https://doi.org/10.5772/37212
8. Abbasimehr, H., Setak, M., Tarokh, M.: A Neuro-Fuzzy Classifier for Customer Churn Prediction. Int. J. Comput. Appl. **19**(8), 35–41 (2011)
9. Melin, P., Castillo, O.: A review on the applications of type-2 fuzzy logic in classification and pattern recognition. Expert Syst. Appl. **40**(13), 5413–5423 (2013). https://doi.org/10.1016/j.eswa.2013.03.020
10. Espin, R., Gómez, J., Téllez, G., González, E.: Compensatory logic: A fuzzy approach to decision making. In: 4th International Symposium on Engineering of Intelligent Systems (EIS'2004). Madeira, Portugal (2004)
11. Zarandi, M., Zarinbal, M., Izadi, M.: Systematic image processing for diagnosing brain tumors: A Type-II fuzzy expert system approach. Appl. Soft Comput. **11**(1), 285–294 (2011). https://doi.org/10.1016/j.asoc.2009.11.019
12. Comas, D., Pastore, J., Bouchet, A., Ballarin, V., Meschino, G.: Type-2 Fuzzy Logic in Decision Support Systems. In: Espin Andrade, R.A., Bello Pérez, R., Cobo, Á., Marx Gómez, J., and Racet Valdés, A. (eds.) Soft Computing for Business Intelli-gence, pp. 267–280. Springer Berlin Heidelberg, Heidelberg, Germany (2014). https://doi.org/10.1007/978-3-642-53737-0_18
13. Comas, D., Meschino, G., Nowé, A., Ballarin, V.: Discovering knowledge from data clustering using automatically-defined interval type-2 fuzzy predicates. Expert Syst. Appl. **68**, 136–150 (2017). https://doi.org/10.1016/j.eswa.2016.10.018

14. Comas, D., Pastore, J., Bouchet, A., Ballarin, V., Meschino, G.: Interpretable in-terval type-2 fuzzy predicates for data clustering: A new automatic generation method based on self-organizing maps. Knowledge-Based Syst. **133**, 234–254 (2017). https://doi.org/10.1016/j.knosys.2017.07.012
15. Bellman, R., Giertz, M.: On the Analytic Formalism of the Theory of Fuzzy Sets. Inf. Sci. **5**, 149–156 (1973). https://doi.org/10.1016/0020-0255(73)90009-1
16. Mendel, J.: Type-2 fuzzy sets and systems: an overview. IEEE Comput. Intell. Mag. **2**(1), 20–29 (2007). https://doi.org/10.1109/MCI.2007.380672
17. Comas, D., Meschino, G., Ballarin, V.: Discovering type-2 fuzzy predicates in da-ta guided by automatic clustering algorithms. Eureka International Virtual Physical Meeting 2014, México (2014)
18. Meschino, G., Comas, D., Ballarin, V., Scandurra, A., Passoni, L.: Automatic design of interpretable fuzzy predicate systems for clustering using self-organizing maps. Neurocomputing **147**, 47–59 (2015). https://doi.org/10.1016/j.neucom.2014.02.059
19. Dubois, D., Prade, H.: A Review of Fuzzy Set Aggregation Connectives. Inf. Sci. **36**(1–2), 85–121 (1985). https://doi.org/10.1016/0020-0255(85)90027-1
20. Zimmermann, H., Zysno, P.: Latent connectives in human decision making. Fuzzy Sets Syst. **4**(1), 37–51 (1980). https://doi.org/10.1016/0165-0114(80)90062-7
21. Actis Di Pasquale, E.: Bienestar Social, resignificación del concepto y de su operacionalización. Un aporte metodológico aplicado al caso argentino, Doctoral Thesis, National University of Quilmes. Quilmes, Argentina (2013)
22. Actis Di Pasquale, E.: Hacia una definición conceptual de bienestar social. El debate desde la Economía del Bienestar hasta Enfoque de las Capacidades. In: R. F. UNICEN (ed.). VI Encuentro Regional de Estudios del Trabajo Pre-Aset 2015, Tandil (2015)
23. Nussbaum, M.: Capabilities as fundamental entitlements: sen and social justice. Fem. Econ. **9**(2–3), 33–59 (2003). https://doi.org/10.1080/1354570022000077926
24. Sen, A.: Inequality Reexamined. Clarendon Press, Oxford, New York (1992)
25. Sen, A., Nussbaum Martha: Capability and Well-being. In: Sen, A., Nussbaum, M. (eds.) Oxford Scholarship Online, pp. 30–53. Clarendon Press, Oxford (1993). https://doi.org/10.1093/0198287976.003.0003
26. Doyal, L., Gough, I.: A theory of human need. Palgrave, London (1991). https://doi.org/10.1007/978-1-349-21500-3
27. INDEC: Instituto Nacional de Estadística y Censos de la República Argentina, https://www.indec.gob.ar/. Accessed 22 February 2019
28. DEIS: Dirección de Estadísticas e Información de Salud, http://www.deis.msal.gov.ar/. Accessed 22 Feb 2019
29. Bouchet, A., Pastore, J., Andrade, R., Brun, M., Ballarin, V.: Arithmetic Mean Based Compensatory Fuzzy Logic. Int. J. Comput. Intell. Appl. **10**(2), 231–243 (2011). https://doi.org/10.1142/S1469026811003070
30. Espin Andrade, R., MazcorroTéllez, G., Fernández González, E., Marx-Gómez, J., Lecich, M.: Compensatory Logic: a fuzzy normative model for decision making. Invest. Oper. **27**, 178–193 (2006)
31. Grinnell, R., Unrau, Y.: Social work research and evaluation: quantitative and qualitative approaches. Oxford University Press, New York (2005)
32. Bostwick, G., Kyte, N.: Measurement social work: research and evaluation quantitative and qualitative approach. Oxford University Press, New York (2005)
33. Thyer, B.: The handbook of social work research methods. Sage, Los Angeles (2010)

A New Plugin to Include FuzzyPred in KNIME

Orenia Lapeira, Ernesto Álvarez, René Cutie, Alejandro Prieto, Alejandro Rosete, and Taymi Ceruto

Abstract Knowledge discovery from databases is a very attractive and challenging task. The elicitation of fuzzy predicates (called FuzzyPred) in conjunctive and disjunctive normal form provides a convenient and effective general way to identify and to represent certain dependencies among items in fuzzy transactions. Konstanz Information Miner (KNIME) is a strong and comprehensive free platform for drag-and-drop analytics, machine learning, statistics, and data processing. It already offers a large variety of nodes, which enables easy execution of data pipelines. This paper presents a new plug-in that integrates FuzzyPred into KNIME. It allows reducing the amount of knowledge and experience required by users to use the method. A case of study is given to illustrate the use of the proposal.

Keywords Data mining · Fuzzy mining · FuzzyPred · KNIME

1 Introduction

Nowadays, modern societies generate a huge volume of data in many ways by using millions of computers, devices, sensors, etc. This large amount of data can be transformed and analyzed to produce valuable insights [1]. In order to do that, data is analyzed and characterized by different points of view and with several objectives. With such amount of data, there is a need for powerful techniques for better interpretation of these data that exceeds the human's ability for comprehension and decision making [2]. Therefore, humans need data mining (data science), and there are several examples of successful applications in several fields such as marketing, business, science, engineering, games, and bioinformatics [3, 4].

Today the use of data mining for solving real-world problems is an extended practice. However, their use requires certain expertise, considerable time, and effort according to the user needs. To reduce the complexity of this task, in the last years,

O. Lapeira · E. Álvarez · R. Cutie · A. Prieto · A. Rosete (✉) · T. Ceruto
Universidad Tecnológica de La Habana José Antonio Echeverría (Cujae), 19390 Marianao, La Habana, Cuba
e-mail: rosete@ceis.cujae.edu.cu

W. Pedrycz et al. (eds.), *Computational Intelligence for Business Analytics*, Studies in Computational Intelligence 953,
https://doi.org/10.1007/978-3-030-73819-8_23

many software tools have been developed to support data management and experimentation [5, 6]. These tools help the researchers to make their methods easily accessible to others.

A lot of data mining tools [7, 8] are available either as commercial tools (e.g., SPSS Clementine, Oracle Data Mining) or as open source software (e.g., WEKA: Waikato Environment for Knowledge Analysis, Orange, KNIME). KNIME (Konstanz Information Miner) [9] is a particularly interesting option due to its graphical interface, its reliability, and the huge amount of options that cover all the steps of any data-mining project [9]. KNIME is a modular environment written in Java, and its graphical workflow editor is implemented as an Eclipse plug-in. This simplicity facilitates its wide acceptance in diverse fields [10, 11]. It is easy to extend through an open Application Programming Interface (API) and a data abstraction framework, which allows the easy addition of nodes and workflows [10, 11].

In this paper, we propose a new plug-in (with several nodes) to integrate FuzzyPred into KNIME. FuzzyPred [12] is a novel unsupervised data mining method that searches for good fuzzy predicates (often expressed in conjunctive or disjunctive normal forms) to describe a fuzzy database. It combines fuzzy set concepts and metaheuristic algorithms to search for logic predicates to describe a given dataset [12]. This approach to data mining is connected with the methodology of Logical Analysis of Data based on a combinatorial search [13]. The search for good fuzzy predicates in normal forms implies a combinatorial optimization problem given by the number of possible combinations of logical operators and fuzzy variables; thus metaheuristics have been used in the search for good fuzzy predicates [12]. FuzzyPred allows obtaining patterns (with the presence of different connectives to combine variables) that the classic data mining methods cannot obtain. Besides, FuzzyPred can be used to generate predicates that are related to several types of patterns, such as rules or clusters [12]. The integration of FuzzyPred into KNIME allows reducing the level of knowledge and experience required by users to use the method.

The rest of the paper is organized as follows. The next section presents the main concepts related to predicates and FuzzyPred. Section 3 introduces several basic characteristics of KNIME, including its advantages. Section 4 explains the proposal, i.e., the new plug-into integrate FuzzyPred into KNIME. In Sect. 5, one example illustrates how the proposal may be used. Finally, Sect. 6 points out some conclusions and future work.

2 Fuzzy Predicates

Fuzzy predicates, as a way of expressing complex fuzzy sets, are commonly used to talk about the properties of objects by defining the set of all objects that have some property in common. In general, a predicate is a statement that may be either true, false, or an intermediate value depending on the truth values of its variables. In fuzzy logic, the strict true/false valuation of the predicates in classical logic is replaced by a quantity interpreted as the degree of membership (which may also be interpreted as a truth value) [12, 14].

A fuzzy predicate may be interpreted as a complex fuzzy concept or class with a certain degree of membership for each object. Each fuzzy predicate is expressed as a combination of fuzzy concepts (corresponding to variables or columns in a fuzzy database) and operators (such as intersection/conjunction, union/disjunction, and complement/negation). For instances, in a fuzzy database where *healthy*, *old*, and *educated* are fuzzy variables, some examples of fuzzy predicates may be:

- *(healthy* and not *old)* or *educated*
- *healthy* and *old* and *educated*
- *healthy* or *old*
- not *healthy* or *(old* and *educated).*

Although several measures are available to assess the quality of a fuzzy predicate [15], we resort here to the so-called Fuzzy Predicate Truth Value (FPTV). This measure is interpreted as the truth value of the fuzzy predicate over all the examples in a database. The measure FPTV has values in the interval [0; 1] being 1 the maximum quality and 0, the lowest one. Other measures were described in [15].

A fuzzy predicate may also be viewed as a tree where each internal node is a fuzzy operator (conjunction, disjunction, and negation), and each leaf is a fuzzy variable of the database. Each fuzzy variable may also be associated with adverbs called hedges. Hedges are terms that modify the shape of fuzzy sets. It has two main behaviors: reinforcement (such as "very") or weakening (such as "little") [16].

A formula is in conjunctive normal form (CNF) if it is a conjunction of clauses, where a clause is a disjunction of literals [17], i.e., it is a conjunction (AND) of disjunctions (OR). A formula is in disjunctive normal form (DNF) if it is a disjunction of clauses, where a clause is a conjunction of literals [17], i.e., it is a disjunction (OR) of conjunctions (AND). Thus, CNF and DNF are only composed of AND, OR, NOT. The NOT operator can only be used as part of a literal.

It may be observed that fuzzy predicates in CNF and DNF have some grade of generality because, in classical logic, different patterns (such as rules or equivalences) may be transformed to CNF and DNF. For example, a pattern/predicate "NOT *healthy* OR (*old* and *educated*)" may be interpreted as a general description of a database, where some examples are "NOT *healthy*" and the other examples are "*old* and *educated*", i.e., this is similar to the meaning associated to a clustering [2, 4]. However, this predicate may also be regarded as a rule IF *healthy* THEN *old* and *educated* based on a classic logic transformation (NOT A OR B $\equiv$ A $\rightarrow$ B). This syntactic transformation is based on rules associated with logical equivalences [17]. Even when these equivalences are not perfectly true in multivalued logic [18, 19], fuzzy predicates in CNF and DNF seems to be a good representation schema to obtain knowledge with a general scope. There are several papers addressing the extension of normal forms (and the transformation of logical expressions) to the context of Fuzzy Logic (e.g. [20, 21]), and it is an active field of research [22, 23].

It is worth noting that these predicates may be created by human experts. However, they can also be discovered via algorithms that learn them from data [13, 24]. Predicate mining is a task that can be faced as an optimization problem, and it can be solved by using metaheuristics [12, 15]. Metaheuristics are a good alternative to face this task because of their recognized ability to search large search spaces in a robust way [25–27].

The degree of membership (equivalent to a truth value in this case) of complex predicates (e.g., the four predicates listed in the previous page) may be calculated based on the membership of each example to the elementary predicates (i.e., the membership of the persons to the sets *healthy*, *old* or *educated*). This can be done by using the definition of union/disjunction and interception/conjunction of fuzzy sets based on t-norms. For example, based on the Zadeh min/max function, a person that belongs to *healthy* with 0.8, to *old* with 0.5, and to *educated* with 0.9 has a truth value (degree of membership) of 0.8 to the set *healthy or old* (because max(0.8, 0.5) = 0.8), and 0.5 to the set *not healthy or (old and educated)* (because max(1–0.8, min(0.5, 0.9)) = max(0.2, 0.5) = 0.5). The overall truth value of a fuzzy predicate in a database may be obtained by the intersection/conjunction of the values of membership of each example [15].

It is worth noting that the truth value of a fuzzy predicate depends on the database. Thus, a predicate with high truth value in a database describes it, in general terms, i.e., it generalizes the information contained in the database. In this sense, obtaining fuzzy predicates is connected with the field of data summarization [28]. The predicates obtained by FuzzyPred has demonstrated its potential value as part of a data mining processes in diverse fields such as analysis of algorithms performance [29] or information technologies national indices [30].

As in any process of knowledge discovery in databases [31], FuzzyPred involves the application of different steps. In the original version of FuzzyPred [15], some steps require a major effort from the user because some steps are not always intuitive, and they were not available in a unique environment. In addition, this implied that users could not focus on the issues that they should be really working on. This motivates our effort to obtain a simpler and integrated environment to obtain fuzzy predicates. This is in line with similar recent efforts to develop data mining tools with this easy-to-use and integrated approach [7, 32], and particularly within the KNIME framework [10, 33].

This is a simple and scalable way to face this challenge, i.e., to resort to integrate particular algorithms (as it is the case of FuzzyPred) into recognized data mining tools (such as Knime) that provide and support many basic features required in a knowledge discovery process (such as preprocessing and visualization which are unavoidable steps for obtaining useful knowledge [5, 34]). In spite of the importance of fuzzy approaches in data mining [35], this algorithmic development is not reflected in a remarkable presence in the most known data mining tools [5, 7].

3 KNIME

Based on the requirements explained in the previous section, we believe that it is convenient to integrate FuzzyPred into a professional data mining tool. There are many commercial and non-commercial DM tools and libraries [5, 7]; see KDnuggets software directory (http://www.kdnuggets.com/software) and The-Data-Mine site (http://the-data-mine.com/bin/view/Software) for an updated comparison.

Here, we focus on free software dedicated to the whole range of the data mining field. Moreover, we are interested in tools within the open source paradigm in order to extend it. Weka [36] is a clear option because it provides implementations of learning algorithms that can be easily applied to the dataset. Weka is a widely used data mining tool that supports all phases of the mining process, encapsulates well tested implementations of many popular mining methods, offers an interface that supports interactive mining and result visualization, and automatically produces statistics to assist result evaluation. In addition, there are other good alternatives (e.g., Orange, R, Rapid Miner, Scikit-learn) [5–7].

However, for our purposes, we prefer KNIME because of its visual approach for data mining that allows a very simplified learning curve for non-advanced users [10, 11]. In addition, KNIME allows a simple way to integrate different programming languages in the same visual workflow environment, and it is supported by an extensive community of users and developers. Since KNIME is built on top of Eclipse, it shares the benefit of a plug-in architecture that makes it easily extensible; many custom-built nodes are available and easily accessible. Another strength is its open source collaborative ecosystem, where contributors are free to develop new algorithms, tools as well as data manipulation or visualization methods. An active community of KNIME users and developers supports constant software upgrades, which is one of the key advantages of free open source software. Some of the most widely used programming languages (Java, Python, R, Octave, and Matlab) are found in the KNIME library as snippet nodes. Besides, its visual interface allows an easy and interactive analysis [37, 38] that it is needed in order to enable the user to explore the results. Next, we present an extended description of KNIME.

KNIME is a modular environment that enables easy integration of new algorithms, data manipulation, and visualization methods as models. It is compatible with Weka, and it also includes statistical methods via the embedded usage of R [9]. It is an open source predictive analytics platform (released under the GNU General Public License v3) suited to process a variety of data formats, from basic CSV or XLSX files to more complex data structures such as XML, URL, and relational databases (e.g., db2, Oracle, MySQL). Information on KNIME is available through the web in several ways (e.g., a dedicated YouTube channel: KNIME-TV; the KNIME web site: http://www.KNIME.org/ and other independent communities such as Stack-Overflow, http://stackoverflow.com/). More information, as well as downloads, can be found at http://www.KNIME.org.

The architecture of KNIME was designed with three main principles in mind:

- Visual interactive framework: data flows should be combined by simple drag and drop from a variety of processing units. Customized applications can be modeled through individual data pipelines.
- Modularity: processing units and data containers should not depend on each other in order to enable easy distribution of computation and allow for independent development of different algorithms.
- Easy expandability: it should be easy to add new processing nodes or views and distribute them through a simple plug and play principle without the need for complicated install/uninstall procedures.

A data analysis process in KNIME consists of a pipeline of nodes. Pipelines are based on edges that transport data and models. Each node processes the input data and/or model(s) and produces results on its outputs. The Workflow Manager allows to insert new nodes and to add directed edges (connections) between two nodes. It also keeps track of the status of nodes (not configured, configured, executed) and returns, on demand, a pool of executable nodes. Each node can have an arbitrary number of views associated with it. Views can range from simple table views to more complex views of the underlying data or the generated model.

To add new nodes to KNIME, it is necessary to extend three abstract classes:

- NodeModel: this class is responsible for the main computations. It requires to overwrite three main methods: configure (), execute (), and reset (). The first takes the meta information of the input tables and creates the definition of the output specification. The execute-function performs the actual creation of the output data or models, reset discards all intermediate results.
- NodeDialog: this class is used to specify the dialog that enables the user to adjust individual settings that affect the node's execution. A standardized set of Default-DialogComponent objects allows to very quickly create dialogs where only a few standard settings are needed.
- NodeView: this class can be overwritten multiple times to allow for different views onto the underlying model.

Figure 1 shows a diagram of this structure. This schema follows the well-known Model-View-Controller design pattern.

In addition to the model, dialog, and view classes, the programmer also needs to provide a NodeFactory, to create new instances. The factory also provides names and other details such as the number of available views or a flag indicating the absence or presence of a dialog.

The approach described above enables the user to build a workflow for different types of DM problems. In this case, we add a new method called FuzzyPred to offer several advantages. One of the advantages is to increase the range of possible users requesting FuzzyPred. The next section describes the proposal.

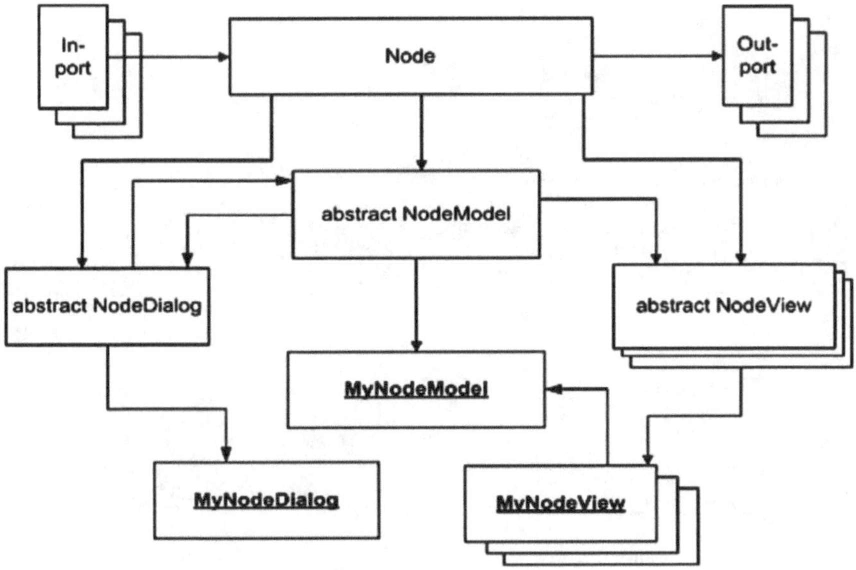

Fig. 1 A diagram depicting the main classes of a KNIME [9]

4 A New Plug-into Include FuzzyPred in KNIME

KNIME offers a wide range of nodes with different purposes. But in this case, we needed to expand the functionality of KNIME by implementing our own nodes with the components corresponding to the FuzzyPred aspects. This also contributes to the Repository Nodes of KNIME.

Specifically, in order to integrate FuzzyPred into KNIME, five nodes were developed: FuzzyTransformation, FuzzyPredAlgorithm, FuzzyPMMLWrite, FuzzyPMMLExtension, and DiversityPredicates. Each node was implemented based on the KNIME architecture of classes, where three abstract classes were extended. Figure 2 shows a class's diagram of the node FuzzyPredAlgorithm as an example of the style used in the implementation of the plug-in. In the next pages, each developed node is described.

4.1 FuzzyTransformation Node

The FuzzyTransformation node was implemented because FuzzyPred is an algorithm that works with fuzzy values instead of using the original data, i.e., it depends on particular preprocessing steps. This node has two configurations: automatic and personalize. The automatic configuration was designed for non-experts, based on our empirical experiences. This automatic configuration is based specifically on three

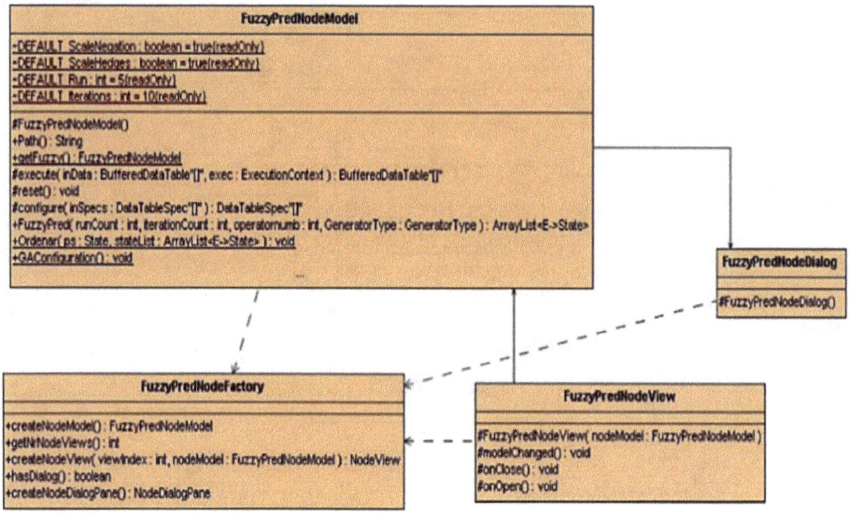

Fig. 2 Architecture of FuzzyPredAlgorithm Node

linguistic terms (little, medium, high), and the universe of discourse is divided into three equally sized regions. The user can personalize this configuration in order to reflect its knowledge and insights. This node supports the transformation of nominal (age: young, adult, old), numeric (height in centimeters: 175, 165, 180), and ordinal attributes (educational level: primary, high school, pre-university studies, and university studies). It is worth noting that some of these transformations may be obtained by combining other nodes available in KNIME, but we include this particular node in order to ease this kind of transformation that is needed for the main algorithm of FuzzyPred.

To create linguistic variables, it is necessary that the user define three parameters: the universe of discourse (minimum and maximum), the type of membership function, and the number of membership functions (labels). The configuration window has some default function types of membership functions, for instance: Gaussian, Sigmoidal, Bell curves.

For example, *age* is a variable that may be used to derive linguistic labels such as *young*, *adult,* and *old*. Figure 3 shows the settings of this variable, using three membership functions (2 trapezoidal function and 1 triangular) and the universe of discourse between 0 and 100.

The value for each linguistic term may be adjusted, as it is shown in Fig. 4.

Membership functions for fuzzy sets can be defined in several ways [39, 40]. The decision on which type of function used depends on the particular context and interest. Triangular or trapezoidal shapes (like in the previous examples) are simple to implement and fast for computation.

In addition to numeric attributes such as age, there are different types of attributes, for example, nominal (eye color) and ordinal (rankings like grades). The type of

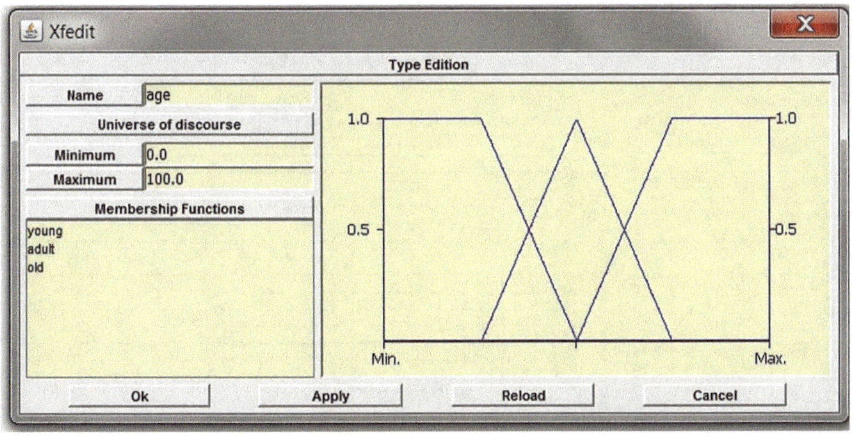

Fig. 3 Setting up the linguistic variable age

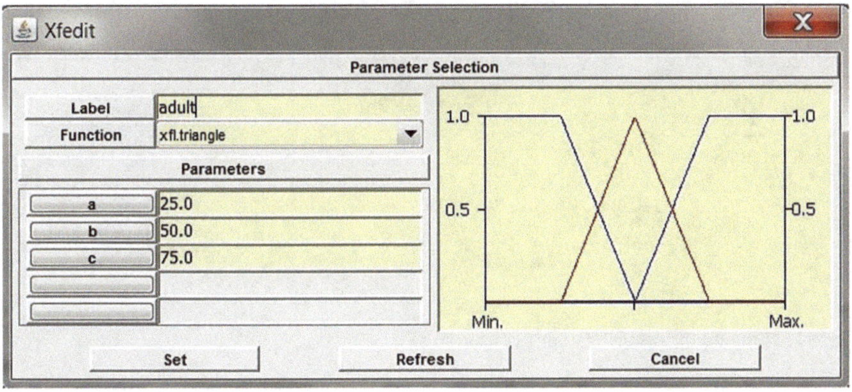

Fig. 4 Setting up the linguistic terms adult

attribute depends on its properties. For example, nominal possesses distinctness ($= \neq$) and ordinals, in addition to the distinction, it has order ($< >$). Before constructing a model in FuzzyPred, all nominal and ordinal attributes that are to be used in the data mining process should be transformed into fuzzy variables in the interval [0, 1]. Thus, if the nominal attribute has k possible values, it is replaced by k linguistic binary variables (being 1 only for the new variable associated with each value). For example, Fig. 5 shows the variable workclass and its three possible values: private, state-grov, and self-emp-not-inc. It is worth clarifying that if the attribute can take too much different values, it might not provide valuable information to the model because a lot of particular details are hard to be generalized.

If the user identifies that a particular nominal attribute has an ordinal semantics, then the user has the possibility to define the correct order as it is shown in Fig. 6.

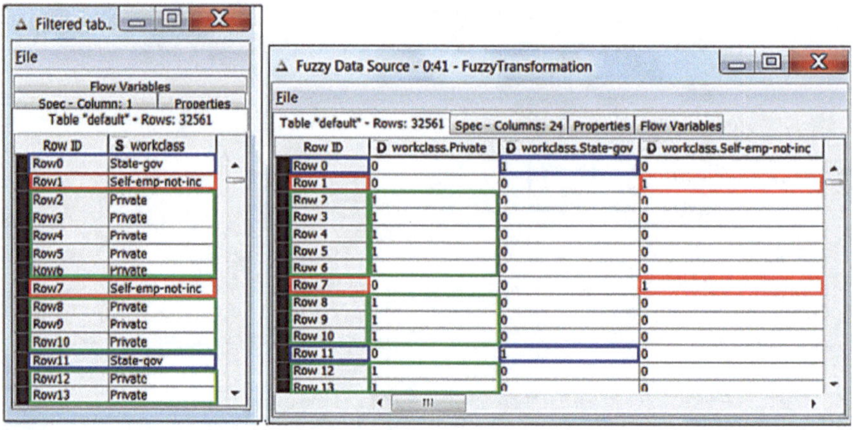

Fig. 5 Setting up the nominal variable workclass

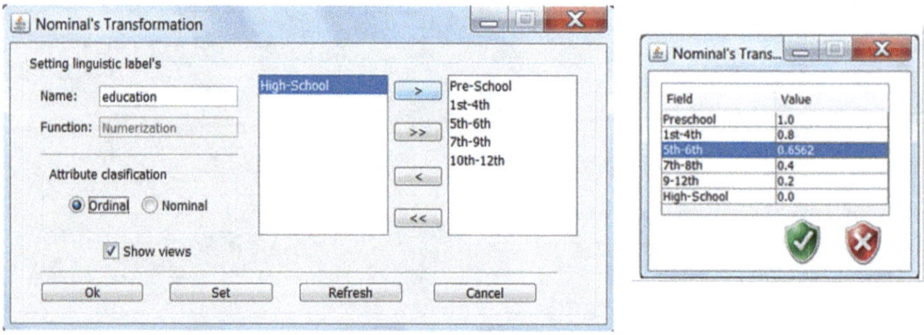

Fig. 6 Setting up the ordinal variable education

The extreme values of the list take 0 and 1 respectively, and the rest are calculated according to the size of the list.

The status of the transformation of each attribute is indicated to the user in separate windows. The possible values are Pending, Processing, Completed (if the attribute was converted successfully), and Not Completed (if the user decides to cancel a particular attribute). At the end of the process, it is really important to have the whole set of relevant features. The portion of the database to be mined is called the minable view. This view is the input to the next node called FuzzyPredAlgorithm. If the user needs to save the fuzzy dataset to an external file, KNIME gives the possibility to download the database in different formats. These nodes are in the I/O Category of KNIME software, and its use is an example of the advantage of the integration.

4.2 FuzzyPredAlgorithm Node

FuzzyPredAlgorithm is the central stage in the process of knowledge discovery for obtaining fuzzy predicated. This stage transforms the minable view into some kind of model (in this case, fuzzy predicates in CNF or DNF).

Finding good values for the parameters of the algorithm is a non-trivial task [25, 41]. Indeed, it is difficult to find the best compromise between the stopping criterion and the algorithm performance. In some cases, the algorithm may perform a huge and unnecessary number of iterations when the optimal solution is quickly found. In other situations, the algorithm may stop just before the iteration when it could find a better solution. Particularly, if your search space is very large, you should think about the percent of the space that you want to explore. How well it will work depends on the experience of the user.

The node FuzzyPredAlgorithm needs a previous configuration (see Fig. 7). This configuration varies depending on the number of fitness functions to be optimized. If there is only one fitness function, the problem is considered mono-objective. However, if two or more objectives are defined, the problem becomes a multi-objective problem.

The parameters to be defined are the following:

- Data: database to be mined.
- Fuzzy logic operator to be used in the computation of the truth value [15, 18]: Zadeh (Min-Max), Probabilistic (Algebraic product and sum), Compensatory Fuzzy Logic (Geometric Mean-Dual).
- Connectives: this option allows defining the number of clauses include n the predicate and the type of normal form (CNF, DNF). It is also possible to allow an

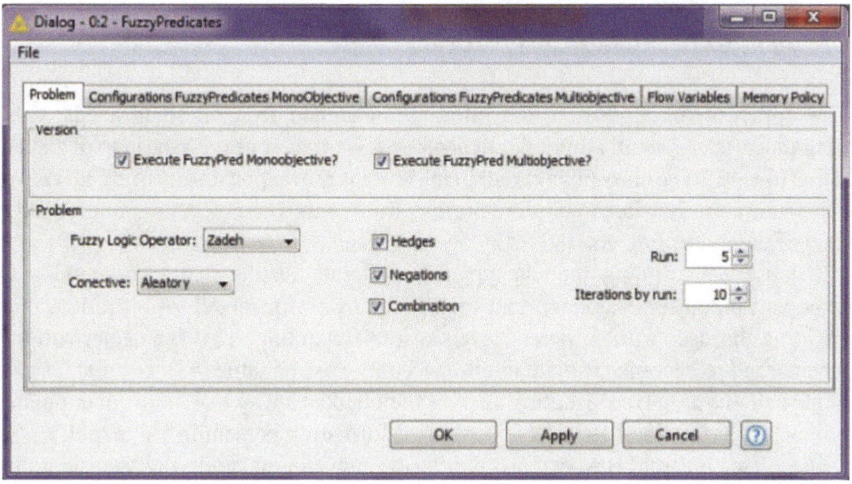

Fig. 7 Setting up the node FuzzyPredAlgorithm

unrestricted search where different predicates may be obtained in terms of their size and normal form. More than 3 clauses may result in a huge search space and hard to understand predicated; thus, this option may be adjusted carefully.
- Hedges: this allows to use modifiers of a predicate “a”, such as negation (NOT a), very (a^2), extremely (a^3), a little ($a^{0.5}$).
- Fitness function: this option defines the quality that it is desired in the obtained predicates [15], such as Fuzzy Predicate Truth Value (FPTV), Fuzzy Predicate Support (FPS), Fuzzy Predicate Binary Support (FPBS(α)), Fuzzy Predicate Central Pruning Average (FPCPA), Fuzzy Predicate Low Pruning Average (FPLPA), Fuzzy Predicate High Average Pruning (FPHPA), Fuzzy Predicate Comprehensibility (FPC).
- Stop condition: number of iterations and the number of repetitions for each iteration.
- Metaheuristic for mono-objective problems [25]: Random search (RS), Hill Climbing (HC), Simulated Annealing (SA), Tabu Search (TS), Genetic Algorithm (GA), Evolution strategy (ES), Estimation of Distribution Algorithm (EDA).
- Metaheuristics for multi-objectives problem [25]: Multiobjective Stochastic Hill Climbing, Multiobjective Tabu Search, Multicase Simulated Annealing, Non dominated Sorting Genetic Algorithm (NSGA II), Multiobjective Evolutionary Algorithm based on Decomposition (MOEADDE), Multiobjective Genetic Algorithm (MOGA).

This node also contains an important post-processing phase. These post-processing functions must be performed in a particular order. For that reason, we don’t give these four nodes as an option for the user, and we have integrated all of them into a single node making this process transparent to the user.

4.3 SpaceTreeVisualization Node

After applying the algorithm, the information gained from the data needs to be communicated. Normally this analytical process is often done with the help of visualization [33, 42]. The class FuzzyPredNodeView has the responsibility to create views. In this view, it is possible to observe the encoding of the predicate, its evaluation, the name of the algorithm, and the fuzzy operator used.

Because data mining models typically generate results that were previously unknown to the user, it is important to provide a way for model visualization, thus providing the user with sufficient levels of understanding. This is in line with the current trend to become artificial intelligence more explainable [43]. For that reason, a node called SpaceTreeVisualization was included as a way to display, manipulate, and visualize trees in order to reveal all the information present in the model.

Space Tree is a tree browser that builds on the conventional node-link tree diagrams [44]. It adds dynamic rescaling of branches of the tree to best fit the available screen space. Branches that do not fit on the screen are summarized by a triangular preview.

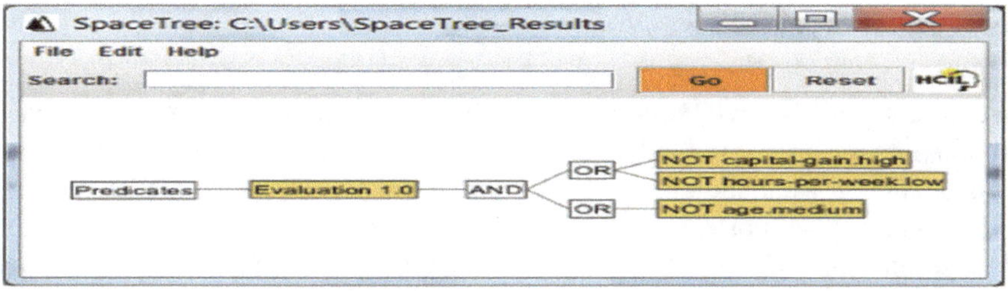

Fig. 8 Example of the SpaceTreeVisualization

When users select a node to change the focus of the layout, the number of levels opened is maximized. In addition, it includes an integrated search and filter functions.

This node does not need any settings; it simply must be related to the output port of the node FuzzyPredAlgorithm. An example of the results of this node is shown in Fig. 8. Users can navigate the tree by clicking on the nodes to open branches, or by using the arrow keys to navigate through ancestors or descendants.

4.4 FuzzyPMMLGenerator and FuzzyPMMLWriter Nodes

For the data mining model, the visualization is important for the user, but it is also very important to allow other tools to load and show the obtained models. For this reason, it is possible to save the model in the format described in Predictive Model Markup Language (PMML), which is extensively used in other KNIME nodes [39]. It provides a convenient mechanism for working with different types of models in data mining. However, FuzzyPred has a particular model to represent the knowledge obtained by the data mining process; that's the reason why it was necessary to define two nodes: FuzzyPMMLWriter and FuzzyPMMLGenerator. The Predictive Model Markup Language (PMML) is an open standard for storing and exchanging models in XML format [45]. Its structure follows a set of predefined elements and attributes which reflect the inner structure of one or more models. Ideally, a model trained by KNIME (or by any other tool supporting PMML) and stored as PMML can be used in other different statistical and data mining tools [46].

PMML defines specific elements for several techniques, including neural networks, decision trees, and clustering models, to name just a few. But FuzzyPred is a new technique that is not supported yet. However, PMML has an extensible model (rules) that it is close to our interest [45, 46].

For FuzzyPred, FuzzyPMMLWriter and FuzzyPMMLGenerator nodes have been added as well. FuzzyPMMLGenerator has a configuration interface where the user can select between two options: generate the model as an extension of standard (RuleSetModel) or as a complete new model. FuzzyPMMLWriter writes the XML file in the location that the user has been configured.

4.5 *DiversityPredicate Node*

In classical logic, Normal Forms are very useful to normalize the knowledge in order to found equivalences and to simplify other processes [17]. However, in the case of data mining, we think that it is interesting to present several alternative ways of similar knowledge. For example, given a database describing persons in terms of three features (*healthy, old, and educated*), a possible predicate that may be obtained in DNF is:

- Not *healthy* or *(old* and *educated)*

If this predicate has a great truth value in the database, it is possible to understand that the persons in the database may be described by using two complementary (not excluding) descriptions: some personas are "not *healthy*", while others are "*old and educated*". In a general sense, this can be interpreted similarly to fuzzy clustering results [47]. However, this pattern in CNF may be transformed to a rule based on the equivalence (NOT A OR B $\equiv$ A $\rightarrow$ B), similar to those obtained through fuzzy association rule mining [48] in the form:

- If *healthy* then (*old* and *educated*)

The purpose of the DiversityPredicate node is to obtain several patterns based on this type of classical transformation to ease the interpretation of the results [43]. Figure 9 present a KNIME workflow, including the proposed nodes to obtain several predicates from a database, while Fig. 10 illustrates an example of how several predicates in normal forms may be diversified to obtain several patterns.

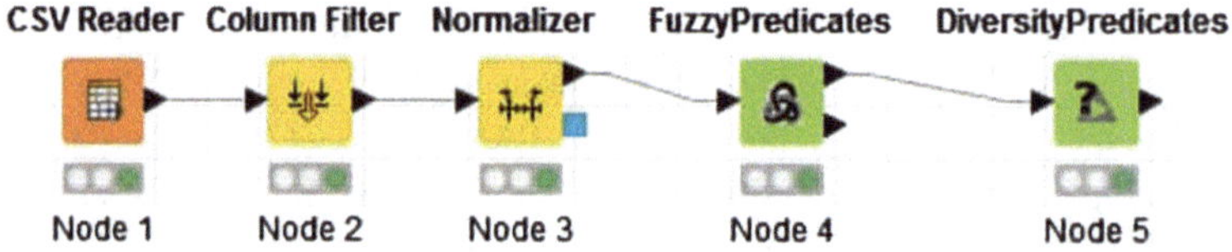

Fig. 9 A complete workflow from data to several patterns

Out-Port name - 0:6 - DiversityPredicates

File

Table "default" - Rows: 11 Spec - Columns: 5 Properties Flow Variables

Row ID	I Iteración	S Predicado	S Modelo equivalente	D Evaluación	S Regla aplicada
Row 0	1	(price) or (width)	(width) or (price)	0.32	Conmutativa
Row 1	1	(price) or (width)	(not (price)) then (width)	0.32	Definición de implicación
Row 2	1	(not (width) and height) or (not (price))	(not (price)) or (not (width) and height)	0.25	Conmutativa
Row 3	1	(height) or (width and not (price))	(width and not (price)) or (height)	0.25	Conmutativa
Row 4	1	(height) or (width and not (price))	(not (height)) then (width and not (price))	0.25	Definición de implicación
Row 5	1	(not (height)) or (width and not (price))	(width and not (price)) or (not (height))	0.28	Conmutativa
Row 6	1	(not (width) and not (height)) or (not (price))	(not (price)) or (not (width) and not (height))	0.34	Conmutativa
Row 7	1	(width) or (not (height) and price)	(not (height) and price) or (width)	0.28	Conmutativa
Row 8	1	(width) or (not (height) and price)	(not (width)) then (not (height) and price)	0.28	Definición de implicación
Row 9	1	(not (price)) or (not (height))	(not (height)) or (not (price))	0.34	Conmutativa
Row 10	1	(not (price)) or (not (width))	(not (width)) or (not (price))	0.45	Conmutativa

Fig. 10 Example of the SpaceTreeVisualization

5 An Illustrative Example

This section presents a case study as an example of the functionality and process of creating an experiment with FuzzyPred in KNIME. We will show it, step by step, through the process of building a small and simple workflow.

To evaluate the usefulness of the proposed approach, some experiments have been carried on a real-world database extracted from the UC Irvine Machine Learning Repository (http://archive.ics.uci.edu/ml/). The first step of the experiment is to choose the data sets to be used. This example uses the Basketball dataset (5 attributes (assist per minute (numeric attribute (real)), point per minute (numeric attribute (real)), age (numeric attribute (integer)), height (numeric attribute (integer), time played (numeric attribute (real))) and 96 records).

A typical workflow for applying the proposed nodes for using FuzzyPred in KNIME is illustrated in Fig. 11.

To create this workflow, first of all, we need to expand "IO" in the Node Repository and the contained category "Read" to drag&drop the File Reader icon into the workflow editor window. After this, some preprocessing (for example, those that include statistical processing) nodes are included in the Preprocessing meta-node. This node calculates statistical measures such as minimum, maximum, mean, standard deviation, variance, median, overall sum, number of missing values, and row count across all numeric columns, and counts all nominal values together with their occurrences. Figure 11 presents an example of a workflow connecting several of the nodes presented in this paper. This flow includes the automatic configuration of the fuzzy sets and some transformations of the loaded data. The result of this process is saved (XLS Writer) for further analysis or processing.

In this workflow, the output of the metanode were 15 linguistic variables (3 for each attribute). The filter node (Column Filter) offer a user-friendly way to filter the linguistic variables that will be part of the minable view. For which, applied the

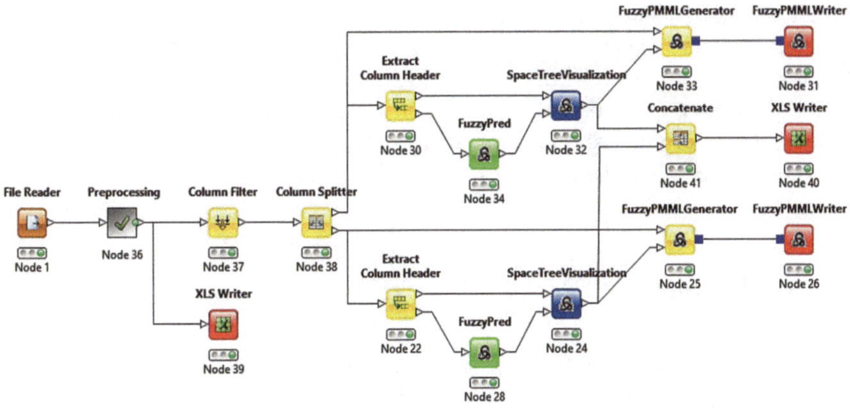

Fig. 11 Workflow created in KNIME using the proposed nodes

```xml
<?xml version='1.0' encoding='UTF-8'?>
<PMML version='4.1' xmlns:xsi='http://www.w3.org/2001/XMLSchema-instance'>
<Header copyright='None'>
  <Aplication name='FuzzyPred System' version='1.0'/>
  <Annotation>Exported whith PMML format. Form data source named 'null'.</Annotation>
</Header>
<DataDictionary numberOfFields='3'>
  <DataField name='basketball.assists per minuteReal.AsisstMinBaja' optype='continuous' dataType='double'>
      <Interval closure='openOpen' leftMargin='0' rightMargin='1'>
      </Interval>
  </DataField>
  <DataField name='basketball.height.HeightBajo' optype='continuous' dataType='double'>
      <Interval closure='openOpen' leftMargin='0' rightMargin='1'>
      </Interval>
  </DataField>
  <DataField name='basketball.points per minuteReal.PointsMinMucho' optype='continuous' dataType='double'>
      <Interval closure='openOpen' leftMargin='0' rightMargin='1'>
      </Interval>
  </DataField>
</DataDictionary>
  <FuzzyPredModel name='FuzzyPredicateModel' funtionName='fuzzyPredicate' algorithmName='Hill Climbing'>
  <MiningSchema>
```


Fig. 12 FuzzyPred model (PMML)

node Column Filter, were selected 9 linguistics variables, for the attributes assist per minute (low, mid, high), points per minute (low, mid, high), and age (young, adult, old). Later, the node Column Splitter was used, with the purpose of analyzing each linguistic variable filtered, for the fuzzy predicates obtained. It is for this that Fig. 11 shows two parallel workflows (very similar). This view is the input of FuzzyPred (the core of the algorithm). In the end, the results are displayed (SpaceTreeVisualization) and saved (FuzzyPMMLWriter, XLS Writer). They may also be derived to obtain alternative patterns through DiversityPredicate node, in a similar way.

As it is shown in Fig. 11, every node is connected to the next in order to get the data flow. In this workflow, each node shows a green status, because they were previously configured and executed. In order to examine the data and the results, it is only necessary to open the nodes' views. Figure 12 shows one of the generated PMML models.

This tool relieves researchers of much technical work and allows them to focus on the analysis of this new learning model. The user can start, suspend/pause and

stop the experiment at any moment in order to see step by step partial reports of the execution. The experiment set up is not complex, and the interface is intuitive. Any researcher can use KNIME on their computers with Java, independently of the operating system. It is worth noting that this process may be easily repeated with a different database with a similar structure by only changing the name of the database that is the source of all the processes.

6 Conclusions

An extensive library of algorithms, together with easy to-use software, considerably reduces the experience required by users to discover new and useful knowledge from available data. As a result of the integration of FuzzyPred into KNIME, researchers with less knowledge would be able to apply this new method successfully to their problems. In this paper, we have described a plug-in with five nodes (FuzzyTransformation, FuzzyPredAlgorithm, Space Tree Visualization, FuzzyPMMLGenerator, FuzzyPMMLWriter, DiversityPredicate), that together with the original nodes in KNIME, support all phases of the KDD process to discover fuzzy predicates, and also to transform the obtained predicates in normal forms to other patterns. A workflow with an example of use was included.

References

1. Sunhare, P., Chowdhary, R.R., Chattopadhyay, M.K.: Internet of things and data mining: An application oriented survey, Journal of King Saud University—Computer and Information Sciences (2020). https://doi.org/10.1016/j.jksuci.2020.07.002
2. Skiena, S.S.: The Data Science Design Manual. Springer (2017)
3. Oliveira, C., Guimarães, T., Portela, F., Santos, M.: Benchmarking Business Analytics Techniques in Big Data. Procedia Computer Science **160**, 690–695 (2019). https://doi.org/10.1016/j.procs.2019.11.026
4. Zain, M.S.I.M., Rahman, S.A.: Challenges of Applying Data Mining in Knowledge Management towards Organization. Int. J. Acad. Res. Bussines Soc. Sci. **7**(12), 405–412 (2017). https://doi.org/10.6007/IJARBSS/v7-i12/3621
5. Verma, K., Bhardwaj, S., Arya, R., Salim, M., Islam, U., Bhushan, M., Kumar, A., Samant, P.: Lastet Tools for Data Mining and Machine Learning. Int. J. Innov. Technol. Explor. Eng. **8**(9S), 18–23 (2019). https://doi.org/10.35940/ijitee.I1003.0789S19
6. Ranjan, R., Agarwal, S., Venkatesan, S.: Detailed Analysis of Data Mining Tools. Int. J. Eng. Res. & Technol. **6**(5), 785–789 (2017)
7. Pynam, V., Spanadna, R.R., Srikanth, K.: An extensive study of Data Analysis Tools (Rapid Miner, Weka, R Tool, KNIME, Orange). Int. J. Comput. Sci. Eng. **5**(9), 4–11 (2018). https://doi.org/10.14445/23488387/IJCSE-V5I9P102
8. Naik, A., Samant, L.: Correlation review of classification algorithm using data min-ing tool: WEKA, Rapidminer, Tanagra. Orange KNIME. Procedia Comput. Sci. **85**, 662–668 (2016). https://doi.org/10.1016/j.procs.2016.05.251
9. Berthold, A., Cebron, N.: KNIME—the Konstanz information miner: version 2.0 and beyond. ACM SIGKDD exploration Newsletter, 11(1), 26–31 (2009). https://doi.org/10.1145/1656274.1656280
10. Radosevic, N., Duckham, M., Liu, G.-J., Sun, Q.: Solar radiation modeling with KNIME and Solar Analyst: Increasing environmental model reproducibility using scientific workflows. Environ. Model Softw. **132**, (2020). https://doi.org/10.1016/j.envsoft.2020.104780
11. Ogungbemi, A.O., Teixido, E., Massei, R., Scholz, S., Küste, E.: Optimization of the spontaneous tail coiling test for fast assessment of neurotoxic effects in the zebrafish embryo using an automated workflow in KNIME. Neurotoxicol. Teratol. **81**, (2020). https://doi.org/10.1016/j.ntt.2020.106918
12. Ceruto, T., Rosete, A., Espin, R.: Knowledge Discovery by Fuzzy Predicates. In: Espin, R., Pérez, R., Cobo, A., Marx, J., Valdés, A. (eds.) Soft Computing for Business Intelligence, 537, pp. 187–196. Springer, Heidelberg (2014). https://doi.org/10.1007/978-3-642-53737-0_13
13. Lejeune, M., Lozin, V., Lozina, I., Ragab, A., Yacout, S.: Recent advances in the theory and practice of Logical Analysis of Data. Eur. J. Oper. Res. **275**(1), 1–15 (2019). https://doi.org/10.1016/j.ejor.2018.06.011
14. Atanassov, K.T.: Intuitionistic Fuzzy Predicate Logic. Studies in Fuzziness and Soft Computing, vol. 351, pp. 65–77. Springer, Cham (2019)
15. Ceruto, T., Lapeira, O., Rosete, A.: Quality measures for fuzzy predicates in conjuctive and disjunctive normal form. Ingeniería e Investigaci6n **34**(3), 63–69 (2014). https://doi.org/10.15446/ing.investig.v34n3.41638

16. Le, V.: Fuzzy Logic in Narrow Sense with Hedges. Int. J. Comput. Sci. & Inf. Technol. **8**(3), 133–143 (2016). https://doi.org/10.5121/ijcsit.2016.8310
17. Bruno, A.: Normal Forms. Math. Comput. Simul. **45**(5–6), 413–427 (1998)
18. Espín, R., Pedrycz, W., González, E., Fernández, E.: Archimedean-Compensatory Fuzzy Logic Systems. International Journal of Computational Intelligence Systems, 8(Sup 2), 54–62, 2015. https://doi.org/10.1080/18756891.2015.1129591
19. Espín, R., Pedrycz, W., Gonzalez, E., Fernandez, E.: An Interpretable Logical Theory: The case of Compensatory Fuzzy Logic. Int. J. Comput. Intell. Syst. **9**(4), 612–626 (2016). https://doi.org/10.1080/18756891.2016.1204111
20. Turksen, I.B.: Fuzzy normal forms. Fuzzy Sets Syst. **69**(3), 319–346 (1995). https://doi.org/10.1016/0165-0114(94)00166-5
21. Gehrke, M., Walker, C.L., Walker, E.A.: Normal forms and truth tables for fuzzy logics. Fuzzy Sets Syst. **138**(1), 25–51 (2003). https://doi.org/10.1016/S0165-0114(02)00566-3
22. Gerla, G.: Fuzzy Turing machines: Normal form and limitative theorems. Fuzzy Sets Syst. **333**, 87–105 (2018). https://doi.org/10.1016/j.fss.2017.01.008
23. Zeinali, M., Alikhani, R., Shahmorad, S., Bahrami, F., Perfilieva, I.: On the structural properties of Fm-transform with applications. Fuzzy Sets Syst. **342**(1), 32–52 (2018). https://doi.org/10.1016/j.fss.2017.12.008
24. Chen, T., Shang, C., Su, P., Shen, Q.: Induction of accurate and interpretable fuzzy rules from preliminary crisp representation. Knowl.-Based Syst. **146**, 152–166 (2018). https://doi.org/10.1016/j.knosys.2018.02.003
25. Talbi, E.G.: Metaheuristics from Design to Implementation. Wiley, London (2009)
26. Cuevas, E., Gálvez, J., Camarena, O., Díaz-Cortés, M.A.: A new metaheuristics optimization methodology based on Fuzzy Logic. Appl. Soft Comput. **61**, 549–569 (2017). https://doi.org/10.1016/j.asoc.2017.08.038
27. Ochoa, A., Rivera, G., Gómez-Santillán, C., Sánchez, B.: Handbook of Research on Metaheuristics for Order Picking Optimization in Warehouses to Smart Cities. IGI Global (2019). https://doi.org/10.4018/978-1-5225-8131-4
28. Boran, F.E., Akay, D., Yager, R.R.: An overview of methods for linguistic summa-rization with fuzzy sets. Expert Syst. Appl. **61**(1), 356–377 (2016). https://doi.org/10.1016/j.eswa.2016.05.044
29. Aledo, J.A., Gámez, J.A., Lapeira, O., Rosete, A.: Characterization of the Optimal Bucket Order Problem Instances and Algorithms by Using Fuzzy Logic. Studies in Fuzziness and Soft Computing **377**, 49–70 (2019). https://doi.org/10.1007/978-3-030-10463-4_3
30. Ceruto, T., Lapeira, O., Rosete, A.: Analyzing Information and Communications Technology National Indices by Using Fuzzy Data Mining Techniques. In: Llanes-Santiago O., Cruz-Corona C., Silva-Neto A., Verdegay, J. (eds.) Studies in Computational Intelligence, SCI 872, 255–279 (2020). https://doi.org/10.1007/978-3-030-34409-2_15
31. Alasadi, S.A., Bhaya, W.S.: Review of Data Preprocessing Techniques on Data Mining. J. Eng. Appl. Sci. **12**(16), 4102–4107 (2017)
32. Triguero, I., Gonzalez, S., Moyano, J.M., Garcia, S., Alcala-Fdez, J., Luengo, J., Fernandez, A., del Jesus, M.J., Sanchez, L., Herrera, F.: KEEL 3.0: an Open Source Software for Multi-Stage Analysis in Data Mining. International Journal of Computational Intelligence Systems, 10(1), 1238–1249, (2017). https://doi.org/10.2991/ijcis.1.82
33. Iglesias, A.I., Ilisástigui, L.B., Córdovez, T.C., Rodríguez, D.M.: Nuevos plugins para la herramienta Knime para el uso de sus flujos de trabajo desde otras aplicaciones. Ciencias de la Información **46**(1), 47–52 (2015)
34. Yun, Y., Ma, D., Yang, M.: Human–computer interaction-based Decision Support System with Applications in Data Mining. Futur. Gener. Comput. Syst. **114**, 285–289 (2021). https://doi.org/10.1016/j.future.2020.07.048
35. Mirzakhanov, V.E.: Value of fuzzy logic for data mining and machine learning: A case study. Expert Syst. Appl. **162**, (2020). https://doi.org/10.1016/j.eswa.2020.113781
36. Alam, F., Pachauri, S.: Comparative Study of J48, Naïve Bayes and One—R Classification Technique for Credit Card Fraud Detection using WEKA. Adv. Comput. Sci. Technol. **10**(6), 1731–1743 (2017)

37. Verhoeven, A., Giera, M., Mayboroda, O.A.: KIMBLE: A versatile visual NMR metabolomics workbench in KNIME. Anal. Chim. Acta **1044**, 66–76 (2018). https://doi.org/10.1016/j.aca.2018.07.070
38. Basha, S.M., Rajput, D.S., Poluru, R.K., Bhushan, S.B., Basha, S.A.K.: Evaluating the Performance of Supervised Classification Models: Decision Tree and Naïve Bayes using KNIME. International Journal of Engineering & Technology, 7(4.5), 248–253 (2018). https://doi.org/10.14419/ijet.v7i4.5.20079
39. Zadeh, L.A.: Fuzzy Logic-a personal perspective. Fuzzy Sets Syst. **281**, 4–20 (2015). https://doi.org/10.1016/j.fss.2015.05.009
40. Ebrahimnejad, A., Verdegay, J.L.: Fuzzy Sets-Based Methods and Techniques for Modern Analytics. Studies in Fuzziness and Soft Computing, vol. 364. Springer Nature (2018)
41. Marti, L., Garcia, J.: A stopping criterion for multi-objective optimization evolutio-nary algorithms. Inf. Sci. **367–368**, 700–718 (2016). https://doi.org/10.1016/j.ins.2016.07.025
42. Borhade, M., Mulay, P.: Online Interactive Data Mining Tool. Procedia Computer Science **50**, 335–340 (2015). https://doi.org/10.1016/j.procs.2015.04.039
43. Barredo-Arrieta, A., Díaz-Rodríguez, N., Del-Ser, J., Bennetot, A., Tabik, S., Barbado, A., Garcia, S., Gil-Lopez, S., Molina, D., Benjamins, R., Chatilaf, R., Herrera, F.: Explainable Artificial Intelligence (XAI): Concepts, taxonomies, opportunities and challenges toward responsible AI. Info. Fusion **58**, 82–115 (2020). https://doi.org/10.1016/j.inffus.2019.12.012
44. Plaisant, C., Grosiean, J., Bederson, B.B.: Spacetree: Supporting exploration in large node link tree, design evolution and empirical evaluation. Information Visualization, 57–64 (2002). https://doi.org/10.1109/infvis.2002.1173148
45. Guazzelli, A., Lin, W.-C., Williams, G., Zeller, M.: PMML: An open standard for sharing models. The R J. **1**(1), 60–65 (2009). https://doi.org/10.32614/RJ-2009-010
46. Morent, D., Stathaos, K., Lin, W.-C., Berthold, M.R.: Comprehensive PMML preprocessing in KNIME. Proceedings of the 2011 workshop on Predictive markup language modeling, ACM, pp. 28–31 (2011). https://doi.org/10.1145/2023598.2023602
47. Askari, S.: Fuzzy C-Means clustering algorithm for data with unequal cluster sizes and contaminated with noise and outliers: review and development. Expert Syst. Appl. (2020). https://doi.org/10.1016/j.eswa.2020.113856
48. Zhang, Z., Pedrycz, W., Huang, J.: Efficient mining product-based fuzzy association rules through central limit theorem. Appl. Soft Comput. **63**, 235–248 (2018). https://doi.org/10.1016/j.asoc.2017.11.025